From the Family Farm
to Agribusiness

From the Family Farm to Agribusiness

The Irrigation Crusade in California and the West, 1850–1931

DONALD J. PISANI

UNIVERSITY OF CALIFORNIA PRESS
Berkeley · Los Angeles · London

University of California Press
Berkeley and Los Angeles, California
University of California Press., Ltd.
London, England
Copyright © 1984 by
The Regents of the University of California

Library of Congress Cataloging in Publication Data

Pisani, Donald J.
    From the family farm to agribusiness.

    Bibliography
    Includes index.
        1. Irrigation—Economic aspects—California—History.
    2. Agriculture—Economic aspects—California—History.
    I. Title.
    HD1739.C2P57 1984   338.1'62   83-17928
    ISBN 0-520-05127-0

Printed in the United States of America
1 2 3 4 5 6 7 8 9

For
Joseph J. and Margaret L. Pisani
Lloyd Bruno
Carson P. Sheetz
Robert L. Middlekauff
W. Turrentine Jackson
Lawrence B. Lee

Contents

Preface

As a boy growing up in California's Central Valley, I had no understanding of, or even interest in, how the state's climate and geography helped mold its institutions. The long, hot, dry summers, the gloriously blue skies, the parched brown carpet of grass that covered the foothills (save for a few weeks in the spring), the endless acres of irrigated farmland, the hydraulic web of dams, canals, and ditches, and countless other symptoms of aridity, were just part of life. Unfortunately, my ignorance of agriculture compounded my ignorance of the physical environment. We learned in school that California was a wealthy state, agriculturally, but we did not learn to recognize the plants that created that wealth. Nor did we learn who planted or harvested those crops, or how, or when. America may have been born on a farm, but it grew up in the city, most of my teachers told me, and when we studied politics, economics, culture, or society, urban institutions took precedence. College did not change my perceptions or attitudes. At the University of California, Berkeley, rural America seemed faintly absurd and always anachronistic. The nation's hinterland, I discovered, was saturated with religious fundamentalism and intolerance, plagued by dreary isolation, oppressive conventionality, and homogenized values, infected with a virulent strain of antiintellectualism. The "sturdy yeoman farmer" competed for historical attention with an amazing assortment of hicks and rubes. Agricultural history was not for me.

An odd set of circumstances, but especially the inspiration and guidance provided by professors W. Turrentine Jackson, James

Shideler, and Donald C. Swain at the University of California, Davis, where I received my Ph.D. degree, kindled my interest in natural resources history and cured me of most of my prejudices toward rural America. Nevertheless, I did not begin work on this book until 1977, when I moved to East Texas from San Diego and began teaching at Texas A&M University. In East Texas, humidity, rather than aridity, governs man and nature. There, many trees and shrubs keep their leaves all year long. In a region where rain falls twelve months a year, and often hardest in the spring and summer, the countryside wears many shades of green. The damp, smothering summer heat and an astounding array of formidable insects remind inhabitants that this is a land of too much water, not too little. The contrasts occurred to me time and again during the many summer months I spent working in California libraries and archives, grateful to escape the enervating Texas climate. My experience was the reverse of most Americans. I had migrated from West to East, from a dry environment to a wet one. But the effect was the same. I began to look at what had been taken for granted in a new light; hence this book.

Everybody talks about the importance of water in the American West, but few professional historians write about it. Stories of drought, depleted underground aquifers, conflicts over water rights, pollution, and massive state and federal subsidies to agribusiness, fill columns in newspapers and popular magazines. But historians have focused largely on the romantic West of the nineteenth century: Indians, trappers and explorers, miners and cattlemen, and the railroads. A few, most notably Paul Wallace Gates, have explained patterns of land use in the West, but there is no comprehensive history of arid land reclamation, or even of western agriculture. This book constitutes the first full study of irrigation agriculture in an arid state.

It is not a comprehensive history of irrigation. Such important topics as the changing technology of dam, canal, and pump construction, the development of plant and soil sciences, and the relationship of irrigation to hydroelectric power generation—to name but a few—receive scant attention. Nor will readers find any extensive discussion of the Hetch-Hetchy, Owens Valley, or Colorado River controversies. Those stories have already been well told; my concern is the broader story of the much larger supply of water used in agriculture.

I had two main purposes in writing this book. First, I wanted to show how irrigation contributed to the evolution of California agriculture from the pastoral and wheat boom era (1850–1890), through the horticultural small-farm phase (1880–1920), to the concentration of farms into ever larger "factories in the fields" in the 1920s and after. By the 1930s, if not sooner, irrigation had become the most important feature of California agriculture, as it was in many other arid states. Second, I wanted to show that irrigation was more than just a way to make money from the land. Nineteenth-century California bore little resemblance to the older agricultural states of New England and the Midwest, and social critics worried deeply about the Golden State's future. Irrigation became a tool of social and economic reform, a tool by which the arid West could be made to conform to the familiar, traditional patterns of land tenure "back home." Ironically, by the 1930s irrigation became the ally, instead of the enemy, of land monopoly and concentration. My title suggests this important change, though readers should be forewarned that this book is not *directly* concerned with the nature of the family farm or agribusiness.

Many topics and themes are explored in this volume. The dominant theme, one of importance to all Americans, is the persistent mismanagement and ineffectiveness of both private enterprise and government in regulating the use of water. The process of allocating this precious resource was seldom guided by either wisdom or equity. The California legislature was slow to enact water laws. When it did act, it failed to display much courage, imagination, or foresight. Nor could the state's multitude of different water users coordinate their needs or reconcile their differences by themselves. The quest for a state water plan constitutes a second theme. The need to provide cheap water, move it great distances, and integrate wasteful local water systems stimulated consideration of such a plan. But the persistent public suspicion of government limited the state's role. Other themes include the hostility toward land monopoly in nineteenth-century California; the importance of land speculation; the pervasive sectionalism that blocked most water legislation; the amazing political longevity of the "mining block" in Sacramento, long after the industry's economic power had ebbed; and the inability of the federal government to stimulate irrigation or produce more rational water laws. The book also discusses the institutions associated with irrigation, ranging from private water companies to irrigation districts.

In short, this book does not fit neatly into any single category. It is as much about ideas as institutions, as much about government as agriculture, as much about land as water. Nor is it exclusively concerned with California. Every attempt has been made to link California's experience to the entire arid West.

In writing this book, I incurred debts too many to be acknowledged fully here. Lawrence B. Lee inspired me to undertake the study, and his careful, thorough scholarship has served as a model for my own. Larry, along with Elmo Richardson and Robert Dunbar, read the manuscript and offered many useful suggestions. Other scholars whose work and counsel I have freely drawn upon include, in no particular order, Norris Hundley, Paul W. Gates, Harry Scheiber, Samuel P. Hays, Gerald Nash, David J. Weber, Robert Kelley, Arthur Maass, W. Turrentine Jackson, James Shideler, and Donald C. Swain. My apologies to those whom I have overlooked.

Much of the research was done at the Bancroft Library, whose intelligent, efficient staff sets a high standard. Gerald Giefer and Susan Munkres at the Water Resources Center Archives on the Berkeley campus opened their valuable collections and provided many useful leads. So did Mrs. Bessie Raymond at the California Department of Water Resources Archives in Sacramento and Richard Crawford at the Natural Resources Branch of the National Archives in Washington, D.C. I also want to thank the librarians, curators, archivists, and other helpful workers at the Library of Congress; University of California, Berkeley, Law Library; Huntington Library; Special Collections repositories at UCLA and the University of California, Davis; California State Archives; California Room of the California State Library; and State of California Law Library. The Interlibrary Loan Office at Texas A&M University's Evans Library tracked down many obscure books, documents, and newspapers. Grants from the A&M College of Liberal Arts and the California Water Resources Center provided financial assistance. Professor J. Herbert Synder, director of the Water Resources Center, deserves special recognition for encouraging the historical study of water in California. He has managed to be an extremely efficient administrator without forgetting the needs and concerns of the scholars he serves. Carole Knapp typed the manuscript with her customary speed and precision. Carol Leyba edited the manuscript with care, precision, and patience. The entire staff of the University of California Press, especially Stanley Holwitz, impressed me with its dedication, courtesy, and efficiency. My parents-in-law, Engel and Shirley Sluiter, extended many, many kindnesses. Without their hospitality, this book could not have been written. Similarly, Al Runte graciously shared his Washington, D.C., apartment during the summer of 1979. I also want to acknowledge the friendship of my colleagues Terry Anderson, Walter Buenger, and especially, Lawrence D. Cress. They did not read the manuscript and know little about this book. But they made writing it a lot easier. Mary Alice knows her contribution. Finally, I hope the dedication conveys some sense of my appreciation to a few of the teachers who have contributed so much to my life.

D. J. P.

Bryan, Texas, 1983

1

Introduction: Nineteenth-Century California, A Fragmented Commonwealth

Nineteenth-century California had a dual personality. It was a "western" state, but western with important differences. Like other "frontier" states, its economy passed through several distinct and fairly abrupt stages. A mining and pastoral era gave way in the late 1860s to a wheat boom which, in turn, gave way to horticulture in the late 1880s and 1890s. Sharp economic fluctuations resulted from speculation in Nevada mining stocks, completion of the first transcontinental railroad, droughts, and real estate booms. Despite the state's varied resources, economic diversification—which helped mitigate these swings—was not achieved until well into the twentieth century. Dependence on mining, and later on wheat cultivation, sucked California into an unstable international economy. The state was far removed from large markets for its crops and suffered from inadequate transportation facilities (at least until the 1880s), sectional rivalries, labor shortages, and ethnic and racial conflicts.

Nevertheless, unlike its sister states and territories, California was highly urbanized, with 27 percent of its residents living in San Francisco in 1870. By 1890, San Francisco was the eighth largest community in the nation and the most populous city west of Chicago. In the 1850s, the city exhibited many of the violent and unstable elements of a frontier society. But by the 1870s and 1880s, problems of law and order took second place to the same difficulties encountered by rapidly growing cities on the eastern seaboard—including municipal corruption, the cost of government, relief for the poor, and the need for new water and sanitation systems. By the 1880s, San Francisco sported a wide variety of financial institutions, and while money was often scarce in California—forcing businessmen to rely on European or Eastern capital—the state depended far less on distant institutions than did other western states.

For all the advantages California enjoyed compared with other western states, its population growth lagged behind its enormous economic potential. California grew by 47 percent in the 1860s and by 54 percent in the 1870s, while Kansas grew by 240 percent in the 1860s and by 173 percent in the 1870s, Minnesota by 155 percent and 77 percent, and Nebraska by 355 percent and 270 percent. In 1870, California's population averaged less than 1 person to the square mile while New England's ratio was 49 to 1, the Middle Atlantic states 69 to 1, and the South Coast states 15 to 1. Even sprawling Texas averaged 2 people per mile. During the 1880s, California's population expanded at about the same rate as during the 1860s, despite a land boom in southern California. Then, in the 1890s, the growth rate dropped to 22.4 percent. Of all the arid West, only Nevada, which actually lost population during the 1890s, attracted fewer new residents.[1]

Though California's relative isolation partially explains these figures, more important were the state's dry and unpredictable climate, the limited supply of land available under the Homestead Act, contests over Mexican land grants (which blocked the sale of rich agricultural land along the coast for more than two decades after statehood), high unemployment, and, in some places, land monopolies. Moreover, California developed a reputation as the home of a rootless society of gamblers, speculators, and businessmen with little allegiance to the traditional values promoted by the family farm. Most Eastern and California journals attributed this phenomenon to the nature of the mining and later real estate booms.

1. For population statistics see the U.S. Bureau of Census, *Abstract of the Twelfth Census of the United States, 1900* (Washington, D.C., 1904), 35, 227, 234, 291; *Twelfth Census of the United States, 1900* (Washington, D.C., 1901), 1: lxxxv, xc, 11; and James E. Vance, Jr., "California and the Search for the Ideal," *Annals of the Association of American Geographers* 62 (June 1972): 202.

The *Kern County Weekly Courier* observed that the Californian "despises all certainties and the wilder and more absurd the scheme or speculation . . . the more readily does he go in. . . . Offer to his view any legitimate business such as stock-raising, farming, manufacturing, or the acquisition of fertile lands that offer the absolute certainty of a fortune in, say at most ten years, and he will hardly deem it worthy of notice, if he does not regard it with absolute contempt."[2]

California agriculture was in its infancy in the 1850s and early 1860s, placing third in the economy behind mining and the livestock industry. In 1846, more than 500 ranches in upper California covered hundreds of thousands of acres in the Los Angeles basin and along the coast from San Francisco to Santa Barbara. The lanky Spanish cattle raised on these open ranges were valued for their hides and tallow, and only incidentally as a source of meat and milk. During the Mexican period, there was little attempt to improve the breeding stock, and the "industry" remained isolated. The only substantial overland cattle drive occurred in 1837 when 700 head were sent north to the Willamette Valley. But the gold rush transformed the industry from a pastoral life-style to a speculative business. Miners bought so much meat that for a time cattle runs from southern California to Sacramento and the mining camps rivaled the great Texas drives. Demand so outstripped supply that the price soared from $4 a head in 1846 to over $500 a head delivered in Sacramento in 1849. During the 1850s, cattle were imported from Mexico, Texas, and the Middle West, increasing both the supply and quality of meat. The state's herd increased from 448,796 in 1852 to 1,116,261 in 1869, and the number of milk cows tripled. In 1861, the nation's leading agricultural newspaper, the *Country Gentleman*, counted fifty-five ranchos in Los Angeles County alone. Abel Stearns's twelve ranches covered 230,815 acres on which grazed

2. *Kern County Weekly Courier*, July 18, 1874. Also see Paul W. Gates, "Public Land Disposal in California," *Agricultural History* 49 (January 1975): 177; J. Ross Browne, "Reclamation and Irrigation," *Transactions of the California State Agricultural Society during the Year 1873, Journals of the California Senate and Assembly* (hereafter *JCSA*), 20th sess. (Sacramento, 1874), Appendix, 3: 391; *Memorial and Report of the California Immigrant Union, to the Legislature of the State of California, JCSA,* 19th sess. (Sacramento, 1872), Appendix, 3: 25; "Wants and Advantages of California," *Overland Monthly* 8 (April 1872): 339.

18,000 cattle and 3,000 horses. The other ranches ranged in size from a modest 4,000 acres to 60,000 acres.[3]

Cattle raising was an attractive business in the 1850s. It required little capital to get started, little labor to tend the stock, and no transportation system. Moreover, stockmen could graze their herds on contested land within Spanish and Mexican grants without making improvements (whose value would be lost if the land was finally awarded to a rival claimant). Land was cheap in the 1850s, and the 15 to 25 acres of grassland needed to feed each head of cattle—about the acreage of an irrigated farm around Fresno in the 1870s and 1880s—was easy to acquire.

Overbreeding, high interest rates, and competition from other states weakened the range cattle industry in the late 1850s, and nature crippled it in the early 1860s. A massive flood in 1861–1862 was followed by a severe drought in 1863–1864. Pastures dried up, and cattle carcasses littered the barren countryside. Between 1860 and 1870, the state's cattle herds declined by nearly 50 percent. In Monterey County, the stronghold of the cattle industry in the Mexican period, the county assessor estimated that the size of the herd had fallen from 70,000 to fewer than 13,000 animals. Some ranchers moved their herds to the coastal valleys or to the Central Valley, and the open range survived into the 1870s. But in 1872 the legislature passed the first "no fence" law, tacitly acknowledging the primacy of farming by making stock owners responsible for crop damages caused by free-grazing animals. Many of southern California's huge estates were subdivided during the late 1860s and 1870s, laying the foundation for that region's citrus industry.

Sheep raising expanded during the 1850s along with the range cattle industry, and because sheep could graze on land unsuited

3. On the cattle and sheep industries, see L. T. Burcham, "Cattle and Range Forage in California: 1770–1880," *Agricultural History* 35 (July 1961): 140–149; Paul W. Gates, *California Ranchos and Farms, 1846–1862* (Madison, 1967), 17–40; R. H. Allen, "The Spanish Land-Grant System as an Influence in the Agricultural Development of California," *Agricultural History* 9 (July 1935): 127–142; James M. Jensen, "Cattle Drives from the Ranchos to the Gold Fields of California," *Arizona and the West*, 2 (Winter 1960): 341–352; Robert Glass Cleland, *The Cattle on a Thousand Hills: Southern California, 1850–1880* (San Marino, Calif., 1951); W. H. Hutchinson, *California: Two Centuries of Man, Land, and Growth in the Golden State* (Palo Alto, Calif., 1969), 179–180; and Hubert Howe Bancroft, *History of California, 1860–1890* (San Francisco, 1890), 7: 52–67. On the ranchos of Southern California see the *Country Gentleman* 18 (September 5, 1861): 161.

for cattle, those herds suffered much less from the drought of 1863–1864. The 80,867 sheep in California in 1852 increased to 1,088,002 by 1860. In that year California was the nation's fifth largest sheep-producing state. By 1870 it led the nation as the herd swelled to 2,750,000. Both the quantity and quality of exported wool increased during the late 1860s and early 1870s; the wool clip was 8,000,000 pounds in 1867 but nearly doubled to 15,000,000 pounds in the following year. However, the increasing value of land and a second major drought in 1876–1877 all but destroyed this industry.[4]

The explosive popularity of wheat culture in the middle and late 1860s contributed to the demise of the livestock industry. As early as 1854, a California booster informed the *Country Gentleman* that bountiful wheat fields stretched for eleven miles beyond the outskirts of Sacramento, yielding an astounding sixty to eighty bushels per acre. Midwesterners must have blinked hard at the figure because their farms seldom produced more than fifteen bushels per acre. Not surprisingly, the writer noted: "A man with a good 'ranch,' in the [Sacramento] valley, can make an independent fortune in a few years, not infrequently in one!"[5] Fifteen years later, at the height of the wheat mania of the 1860s, the San Francisco *Bulletin* reported:

> At the beginning of the last rainy season the excitement in favor of wheat growing had only been exceeded by some of the more memorable mining excitements of former days. There was a rush and a furor for wheat lands. Nearly every novice about to try his hand at agriculture, bought or rented lands for wheat culture. Two or three good harvests had turned the heads of thousands. Lands were rented in numerous instances, and a cash rental paid for a single year's occupation which was equal to the entire value of the lands.[6]

4. Gates, *California Ranchos and Farms*, 35; Frank Adams, "The Historical Background of California Agriculture," *California Agriculture*, ed. C. B. Hutchison, (Berkeley, 1946), 34; *Country Gentleman* 29 (February 7, 1867): 97; ibid., 32 (November 26, 1868): 360.

5. *Country Gentleman* 4 (July 27, 1854): 56.

6. San Francisco *Bulletin*, July 9, 1869. Also see the San Francisco *Alta California*, July 12, 1869; *Country Gentleman* 34 (September 2, 1869).

There were several reasons for the boom. First, during the Civil War the Confederate Navy reduced California's trade with the East Coast. This forced San Francisco merchants to increase trade with England, particularly to acquire heavy machinery. Wheat, in effect, offered a medium of exchange, as well as a fleet of empty ships to carry European goods home. Moreover, wheat culture followed the railroad. The Southern Pacific line from San Francisco reached San Jose in 1864 and Salinas in 1872, opening the Santa Clara and Salinas valleys to grain farming. Finally, beginning with the winter of 1866–1867, a string of wet years contributed to the boom as crop yields ran far above normal, particularly where farmers cultivated virgin soil. From 1866 to 1872, the acreage planted to wheat more than tripled.[7]

California newspapers condemned the speculative nature of wheat farming well into the 1890s. The similarities between mining gold and harvesting grain were not lost on the state's social and economic critics. However, wheat farmers had good reasons besides greed to choose that crop. In the 1860s, California was admirably suited to wheat and barley but not to many other crops. They were two of a handful of plants that could be raised in the Central Valley without irrigation, and wheat was the *only* major crop durable enough to export before refrigerated railroad cars became common in the late 1880s. The valley offered rich, flat land which required little preparation, and the would-be farmer needed almost no experience because wheat was easy to plant and virtually took care of itself. He could get by without a house, barn, or even quarters for hired hands. Farm laborers were seldom needed for more than three or four weeks a year, and the mild California climate permitted them to sleep under the stars. Setting up and operating a farm required little capital, and wheat usually returned a good profit from the first year. The farmer who planted vines or orchards had to wait five or ten years for returns on a much larger initial investment. In addition, his plants required much more care, which increased labor costs. Little wonder that wheat won the reputation as a "poor man's crop," even though the industry was dominated by huge farms after the 1860s.

7. Allen, "The Spanish Land-Grant System," 133–134.

There were other, less tangible reasons for raising wheat. Because the crop required little attention, the wheat farmer did not have to endure a dreary rural existence isolated from human society. Many wheat barons lived in Sacramento, Stockton, San Francisco, or Oakland, and left their comfortable urban surroundings only to supervise the planting or harvest. Others simply entrusted this job to lieutenants and rarely visited their farms. During the 1850s and 1860s, many wheat farmers along the coast could not secure clear land titles; the average claim to a Mexican land grant filed with the federal land commission in San Francisco took seventeen years to confirm. Since wheat did not require the farmer to improve his land, the farmer, like the cattle rancher, had little to lose if the commission ruled against him. Then, too, some farmers had no intention of making California a permanent home. Often they came to the state looking for a fortune in gold, only to have their dreams of quick riches crushed by the hard life of the mining camps. They could hope that a few bumper wheat crops would give them wealth and respectability—and allow them to return home in style.[8]

Wheat farmers adapted well to conditions in California, at least during the 1860s and 1870s. Instead of planting crops in the spring, they usually left the land fallow through the summer, waiting for the rainy season in November. The first heavy rains softened up the sun-baked soil to permit plowing and seeding. Crops germinated well in the mild California winters, and the critical phase in the growth cycle did not arrive until mid-February. If several inches of rain fell after that time, the plants matured well. If not,

8. There is no comprehensive history of either agriculture or wheat farming in California. However, in addition to the Adams, Cleland, and Gates works cited above, see Howard S. Reed, "Major Trends in California Agriculture," *Agricultural History* 20 (October 1946): 252–255; E. J. Wickson, *Rural California* (New York, 1923); Gilbert C. Fite, *The Farmers' Frontier, 1865–1900* (New York, 1966); Osgood Hardy, "Agricultural Changes in California, 1860–1900," American Historical Association, Pacific Coast Branch, *Proceedings, 1929* (Eugene, Oreg., 1930), 216–230; Gerald D. Nash, "Stages of California's Economic Growth, 1870-1970," *California Historical Quarterly* 51 (Winter 1972): 315–330; and Rodman Paul, "The Beginnings of Agriculture in California: Innovation vs. Continuity," *California Historical Quarterly* 52 (Spring 1973): 16–27; idem, "The Great California Grain War: The Grangers Challenge the Wheat King," *Pacific Historical Review* 27 (November 1958): 331–349; idem, "The Wheat Trade between California and the United Kingdom," *Mississippi Valley Historical Review* 45 (December 1958): 391–412.

they were stunted. Farmers had to worry about the evenness of rainfall from November through April, not just the total amount, though in the San Joaquin Valley fifteen inches of rain usually produced bumper crops while ten inches yielded only a mediocre harvest. The crop matured very rapidly as temperatures warmed in late March and April. At this time, wind was more of a concern than rain. Every few years, fierce north winds of forty to fifty miles an hour swept down the Central Valley ushering in spring. Particularly in the San Joaquin Valley, where the soil was sandy, crops literally blew away. But barring natural disasters, once wheat matured it could be left in the fields for months as farmers negotiated transportation costs with shippers, holding back their harvest in the hope of getting the best possible price. They had little fear of rust or scale, and even after the harvest, bulging sacks could be left in the field for weeks without danger from mildew.

Many students of the California wheat industry noted another major difference between farming on the Great Plains and in California. Almost from the beginning, wheat farming in California was highly mechanized. In 1860, on his 39,000-acre estate in the Sacramento Valley, John Bidwell used 20 plows, 4 harrows, 2 mowers and reapers, and 2 threshers valued at $7,500. And in 1862, farm machinery dealers in Sacramento and Stockton alone sold 500 mowers, 220 reapers, and 100 threshers.[9] Technology quickly met the challenges of scale in California. Hyde's steam plow, invented in 1871, plowed a strip sixteen feet wide, sowed, and harrowed all in one operation. By the late 1880s, the steam plow had been refined to the point that it could turn 160 acres in twenty-four hours.[10] Harvesting was accomplished by twelve-foot headers drawn by six horses or mules; each machine cut 15 to 25 acres a day. By the 1890s, wheat farmers employed gigantic combines drawn by two or three dozen mules and tended by five to ten men. These reaped, threshed, and sacked grain all in one operation. In the Mexican period, twelve men were needed to raise an acre of wheat, but by

9. Gates, *California Ranchos and Farms,* 44–45.

10. Joseph A. McGowan, *History of the Sacramento Valley* (New York, 1961), 1: 252. On the technology of wheat culture see Effie E. Marten, "The Development of Wheat Culture in the San Joaquin Valley, 1846–1900" (M.A. thesis, University of California, Berkeley, 1924), 45–62. Marten describes planting and harvesting techniques on pp. 30–44.

the early 1890s, mechanization allowed one worker to cultivate 130 acres.[11] Cut off from eastern markets by the high cost of transporting wheat by rail, California farmers then, as now, used technology to improve their competitive position overseas. One of California's leading historians has observed: "California agriculture thus provided a model for commercialized farming throughout the United States."[12]

Crop production statistics measure the effects of mechanization. Though harvests varied dramatically from year to year, demand for California wheat remained high into the 1890s. In 1852, the state's farmers produced only 271,762 bushels, but by 1860 production reached 7,500,000 bushels, and California ranked ninth in the nation.[13] By 1867, the state agricultural society's report indicated that 882,888 acres were planted to wheat and yielded 14,432,883 bushels.[14] Already over 98 pecent of the export crop went to Great Britain.[15] Four years later, 2,128,165 acres bore 28,784,571 bushels.[16] The richest harvests came in the early and middle 1880s when California led the nation. For example, in 1884, 3,587,864 acres provided 57,420,188 bushels.[17] Though the census of 1890 showed California as the second largest wheat-producing state, by 1900 it had fallen to sixth place. In 1916, the state produced

11. *Fifth Biennial Report of the Bureau of Labor Statistics of the State of California for the Years 1891–1892, JCSA,* 30th sess. (Sacramento, 1893), Appendix, 2: 16.

12. Nash, "Stages of California's Economic Growth," 318–319.

13. Gates, *California Ranchos and Farms,* 52. Agricultural statistics for this period are unreliable. For example, Gerald Nash, in *State Government and Economic Development: A History of Administrative Policies in California, 1849–1933* (Berkeley, 1964), 63, estimates the 1860 wheat crop at only 6,000,000 bushels. On December 28, 1858, the *Sacramento Daily Union* noted that for the first time in the state's history, all the county assessors had complied with the law and submitted agricultural statistics to the state surveyor-general's office in Sacramento. But during the nineteenth century, such years were rare. Moreover, even the statistics submitted are open to serious questions.

14. *Transactions of the California State Agricultural Society for the Year 1868, JCSA,* 18th sess. (Sacramento, 1870), Appendix, 3: 132–133.

15. *Country Gentleman* 34 (November 4, 1869), 352–353.

16. *Biennial Report of the Surveyor-General of the State of California from August 1, 1873, to August 1, 1875, JCSA,* 21st sess. (Sacramento, 1875), Appendix, 2: 26–27.

17. *Transactions of the California State Agricultural Society During the Year 1884, JCSA,* 26th sess. (Sacramento, 1885), Appendix, 2: 206–207.

only 4,000,000 bushels; the wheat industry had long since lost its preeminent place in California agriculture.[18]

The wheat industry's sharp decline resulted from Canadian and Russian competition, the introduction of new varieties of grain superior to California strains, declining yields due to soil exhaustion, and the increasing value of farmland devoted to horticulture. But during its heyday, the industry built a painful legacy. In 1877, the *Sacramento Daily Union* bemoaned the fate of California agriculture:

> We are all but too familiar with the picture: A level plain, stretching out to the horizon all around; for a few months a wavering sea of grain, then unsightly stubble; in the center a wretched shieling [hut] of clapboards, weather-stained, parched, and gaping; no trees, no orchard, no garden, no signs of home[,] . . . on everything alike the tokens of shiftlessness and barbarism. Such farmers buy their vegetables, their butter, their bacon, all they need, at the nearest town or settlement. They never think of raising anything beyond the one staple, wheat. . . . Failing . . . a reformatory movement, we see nothing in prospect but a shiftless drifting backward further and further into barbarism, until, the fertility of the soil being exhausted, the reckless and half-civilized tillers of it shall be compelled to migrate, and shall, like other nomads, seek new camping-grounds in regions not yet destroyed for all purposes of production by methods similar to their own.[19]

Critics of wheat farming repeatedly raised a fundamental question: Could wheat provide a stable agricultural foundation for the state's future economic growth? Beyond the damage to the soil,

18. Mansel G. Blackford, *The Politics of Business in California, 1890–1920* (Columbus, Ohio, 1977). Particularly see Blackford's introductory chapter, pp. 3–12.

19. *Sacramento Daily Union*, May 24, 1877. For similar statements see the *Union* of January 16, 1877; Charles Nordhoff, *California: For Health, Pleasure, and Residence* (New York, 1875), 131, 228; "Overworked Soils," *Overland Monthly* 1 (October 1868): 327–331; "Wants and Advantages of California," ibid., 8 (April 1872): 338–347; *Country Gentleman* 50 (December 10, 1885): 992–993; Stephen Powers, *Afoot and Alone: A Walk from Sea to Sea* (Hartford, Conn., 1872), 308.

beyond the rootlessness of wheat farmers, beyond their get-rich-quick mentality, beyond even the dreary sameness of endless acres of grain, the most distressing aspect of wheat farming was the way it stifled the family farm. Small farms were anathema to the wheat baron. Though they could not compete on equal terms in the international wheat market, by diversifying crops they threatened to reduce profits by driving up property values and taxes. Most grain farmers, especially in the Sacramento Valley, opposed small farming well into the 1890s. In turn, critics of wheat farming also charged that the bonanza farms retarded the development of rural communities and degraded the status of labor. The vast farms were all but deserted except in the late fall and spring. Since field hands were needed only during planting and harvesting, they became a transient work force, establishing a pattern that has persisted to this day. Moreover, the unattractiveness of work on these giant estates contributed to the extensive use of Chinese and other minorities. In 1869, one observer noted: "Labor is scarce and high priced; and were it not for the Chinamen in the state, one-half of our luxuriant harvest would annually rot in the fields for want of hands to gather it."[20] California's agricultural wealth seemed to depend on a permanent class of dispossessed who, in turn, undermined the opportunity of white laborers by working for "starvation wages." As the 1860s gave way to the 1870s and 1880s, the "poor man's crop" helped create a closed agricultural system.

The greatest danger posed by wheat farming was its tendency to perpetuate, if not create, land monopoly. Most nineteenth-century Americans believed that the nature of society derived from its agricultural base. Henry Nash Smith has argued that Americans revered the yeoman farmer as a reminder of a simpler, happier, more virtuous society. Though the "agrarian ideal" was threatened by urbanization and industrialization, the American West offered an opportunity to strengthen and restore this tarnished dream. The ideal rested on certain assumptions. Agriculture, not commerce or industry, was the only source of legitimate wealth, and the keystone of an egalitarian society. Every man had a "natural right" to

20. Bentham Fabian, *The Agricultural Lands of California* (San Francisco, 1869), 6. In *California Ranchos and Farms,* p. 62, Paul Gates notes that in 1860 the ratio of farm owners to laborers was only two to one in California as opposed to three to one in Ohio and Illinois and four to one in Indiana.

own land, and land ownership promoted independence, pride, and dignity. The more freeholders the better; a large "middle class" reduced the number of rich and poor, eliminating the basic source of class conflict. Hence, title to land should derive from use, not abstract property rights. Life close to the soil produced a moral society; luxury and artifice were characteristics of the city. Thus, the Homestead Act took on symbolic significance as an attempt to insure that the West would belong to the yeoman farmer.[21]

Many historians have discussed the effect of federal and state land laws on speculation and monopoly in California. Nearly 9,000,000 acres of Spanish and Mexican land grants were confirmed by a special federal commission (1852–1856) and by the courts, though contested claims took an average of seventeen years to adjudicate. The average grant contained nearly 15,000 acres. Nineteen, containing 728,139 acres, were still intact as late as 1950. The federal government deeded the state 2,193,965 acres of unusually fertile and productive "swamp and overflow" land. Most sold at auction for $1.00 to $1.25 an acre. Similarly, 6,700,000 acres of school lands quickly sold at $1.25 an acre, along with a half million acres granted for "internal improvements" and "public buildings." Land companies and individual speculators acquired millions of additional acres using agricultural college and soldier scrip, the Desert Land Act (1877), and the Timber and Stone Act (1878). Finally, in the 1860s, 1870s, and 1880s, the railroad acquired 11,500,000 acres in federal grants, about 16 percent of California's public domain. The alternate sections remaining under federal control were usually sold at auction. As I will argue later, monopoly and speculation might well have stimulated rather than retarded economic development. Nevertheless, the state's disposal of its prodigal gifts was short-sighted and involved fraud and corruption on a grand scale. The Pre-Emption Act (1841) and the Homestead Act (1862), which played a large part in opening the agricultural lands of the Midwest to small farming, had little effect on the disposal of California's government lands. For example, from 1863–1869, 2,848 claims were

---

21. Henry Nash Smith, *Virgin Land: The American West as Symbol and Myth* (Cambridge, Mass., 1950), 138–150. Also see Arthur A. Ekirch, *Man and Nature in America* (New York, 1963), 10–21.

filed under the Homestead Act, comprising only 414,861 acres. Minnesota counted more homestead entries in 1867 than California did during the entire decade of the 1860s.[22]

Land monopoly, like wheat culture, became a symbol of the failure to reproduce familiar, comfortable institutions and life-styles. In the nineteenth century, many Californians viewed the future with alarm. In 1871, Henry George warned that California might suffer the same fate as ancient Rome:

> In the land policy of Rome may be traced the secret of her rise, the cause of her fall. . . . The [Roman] Senate granted away the public domain in large tracts, just as our Senate is doing now; and the fusion of the little farms into large estates by purchase, by force and by fraud went on, until whole provinces were owned by two or three proprietors, and chained slaves had taken the place of the sturdy peasantry of Italy. The small farmers who had given her strength to Rome were driven to the cities, to swell the ranks of the proletarians, and become clients of the great families, or abroad to perish in the wars. There came to be but two classes—the enormously rich and their dependents as slaves; society thus constituted bred its destroying monsters; the old virtues vanished, population declined, art sank, the old conquering race actually

22. Gates, "Public Land Disposal in California," 160, 168. Gates's brilliant and exhaustive studies of land in California have set a high standard. See idem, "Adjudication of Spanish-Mexican Land Claims in California," *Huntington Library Quarterly* 21 (May 1958): 213–236; idem, "California's Agricultural College Lands," *Pacific Historical Review* 30 (May 1961): 103–122; idem, "California's Embattled Settlers," *California Historical Quarterly* 41 (June 1962): 99–130; idem, "The California Land Act of 1851," *California Historical Quarterly* 50 (December 1971): 395–430; idem, "Pre-Henry George Land Warfare in California," *California Historical Quarterly* 46 (June 1967): 121–148; idem, "The Suscol Principle, Preemption, and California Latifundia," *Pacific Historical Review* 39 (November 1970): 453–472. Gerald Nash has also made a significant contribution. In addition to his *State Government and Economic Development*, see "The California State Land Office, 1858–1898," *Huntington Library Quarterly* 27 (August 1964): 347–356; idem, "Henry George Reexamined: William S. Chapman's Views on Land Speculation in Nineteenth Century California," *Agricultural History* 33 (July 1959): 133–137; idem, "Problems and Projects in the History of Nineteenth-Century California Land Policy," *Arizona and the West* 2 (Winter 1960): 327–340. Gates's and Nash's fine work have largely supplanted the only book-length history of land in California, W. W. Robinson's *Land in California* (Berkeley, 1948).

died out, and Rome perished. . . . Centuries ago this hap-
pened, but the laws of the universe are to-day what they were
then.[23]

Other critics found examples closer to home. The *Sacramento Daily
Union* noted that without its tradition of widespread land ownership, "the United States would, in all probability, to-day contain
less wealth and population, and not much more general intelligence
then Mexico, where the land is all monopolized by the rich. Nothing has more retarded the prosperity of the South and Central
American republics."[24] And a special committee of the California
Legislature, one of several formed in the 1870s to study land problems in California, argued that monopoly created a vast class of
renters similar to conditions of land tenure in Europe. "This chief
curse of civilization, upon which all minor monopolies are founded,
is fast attaining such vast proportions in California, that it promises
to soon become so powerful as to defy opposition, just as it defies
all attempts to curb it in the Old World."[25]

Critics of California and national land policies often resorted
to hyperbole. If San Francisco seemed overcrowded and steeped
in corruption, the reason was land monopoly. If the rich and poor
seemed particularly conspicuous in California, the cause was land
monopoly. If the economy was sluggish, if unemployment was
high, if the Workingmen threatened anarchy and revolution, and
if tramps and vagabonds infested the countryside, monopolists
provided a convenient scapegoat.

Nevertheless, the phenomenon was real, even if yellow jour-
nalists, nostalgic reformers, and those who had failed in the race
for wealth often saw vast conspiracies where none existed. In its
report for 1873 and 1874, the State Board of Equalization noted that
there were 122 farms or ranches in California larger than 20,000
acres; 158 ranging from 10,000 to 20,000 acres; 236 from 5,000 to

23. Henry George, *Our Land and Land Policy* (San Francisco, 1871), 47. Also
see Charles A. Barker, "Henry George and the California Background of Progress
and Poverty," *California Historical Quarterly* 24 (June 1945): 97–115.

24. *Sacramento Daily Union*, December 16, 1874. Also see the *Union* of November 4, 1873.

25. *Report of the Special Committee on Land Monopoly*, JCSA, 20th sess. (Sacramento, 1874), Appendix, 5:195.

10,000 acres; 104 from 4,000 to 5,000 acres; 189 from 3,000 to 4,000 acres; 363 from 2,000 to 3,000 acres; and another 1,126 containing 1,000 to 2,000 acres. In short, 2,298 individuals or companies owned parcels of 1,000 acres or more. In 1872, 28,000 farms contained 100 acres or more, but only 9,500 included less than that amount.[26] In its report for 1870 and 1871, the State Board of Agriculture used tax assessment figures to reveal that in eleven California counties, 100 individuals owned a staggering 5,465,206 acres—an average of 54,652 acres per person. In San Joaquin County, where land was still used largely for grazing cattle in the early 1870s, the thirteen largest landowners held title to 3,100,035 acres, an average of 238,464 acres per owner. No other county came close, but in Fresno County, seven individuals owned an average 40,088 acres, and in Kern County, nine owned an average of 33,949 acres. Of these eleven counties, the smallest holdings were in Tehama County, where five wheat farmers owned a "modest" average 9,742 acres each.[27] The largest individual landowner in the early 1870s was William S. Chapman of San Francisco, a land scrip speculator who held 350,000 acres. However, the San Joaquin County ranch of Henry Miller and Charles Lux contained 450,000 acres (surrounded by 160 miles of fence), and two former state surveyors-general owned estates nearly as large.[28]

Nineteenth-century Californians often charged that land monopolists had retarded the state's agricultural development. However, without a dependable water supply only a small amount of California land could approach Midwest land in value. Land prices varied enormously from one section of California to another, based not only on agricultural potential but also on what use the land was put to. Irrigation agriculture was limited in the 1870s; as yet the state lacked a dependable transportation system and markets for irrigated crops. Hence, in 1870, particularly in the Central Valley, land that would one day command premium prices went begging at $2.50 an acre. Pasture land could be acquired in huge

26. *Report of the State Board of Equalization of the State of California for the Years 1872 and 1873,* JCSA, 20th sess. (Sacramento, 1874), Appendix, 2: 22–23.

27. *Biennial Report of the State Board of Agriculture for the Years 1870 and 1871,* JCSA, 19th sess. (Sacramento, 1872), Appendix, 3: 15–16.

28. George, *Our Land and Land Policy,* 71–72.

quantities, but returned a low profit per acre. A 50-foot mining claim, or a 200-foot town lot in Oakland, might well provide more income than a 2,000-acre rancho in the San Joaquin Valley. Similarly, vast estates by Midwest standards often returned less profit than a family farm in Kansas or Iowa. The attitude of Californians toward land monopoly derived from values "alien" to the state. Critics of monopoly pleaded for limits on the size of estates and equal taxation for cultivated and uncropped land; but given the different uses of rural land in California—mining, grazing, agriculture and lumbering—how could an equitable tax policy be constructed? Amid the public outcry over monopoly in the early 1870s, the Board of Equalization wisely noted: "When the vast territorial [area] of California is considered, with the fact that by far the greater part of the large tracts held in private ownership are unfit for any other agricultural purpose than that of grazing cattle and sheep, and are wholly incapable of adaptation to the plow, it would appear that the disadvantages which this state labors under from large holdings of valuable lands are not so great as we have generally supposed."[29] During the Mexican period, most land was worth little more than the value of its native grasses, and large holdings were consistent with the pastoral economy. In the 1860s, the wheat industry often competed with stockmen for land, but it did not destroy the assumption that agriculture as practiced in the Midwest could not flourish in California. After all, except on a very limited amount of land, wheat farming represented a year-to-year gamble with nature, and fortunes were lost as well as made.

As noted above, wheat farmers usually opposed diversified agriculture, but they never used more than about 4,000,000 acres of land. The largest corporate landowner was the Central Pacific-Southern Pacific Railroad. By 1882 it had acquired 9,500,000 acres, of which it had already sold nearly 1,000,000 acres.[30] Federal surveys of tracts granted to these companies lagged far behind railroad construction, so that much of the land could not be sold during the 1870s. During the 1880s, middle-management railroad officials

29. *Report of the State Board of Equalization of the State of California for the Years 1872 and 1873, JCSA,* 20th sess. (Sacramento, 1874), Appendix, 2: 24.

30. *Annual Report of the Central Pacific Railroad Company to the Stockholders for the Year ending December 31, 1881* (San Francisco, 1882), 51–52.

promoted diversified agriculture and densely settled rural communities, a policy consistent with the railroad's economic self-interest. These policies helped foster a positive corporate image at a time when the railroad faced mounting criticism over rates and its political influence. The railroad's land agents from 1865 to 1907, B. B. Redding and W. H. Mills, actively encouraged agricultural colonization in their speeches, writings, and support for private booster groups such as the California Immigrant Union, formed in 1869. The railroad sold its land on liberal terms and provided special "excursion" rates and immigrant trains. It also encouraged the formation of local granges and farm cooperatives; gathered and disseminated information on soils, plants, and precipitation; provided free transportation for plants and other materials used in agricultural experiments conducted by the state university; subsidized publicists of the Golden State such as Charles Nordhoff; touted the state through the columns of railroad-owned or -controlled newspapers; and, in 1898, began publishing *Sunset* magazine to glorify life in California. Richard Orsi notes that the railroad became the state's chief proponent of planned, orderly agricultural settlement: "Increasingly, the Southern Pacific's own land development agencies and its subsidiary land corporations rejected haphazard land disposal in favor of founding organized agricultural settlements as stimulants to land sales and freight and passenger traffic."[31]

Nor did the railroad act alone. During the 1880s, many large land companies and individual speculators promoted colonization as an alternative to unplanned, uncoordinated settlement. As early as 1868, William S. Chapman, probably the largest land speculator in the late 1860s and 1870s, challenged the assumption that speculation retarded agricultural development. He pointed out that the federal government had sold land in small parcels since 1859 but had attracted few buyers. Only the speculator could acquire the vast tracts of land needed by wheat farmers; the "deceit" and "fraud"

31. Richard J. Orsi, "*The Octopus* Reconsidered: The Southern Pacific and Agricultural Modernization in California, 1865–1915," *California Historical Quarterly* 54 (Fall 1975): 208. Also see Edna Monck Parker, "The Southern Pacific Railroad and Settlement in Southern California," *Pacific Historical Review* 6 (June 1937): 103–119.

used to acquire land was justified by unrealistic federal land laws. Chapman explained:

> I showed my faith by my works; I vested all the money I had in the puchase of these lands, and all I could borrow. I induced moneyed men to join me. What I bought I sold again at a small advance to actual settlers, whom I induced to farm the land according to my notions. Men who bought of me at $2.50 an acre, payable in one year (with privilege of another year's time, if the crop should fail) have this year harvested a crop which will very nearly ten times over pay back their purchase money. I have entered some hundreds of thousands of [acres of] this land. I have sold it as fast as I could at reasonable prices to actual settlers, who have been induced by me to settle on it; others, seeing what I was doing, and having thus their attention directed to these lands, have pursued a similar course, the public mind has become excited on the subject, and settlement and cultivation have progressed in the San Joaquin Valley at a ten-fold greater rate than if there had been no 'speculation' in the matter.[32]

Chapman argued that soil exhaustion and the increasing value of land would inevitably kill off wheat farming, but that it represented a necessary first step toward agricultural diversification. In 1868, he sold 80,000 acres to a colony of German farmers for $1.80 an acre. In 1871, he helped establish the Fresno Canal and Irrigation Company, and shortly thereafter joined Isaac Friedlander, Charles Lux, William C. Ralston, and other capitalists to form the San Joaquin and Kings River Canal and Irrigation Company. In 1875, he aided the first major colony in the San Joaquin Valley, the Central Colony, by donating 192 twenty-acre tracts, selling land on credit to small farmers, and sending an agent to Spain to select muscatel cuttings. He was also one of the first to promote alfalfa cultivation, and this crop subsequently became the foundation of dairy farming in California. Chapman lost most of his land when the San Joaquin and Kings River Irrigation Company failed in 1875, but he contin-

---

32. William Chapman's letter, originally published in the San Francisco *Bulletin* of August 27, 1868, is reprinted in Nash's "Henry George Reexamined."

ued to believe that federal land laws unsuited to the arid West posed the greatest single obstacle to the expansion of agriculture.[33]

Despite the near-obsession of nineteenth-century California with land monopoly, there were other important reasons for the sluggish pace of agricultural development. The nature of California politics helps explain the state's inability to do more to encourage economic growth. Not only did most Californians mistrust politicians (and all proposals for state-sponsored public works such as irrigation canals) but sectional rivalries made the adoption of any unified water plan impossible.

Fragmentation characterized California politics in the nineteenth century. As in most mining districts in the American West, during the 1850s and 1860s communities rose and fell so rapidly that apportioning political power in the legislature was guesswork at best. For example, the *Kern County Weekly Courier* of April 13, 1872, reported that Calaveras County, with a population of 8,895, claimed two senate and three assembly seats in the legislature, while Fresno, Tulare, and Kern counties, with a population of 13,782, were served by only one senator and two assemblymen. Moreover, political party affiliations were very weak in California. Geographical alignments—such as northern versus southern California, and interior counties versus San Francisco—put community interests above party. Democrats in Sacramento often shared more

33. Nash, "Henry George Reexamined," 134–135. For other promonopoly statements, see J. Ross Browne, "Agricultural Capacity of California: Overflows and Droughts," *Overland Monthly* 20 (April 1873): 313; idem, "Reclamation and Irrigation," in *Transactions of the California State Agricultural Society during the Year 1872, JCSA,* 20th sess. (Sacramento, 1874), Appendix, 3: 390–426; the statement of the former deputy registrar of the U.S. Land Office in San Francisco, W. W. Johnston, in *Report of the Committee on Land Monopoly, JCSA,* 20th sess. (Sacramento, 1874), Appendix, 5: 278; and the address of the President of the California State Agricultural Society in *Transactions of the California State Agricultural Society During the Year 1882, JCSA,* 25th sess. (Sacramento, 1883), Appendix, 4: 18.

There are two distinct interpretations of the land speculator's role in western economic development. For the classic "robber baron" thesis, see Paul W. Gates, "The Role of the Land Speculator in Western Development," *Pennsylvania Magazine of History and Biography* 64 (July 1942): 314–333; and Ray Billington, "The Origin of the Land Speculator as a Frontier Type," *Agricultural History* 19 (April 1945): 204–212. For a view closer to William Chapman's, see Thomas LeDuc, "State Disposal of Agricultural Scrip," *Agricultural History* 28 (July 1954): 99–107; and Allan and Margaret Bogue, "'Profits' and the Frontier Land Speculator," *Journal of Economic History* 17 (March 1957): 1–24.

common objectives with Republicans in their own city than they did with Democrats in San Francisco. Within each community, economic interest groups also transcended party affiliations. For example, wheat growers opposed changes in land tax policies no matter which party championed reform. The result was political fragmentation, with frequent splits in the two major parties and the formation of ephemeral third parties such as the Workingmen's party in the late 1870s and early 1880s, and the Populist party in the 1890s.[34]

In the legislature, there was little continuity in membership from one session to the next. Most lawmakers had limited political experience, a condition many voters considered an asset. In the pre-media, small-district age, political machines put forward affable, reasonably intelligent, reasonably honest men who could be controlled; in urban districts the machines also decided who could (and would) vote. Sometimes, especially in rural districts, state legislators were elected to champion specific legislation. But most arrived in Sacramento without a legislative agenda and with little political vision. They must have returned "to the people" relieved to resume their permanent occupations. The high turnover, particularly in the assembly, did not reflect a vote-the-rascals-out philosophy so much as a crude form of rotation in office. Especially after 1879, when the new constitution limited the legislature to paid sessions of sixty days, this turnover insured that the state's politics could be orchestrated by a handful of short-sighted, sometimes venal men, most of whom worked behind the scenes. For in the absence of a permanent legislative staff, continuity was provided not simply by those few politicians who had served in the legis-

34. There is no comprehensive history of California politics in the nineteenth century. However, important studies include Bancroft's *History of California*, vol. 7; Theodore H. Hittell, *History of California*, vol. 4 (San Francisco, 1898); Earl Pomeroy, "California, 1846–1860: Politics of a Representative Frontier State," *California Historical Quarterly* 32 (December 1953): 391–402; William H. Ellison, *A Self-Governing Dominion: California, 1849–1860* (Berkeley, 1950); Joseph Ellison, *California and the Nation, 1850–1869: A Study of the Relations of a Frontier Community with the Federal Government* (Berkeley, 1927); Carl B. Swisher, *Motivation and Political Technique in the California Constitutional Convention, 1878–1879* (Claremont, Calif., 1930); Alexander Callow, Jr., "San Francisco's Blind Boss," *Pacific Historical Review* 25 (August 1956): 261–280; R. Hal Williams, *The Democratic Party and California Politics, 1880–1896* (Stanford, Calif., 1973); and Donald E. Walters, "Populism in California, 1889–1900" (Ph.D. diss., University of California, Berkeley, 1952).

lature before but by political bosses and well-organized interest groups such as the railroad. This form of "continuity" further undermined the reputation of politics as a profession, insuring that "the best men" rarely sought office or public service.[35]

By the 1880s, complaints against political corruption in the legislature rivaled criticism of the railroad and land monopolies. For example, in April, 1886, San Francisco's *Bulletin* charged that there were actually two Republican parties in San Francisco. "The one is an appanage of the Street Department—the other, in a lesser sense, of the License Collector's Office. . . . The conventions appointed the county committees and the county committees the conventions. A small group of Bosses divided between them the spoils of city and State."[36] Later in the year, the *Bulletin* described the San Francisco delegation as "the offscourings of public offices." "All the great local bureaus had their representatives. The Federal offices also had their agents. The Democrats went to the City Hall for their Senators and Assemblymen, the Republicans to the Custom House or the Mint."[37] Stories of legislative sinecures make amusing reading today, but they were not taken lightly at the time. In 1889, Democratic Boss Christopher Buckley dominated the legislature. The Democrats created a multitude of petty jobs, including a cuspidor inspector and clerks to aid the "Deputy-Assistant Sergeants-at-Arms." The *San Francisco Chronicle* bitterly noted that although the clerks had not been needed to begin with, they were needed even less after the appointment of four "gate-keepers." Moreover, at the beginning of the session, an "enrolling clerk" and an assistant were appointed, even though the legislature already had a permanent staff of clerks to keep the journals. When the journal clerks fell behind in their work, and the assistant enrolling clerk was asked to help, the latter had to refuse—he could not write! His offer to sacrifice *half* his salary so that another clerk could

35. Legislative rosters indicate that only two assemblymen who served in the 1881 legislature returned to Sacramento two years later. And only one assemblyman and five senators sent to the 1887 legislature had served at the previous session. These are not isolated examples. Turnover seems to have been great throughout the nineteenth century.

36. *Bulletin,* April 6, 1886.

37. Ibid., September 4, 1886. Also see the issue of September 18, 1886.

be hired did not impress the *Chronicle*. Though the California Constitution specified the number and responsibilities of legislative employees, Buckley used the contingency fund to reward his friends.[38]

Sinecures were a relatively mild form of corruption compared with the "grand larceny" of special interest legislation. In 1862, the *Sacramento Daily Union* commented:

> The great legislative evil of the State is the introduction and passage of local and special bills. At least three-fourths of the time of the legislature is consumed each session in the consideration of local bills, which refer either to individuals or counties. If the expenditures of the legislature reach $275,000 annually, about $200,000 may be charged to the account of special legislation in which the state has no particular interest.

During the 1862 legislative session, the Senate adopted a resolution asking the Judiciary Committee to investigate "the best method of reducing to the smallest possible limit . . . the number of local and private bills." The resolution suggested that private bills might be submitted to a special legislative committee to decide whether the purpose of the bill could be achieved by a general law or by amending a general law. It also proposed that no bill be introduced unless first printed and distributed to each member of the legislature, and that no bill be brought up for debate before it had been published at least twice in a newspaper that served the region affected. The resolution noted that because of the preoccupation with "pet" bills, legislation of statewide interest rarely received adequate attention. The most important bills were frequently buried in a pile of legislation left over after the legislature adjourned.[39]

Nothing came of the resolution, and the controversy continued. One of the most brazen pieces of special interest legislation came before the legislature in 1870 and 1872. A group of San Fran-

---

38. *San Francisco Chronicle*, January 11, 12, 30, February 10, 1889. The *Visalia Weekly Delta* of March 21, 1889, described Buckley as "the boodle chief," claiming he had "caused greater loss to the people of this state than all the highwaymen and burglars who have infested it from the earliest times to date."

39. *Sacramento Daily Union*, March 11, 1862.

cisco businessmen, headed by the flamboyant engineer Alexis Von Schmidt, wanted to tap Lake Tahoe as a water supply for San Francisco and other northern California cities. They intended to build a huge aqueduct 163 miles across California at a cost of $6,000,000 to $12,000,000. In 1870, even before construction plans had been published, the group bypassed the San Francisco Board of Supervisors and carried the scheme directly to the legislature. There it joined a half dozen other private bills, all involving private franchises or construction jobs for San Francisco, which attracted heavy fire from the city's newspapers. The *San Francisco Chronicle* commented that these bills would bankrupt the city and asked: ". . . might it not be a good plan to confiscate the city altogether—sell her off at a tax sale—give Sacramento and Oakland their just proportion of the proceeds, and hand the balance to the Tahoe Water Company? This would relieve many persons of anxiety upon the question, how to pay their taxes." Though many critics doubted the scheme's engineering feasibility and others feared it would cost much more than estimated, the bill actually passed the assembly in 1872 by the lopsided vote of 49 to 27. Interior counties lined up behind the plan hoping it would provide Sacramento and hydraulic mines in the foothills with cheap water. Though the San Francisco delegation managed to block the bill in the Senate, the Tahoe project came perilously close to success.[40] At the adjournment of the 1872 session, the *Stockton Daily Independent* commented: "When the Legislature convenes it is usually pronounced a superior body of men, and when it adjourns it is most generally denounced as excelling all of its predecessors in incompetency and corruption. It is to be presumed that the Legislature just adjourned will not be an exception."[41]

Despite criticism from the press, special interest legislation frequently found its way into law. The constitutions of 1849 and

40. See Donald J. Pisani, "'Why Shouldn't California Have the Grandest Aqueduct in the World?': Alexis Von Schmidt's Lake Tahoe Scheme," *California Historical Quarterly* 53 (Winter 1974): 347–360. The *Chronicle* quote is reprinted on p. 351.

41. *Stockton Daily Independent*, April 2, 1872. Newspapers expressed similar judgments at the close of virtually every legislative session during the 1870s and 1880s. As typical examples see the *Fresno Expositor*, April 3, 1872; *Bakersfield Californian*, March 28, 1878.

1879 required that bills cover only one subject, and that the subject be clearly identified in the title. But this requirement could be dodged if the author simply described his bill as an amendment to a particular section of the civil code. Moreover, before the second constitution took effect, there was no legal requirement that bills be printed or read three times prior to a final vote. Three readings were customary, but one or more were frequently omitted in the rush of the closing days of a session. Sometimes only the title of a bill was read; the lawmakers trusted the sponsor's description of its contents. On other occasions, bills were read twice and passed without a final reading. This omission permitted major changes in legislation prior to its consideration by the governor.[42]

Framers of the 1879 constitution tried to reform the lawmaking process. In submitting the document to the public, they explained: "The power of the legislature has been restricted in every case where it would be safe to do so, in respect to the enacting of local or special laws." The new constitution required that all bills be printed and read in full three times on three different days. It also required a majority of *all* members of the legislature to pass a bill, not just those present on any given day of a session, and prohibited the introduction of bills within ten days of the close of a session without a two-thirds vote. Lobbying became a felony, and the legislature could not "appropriate money for any purpose besides the support of the State Government and institutions exclusively under the control of the State." Nor could it lend its credit either to municipal or private corporations, or make gifts of land or money. The first constitution's $300,000 debt limit was maintained, and state debts were limited to twenty years. Lawmakers could meet as long as they wished but could be paid for no more than sixty days.[43]

The new constitution failed to reduce political corruption or improve the legislature's reputation. But the legislature's inefficiency derived as much from sectional differences as from inex-

---

42. *Sacramento Daily Union*, December 25, 1877; January 8, 9, 15, 1880.

43. The quotes are from an "Address to the People of California" by the Constitutional Convention of 1878–1879 reprinted in the *Pacific Rural Press* 17 (March 8, 1879): 156. The best description of the new constitution is in Swisher, *Motivation and Political Technique*.

perience, incompetence, and greed. Sharp differences between northern and southern California emerged as early as the first constitutional convention. Since virtually all the mines were confined to the public domain, and since the federal government refused to recognize the validity of mining claims until 1866, taxes fell disproportionately hard on southern California where much of the land was in private ownership. In 1852, the six "cow" counties south of the Tehachapis paid over twice the property tax collected in northern California even though their population was only 5 percent of that in the twelve mining counties. And while the mining counties insisted on counting their transient populations for purposes of legislative apportionment, northern California contributed less than half the revenue collected from the state poll tax. In 1859 the legislature actually approved a plan whereby the counties from San Luis Obispo south would have become the "Territory of Colorado." However, the Civil War intervened before Congress could act and assured that the "secession" of southern California would never again receive substantial support in the legislature.[44]

After 1876, when the Southern Pacific completed its line into Los Angeles from northern California, southern Californians began to complain about rate discrimination. They hoped mining in Arizona would provide a foundation for Los Angeles's commercial empire, just as San Francisco's prosperity had been built largely on the mines of northern California and western Nevada. In particular, the fledgling citrus industry required new markets to realize its potential. However, by the late 1870s, San Francisco had cornered much of the Arizona trade. The S.P. charged 6½¢ per mile to carry certain goods from San Francisco to Yuma, on the California-Arizona border, while it charged 14¼¢ per mile from Los Angeles to Yuma.[45] Southern California's attacks on the railroad were only partly justified. The cost of transporting goods over short

44. Ellison, *A Self-Governing Dominion*, 167–197; Roberta M. McDow, "State Separation Schemes, 1907–1921," *California Historical Quarterly* 49 (March 1970); 39–46.

45. The *Sacramento Daily Union*, August 4, 1877. Swisher, *Motivation and Political Technique*, 47, 102. Swisher also notes that the Southern Pacific discriminated against Stockton. The shipment of goods cost little more from San Francisco to the southern end of the San Joaquin Valley than to the same point from Stockton (p. 49).

runs often exceeded the cost of long hauls, depending on the kind of cargo, the total volume of goods carried, service to intermediate points, the number of empty cars, the frequency of scheduled runs, and other considerations. Moreover, in northern California the railroad competed with river transportation, which helped hold down rates.

Nevertheless, after 1870 southern California's growth rate exceeded that in the north and provided a burgeoning local market for the region's agricultural products. Los Angeles did not rival San Francisco's population until the second or third decade of the twentieth century, but its businessmen started to open their own markets much earlier. For example, after the turn of the century, Los Angeles began to extend its financial influence over the San Joaquin Valley, formerly an economic province of San Francisco. The conversion of that valley from wheat farming to diversified, intensive horticulture stimulated investment from the East as well as from Los Angeles, and the discovery of oil in 1895 near the Kern River helped fuel the boom. One historian has written: "The struggle for control of the valley, which had been developing since the 1880s, intensified after 1906. The merchants of Los Angeles took advantage of the confusion following the San Francisco earthquake and fire to penetrate north of Bakersfield into substantially the entire Valley."[46]

Boosters in both parts of the state actively worked to lure immigrants to their section, often by belittling the agricultural potential of the rival section. For example, on April 26, 1875, the *Sacramento Daily Union* bitterly reported that southern California land companies had sent agents to Ogden, Utah, to lure migrants away from northern California. "The audacity with which this rascally business is carried on may be gathered from the fact that one of the touters was yesterday convicted [?] of telling immigrants who were bound for San Jose that no grain was or ever could be grown in that neighborhood; that the Santa Clara Valley (really the richest wheat section in the State) was a barren desert; and that all the wheat was grown in Southern counties." Yet northern Cali-

46. Blackford, *The Politics of Business*, 9. Also see the *Pacific Rural Press* 75 (April 4, 1908): 210.

fornia newspapers went out of their way to attack southern California's attractive irrigation colonies. These experiments in desert farming marked a radical departure from agriculture in the Midwest and offered great opportunities to unscrupulous land companies. Northern California boosters portrayed such ventures in the worst possible light. The *Oroville Register,* in the heart of the Sacramento Valley's wheat-growing district, warned that those who bought land in agricultural communities would be "badly swindled." "In our opinion no bunko dealer in the 'dives' of San Francisco ever perpetrated a grosser fraud . . . than is sought to be perpetrated by the promoters of these colony schemes, in the poverty breeding counties of Southern California, when trying to induce our farmers of small means to sell out and invest their little all in a twenty-acre interest in a wild-cat colony in Fresno, Los Angeles, San Diego, Santa Barbara, or other southern counties."[47]

The contest between northern and southern California, first to attract immigrants and later to dominate trade, masked deep sectional rivalries within the north itself. Both Oakland and San Francisco tried to capture as much of the oceanic trade as possible by developing port facilities. Chico, Sacramento, and Stockton competed over trade within the Central Valley, as well as for the lucrative markets offered by mining communities in the Sierra Nevada foothills. Sacramento and San Francisco competed for the state capital, the state fair, state prisons, and the terminus of the first transcontinental railroad. Sacramento newspapers proudly claimed that their city held a competitive advantage over San Francisco. For example, the *Sacramento Daily Union* noted in 1881:

> Sacramento is abundantly able to supply Northern California with everything. Her transportation facilities are in all respects as great as those of San Francisco. She can sell goods cheaper than the latter, because the business expenses of her merchants are much lighter. She can reach the mining counties and the whole northern section of the State from twelve to twenty-four hours quicker than San Francisco can. She can

47. The *Register* editoral is reprinted in the *Sacramento Daily Union,* October 15, 1879. Also see the *Union* of October 20, 30, 1879.

import goods, machinery, everything, from the East as speed-
ily as San Francisco. . . . There is therefore no necessity for
the mining counties or the people of the upper valley to trade
with San Francisco at all.[48]

Yet the *Union* persistently complained that San Francisco had retarded the economic development of the interior counties, first by
monopolizing transportation, and favoring the mining industry (in
which the city's businessmen had invested heavily) over agriculture; and, later, by refusing to sell off the large landed estates
owned by businessmen, speculators, and nonresident farmers in
the metropolis. The *Stockton Daily Independent* charged in 1878: "The
course pursued by the metropolis for more than a quarter of a
century has been so signally selfish that the people of the interior
feel little or no interest in any of its local concerns. It has sought
to build itself up by absorbing the wealth of the state rather than
by extending generous assistance to interior enterprises and fos-
tering general prosperity, hence its wants and desires fail to arouse
any particular interest in the minds of the rural population."[49]

Nineteenth-century California was a state characterized by
extremes. It contained arid and humid regions, mountains and
plains, benign coastal valleys and torrid deserts, mining camps and
large cities, thousand-acre wheat farms and ten-acre orange orchards, Orientals and Occidentals, railroad barons and "tramps."
It was a state neither "eastern" nor "western," nor even a blend
of the two. Rather it embraced many of the qualities of both, the
lawlessness of the frontier no less than the urbanity of the metropolis. To use a contemporary metaphor taken from another con-
text, California was a tossed green salad rather than a melting pot.
Its extremes forced Californians to adapt to a life-style far different
from that found in other parts of the nation. California's fragmen-

---

48. *Sacramento Daily Union*, February 11, 1881.

49. *Stockton Daily Independent*, April 4, 1878. On the same theme see the
Sacramento Daily Union of September 18, 1860; April 6, 7, 1864; September 9, 1864;
November 7, 1864; December 3, 1864; February 9, 13, 1865; March 20, 1865; October
28, 1865; April 30, 1875; August 26, 1876; March 16, 1878; July 10, 1878; March 26,
1879.

tation or polarization, added to its political corruption and the power of special interests in Sacramento, severely limited the state government's ability to deal with important issues. These included land monopoly, flood control, the need for a more equitable tax structure, railroad rate regulation—and the use of the state's vast water supply. The inability or unwillingness of Californians to solve important social and economic problems prevented the implementation of a rational water code and comprehensive irrigation plan until the twentieth century. Meanwhile, private interests, with substantial assistance from the courts and legislature, moved to claim most of the state's flowing water.

2

The Crucible of Western Water Law, 1850–1872

The story of the evolution of water law in California makes difficult and often tedious reading. Except for the titanic legal battle between Henry Miller, James Ben-Ali Haggin, and their armies of paid and unpaid supporters, which culminated in the California Supreme Court's famous 1886 ruling in *Lux* v. *Haggin,* the history lacks dramatic turning points and colorful personalities. Yet the law profoundly influenced the development of irrigation in California. In the 1850s and 1860s, four distinct systems of water law emerged in the state, providing a legal foundation for the entire arid West. Most historians of the West recognize the significance of the conflict between riparian rights and prior appropriation. But pueblo rights and a system of community control over irrigation rights that extended from Los Angeles to most of the state's agricultural counties in the years between 1854 and 1868 were equally important.

Water law evolved slowly in both California and the West, constructed piece by piece, like a quilt, rather than from whole cloth. Individual court cases and statutes were piled layer on layer, not welded together like links in a chain. In creating and interpreting water law, the courts and legislature rarely looked beyond immediate economic needs; neither institution was motivated by any grand design or system, and decisions often ignored the dictates of logic and consistency. Water law was more a reflection of unsettled, often chaotic frontier conditions than the product of legal precedents, dicta, or the philosophical assumptions of particular jurists.[1]

As is well known, the doctrine of appropriation developed in response to the needs of California miners. The argonauts diverted streams to get at placer deposits in river beds and later constructed dams and elaborate networks of flumes to wash away topsoil in search of ore-bearing quartz. In 1851, the legislature, at the behest of Stephen J. Field, then a legislator from Yuba County, declared that in court tests concerning mining claims, "proof shall be admitted of the customs, usages, or regulations established and in force at the bar, or diggings, embracing such claim; and such customs, usages, or regulations, when not in conflict with the Constitution and laws of this State, shall govern the decision of the action."[2] At about the same time, one lower court, the District Court for Nevada County, ruled that "the party who first uses the water of a stream, is by virtue of priority of occupation entitled to hold the same . . . as their property, to the extent of the capacity of the ditch to hold and convey water." This decision advanced three important principles: that priority was the first measure of a water right (rather than, say, the use to which water was put); that ap-

---

1. There is no comprehensive historical survey of the evolution of water law in California. The best study is Gordon R. Miller, "Shaping California Water Law, 1781–1928," *Southern California Quarterly* 55 (Spring 1973): 9–42. Unfortunately, Miller completely ignores the water commission laws passed by the legislature in the 1850s and 1860s, discussed later in this chapter. Also see the appropriate sections of C. S. Kinney's four-volume legal masterpiece, *A Treatise on the Law of Irrigation and Water Rights and the Arid Region Doctrine of the Appropriation of Water* (San Francisco, 1912); S. C. Wiel, *Water Rights in the Western States* (San Francisco, 1905); A. E. Chandler, *Elements in Western Water Law* (San Francisco, 1913); Wells Hutchins, *California Law of Water Rights* (Sacramento, 1956); John D. Works, "Irrigation Laws and Decisions of California," in *History of the Bench and Bar of California*, by Oscar T. Shuck (Los Angeles, 1901), 101–172; Elwood Mead, *Irrigation Institutions* (New York, 1903), 180–219; A. E. Chandler, "Appropriation of Water in California," *California Law Review* 4 (March 1916): 206–216; Lucien Shaw, "The Development of Water Law in California," California Bar Association, *Proceedings of the Thirteenth Annual Convention* 13 (1913): 154–173. Two recent legal surveys, prepared by the staff of a special commission appointed by Governor Edmund G. Brown, Jr., during the California drought of 1976–1977, also merit attention. See David B. Anderson, *Riparian Water Rights in California* (Sacramento, 1977); Marybelle D. Archibald, *Appropriative Rights in California* (Sacramento, 1977).

2. *California Statutes*, 1851, 149.

propriative rights depended on the continued use of water and were not absolute, irrevocable grants; and that one measure of an appropriative right was the capacity of the diversion ditch.[3]

The California Supreme Court rejected the lower court's decision as "impracticable in its application" and fell back on the precepts of "well-settled law." At the time, the issue of federal sovereignty posed a much more immediate problem. The U.S. Constitution's "commerce clause" granted the nation sovereignty over navigable rivers, but most streams flowing over mining lands had no value as avenues of trade. Still, if federal sovereignty over the public lands *implied* ownership of the water flowing over those lands, then California had no more right to guarantee the ownership or use of water than it did to confer titles to the mineral lands themselves. This raised an enormous question: How could hydraulic mining companies raise the substantial sums of money needed to carry on their work if they could not secure clear title to a fixed quantity of water?

Neither Congress nor the federal courts provided an answer to that question, so the California Supreme Court acted on its own in 1855, six years after the beginning of the state's mining boom. *Irwin* v. *Phillips* dodged the question of federal sovereignty over water and justified appropriation on grounds of economic necessity. The expectation that Congress would soon enact laws regulating the use of water and land within the public domain may explain the court's failure to issue a broad philosophical statement on the nature of water rights. Its opinion noted that "a system has been permitted to grow up by the voluntary *occupation* of the mineral region [which] has been *tacitly assented to* by the [federal] government, and heartily *encouraged* by the expressed legislative policy of the other [i.e., state government]." In short, the federal government's refusal to act constituted tacit acceptance of the claims of appropriators. The judges admitted that California water law was as yet "crude and undigested," subject to "dispute," but reasoned that "necessity and propriety" demanded not just recognition of mining claims but also clear title to the water needed to work those claims.[4]

---

3. The lower court's ruling is reprinted in Eddy v. Simpson, 3 Cal. 249 (1853).

4. Irwin v. Phillips, 5 Cal. 146, 147 (1855).

Subsequent cases built on the principles enunciated in *Irwin* v. *Phillips*. Water rights were made effective from the beginning of construction on diversion works, rather than from the date of completion or actual diversion of water. Farmers, lumbermen, millers, and other water users, as well as miners, could claim water under prior appropriation. The priority of a water claim took precedence over where the water was used, how much was used, the purpose for which it was used, or the value of that use. The first appropriator could change the point of his diversion, but not if the new diversion destroyed the rights of subsequent claimants. An appropriator upstream could not use water if that use (e.g., motive power for a sawmill) destroyed the value of the water to an earlier appropriator downstream.

By 1872, the broad contours of the doctrine of appropriation were clear. The first claimant enjoyed an exclusive right, no matter how much water he used. His was the only "absolute" grant; all others depended on the extent and priority of earlier rights. All users were limited to the amount of water put to "beneficial use," regardless of the size of their claims. The water could be used on any lands, any distance from the source of water, not just on land adjoining the water supply. Moreover, the water could be "used up," even if a diversion entirely dried up a supply source. Appropriative rights could be bought and sold in part or whole. They could also be lost or circumscribed in a variety of ways ranging from "disuse" to "adverse use."[5]

The origins of riparian water rights have been subject to debate. Samuel C. Wiel, one of the arid West's leading water lawyers, claimed that the doctrine originated in articles 644 and 645 of the 1804 Napoleonic Code, then found its way into English common law through the commentaries of American jurists Story and Kent beginning in 1833. English lawyers and judges, according to Wiel, did not test the doctrine in court until 1849. Before 1833, English

5. The most important cases defining appropriative water rights are Hill v. Newman, 5 Cal. 445 (1855); Tartar v. The Spring Creek Water and Mining Company, 5 Cal. 395 (1855); Conger v. Weaver, 6 Cal. 548 (1856); Kelly v. The Natoma Water Company, 6 Cal. 105 (1856); Hill v. King, 8 Cal. 336 (1857); Maeris v. Bicknell, 7 Cal. 261 (1857); Butte Canal and Ditch Company v. Vaughn, 11 Cal. 143 (1858); Ortman v. Dixon, 13 Cal. 33 (1859); Rupley v. Welch, 23 Cal. 453 (1863). "Adverse use," or "prescription," applied to water rights established by use, rather than through formal legal claims.

water rights were based on "ancient customs" or "ancient enjoy-ment," principles with a close affinity to the doctrine of appropri-ation. In short, Wiel offered two revolutionary propositions: that the doctrine of appropriation antedated riparian rights, and that the latter was "alien" to Anglo-American law, derived as it was from the Continent.

Arthur Maass and Hiller B. Zobel have challenged Wiel's the-ory in a closely reasoned exposition on the origins of riparian rights.[6] Yet the California experience sustains Wiel on one point: the ri-parian doctrine was not fully formed in English-speaking countries at the time of the California Gold Rush. It was not a set of fixed, immutable principles, or a body of dogma. On April 13, 1850, the California legislature declared that "the Common Law of England, so far as it is not repugnant or inconsistent with the Constitution of the United States, or the Constitution or laws of the State of California, shall be made the rule of decision in all the Courts of this State." This innocent declaration has been seen as inadver-tently fastening riparian rights on California. But formal acceptance of the common law did not preclude recognition of appropriative water rights. That the law should bend to new circumstances was the basic premise of common law, and the English common law had been modified repeatedly to conform to conditions in the east-ern United States. A good example was "prescription," the acqui-sition of a title or right by exercising that right over time, even in violation of established claims. In England this process took twenty years, but in many eastern states only ten.[7]

When the California Supreme Court recognized appropria-tion, it acknowledged economic conditions as they existed in Cali-fornia and bent the law to suit them; it did not adopt a new system of water law. It drew clear distinctions between public and private

---

6. Samuel C. Wiel, "Waters: American Law and French Authority," *Harvard Law Review* 33 (December 1919); 133–167; idem, "Origin and Comparative Devel-opment of the Law of Watercourses in the Common Law and Civil Law," *California Law Review* 6 (May and July 1918): 245–267, 342–371; Arthur Maass and Hiller B. Zobel, "Anglo-American Water Law: Who Appropriated the Riparian Doctrine?" *Public Policy* 10 (1960): 109–156.

7. *Cal. Stats.*, 1850, 219; Arthur Maass and Raymond Anderson, . . . *and the Desert Shall Rejoice: Conflict, Growth, and Justice in Arid Environments* (Cambridge, Mass., 1978), 228–229.

lands as well as between mining and agriculture and had good reasons for doing so. As it explained in *Tartar* v. *Spring Creek Water and Mining Company* in 1855, the rights of miners had to take precedence over those of farmers on the public domain. To gain access to water, farmers invariably settled land adjoining streams first. Had riparian rights been allowed to prevail on the public lands, "the entire gold region might have been enclosed in large tracts, under the pretence of agriculture and grazing and eventually, what would have sufficed as a rich bounty to many thousands, would be reduced to the proprietorship of a few."[8] The court also recognized that riparian rights, which existed only as a corollary to landownership, could not prevail on the public domain, where miners did not enjoy title to the land they worked.[9]

Conditions in California during the 1850s and 1860s made these decisions moot. Outside the public domain, except perhaps in southern California, there were no conflicts over water and no demand for any system to establish water claims or regulate water use. The huge estates established during the Mexican period remained intact. Irrigation was confined to a few thousand acres of riparian land and, since agriculture was still a fledgling industry, virtually all water cases during these decades concerned mining or milling. How the legal principles enunciated in the 1850s and 1860s applied to irrigation remained unclear.

There was always a chance that the California Supreme Court would modify its early decisions in order to meet the needs of farmers, especially after the demise of the mining industry as farmers took up land farther and farther away from streams. But, contrary to the view that the doctrine of riparian rights was alien to

8. Tartar v. The Spring Creek Water and Mining Company, 5 Cal. 397 (1855).

9. Hill v. Newman, 5 Cal. 446 (1855); Irwin v. Phillips, 5 Cal. 145 (1855); Kelly v. The Natoma Water Company, 6 Cal. 108 (1856). Generally, the courts ruled that riparian rights did not apply on the public domain, but there were tantalizing qualifications. For example, in Crandall v. Woods, 8 Cal. 136 (1857), the most extensive early case involving riparian rights, the supreme court suggested that miners who used water to work land immediately adjoining a stream had a right superior to those who diverted the water and carried it to nonriparian lands. Those who established claims adjoining a stream could block the diversion of a subsequent appropriator even though they had never formally posted a claim to the water. This was one of many ways in which riparian and appropriative claims clashed or overlapped in the 1850s and 1860s.

California, or at least took a back seat to appropriation before the 1880s, many early decisions established the supremacy of riparianism. In the 1853 case of *Eddy* v. *Simpson*, the court ruled that "the right of property in water is *usufructuary*, and consists not so much of the fluid itself as the advantage of its use. The owner of land through which a stream flows, merely transmits the water over its surface, having the right to its reasonable use during its passage. The right is not in the *corpus* of water." In *Hill* v. *Newman* (1855), the court declared: "The right to running water is defined to be a corporeal right, or hereditament, which follows or is embraced by ownership of the soil over which it naturally passes. . . . The right to water must be treated in this State as it has always been treated, as a right running with the land . . . and as such, [it] has none of the characteristics of mere personalty." In *Conger* v. *Weaver* (1856), the judges insisted that common law principles had "abundantly sufficed for the determination of all disputes which have come before us; and we claim that we have neither modified its rules, nor have we attempted to legislate upon any pretended ground of their insufficiency." Nine years later the court was even more emphatic when it flatly rejected "the notion, which has become quite prevalent, that the rules of the common law touching water rights have been materially modified in this State upon the theory that they were inapplicable to the conditions found to exist here, and therefore inadequate to a just and fair determination of controversies touching such rights. This notion is without any substantial foundation."[10]

In 1865 the court also tried to clarify the relationship of riparian rights in a suit between two landowners along a stream, one of whom had attempted to divert water to the detriment of a downstream owner who had taken up his land later than the defendant. The court ruled that each riparian owner along a stream had the right to a "reasonable" use of water, but could not "unnecessarily diminish the quantity [of water] in its natural flow." The volume of a stream could be reduced if the water was necessary to meet domestic needs and to water cattle, no matter how much the diversion reduced the supply of water available to downstream ri-

---

10. Eddy v. Simpson, 3 Cal. 249 (1853); Hill v. Newman, 5 Cal. 446 (1855); Conger v. Weaver, 6 Cal. 55 (1856); Hill v. Smith, 27 Cal. 432 (1865).

parian owners. This was "unavoidable." But the upstream riparian owner had no right to alter or obstruct the channel of the stream itself.[11]

By the end of the 1860s, most basic features of riparian rights had been defined. Such rights were not limited in amount; diversions could vary from day to day or year to year. Rights were correlative rather than absolute, definable only when related to other riparian rights. Riverbank owners could use water for any useful purpose. But except for family and stock needs, they could not substantially reduce the flow of a stream if riparian neighbors complained. Riparian owners who needed more water usually purchased the acquiescence of other riverbank owners and filed appropriative claims. However, unchallenged diversions ripened into unassailable "prescriptive rights" after five years, so riparian owners were quick to challenge interlopers in the courts.[12]

Virtually all the important water cases of the 1850s and 1860s pertained to mining in northern California. However, the economy of the "cow counties," as they were contemptuously called by many San Franciscans in those days, was based on stock raising and agriculture. Ironically, at least until 1854 water rights south of the Tehachapis were as uncertain as mining claims on the public domain. Articles VIII and IX of the Treaty of Guadalupe Hidalgo (1848) promised that "property rights of every kind" would be "inviolably respected" under American rule. But while Congress created a land commission to confirm Mexican land titles, it never addressed the issue of water ownership.

Spanish and Mexican law recognized neither riparian water rights nor rights acquired by prior appropriation. Spanish law emphasized that all land, water, and other natural resources belonged to the king. In theory, until 1822, when Mexico won its independence, all property rights were usufructuary rather than absolute. The crown regulated every aspect of colonization in Mexico, in-

11. Ferrea v. Knipe, 28 Cal. 343 (1865).

12. For a discussion of the riparian doctrine in the nineteenth century see John Norton Pomeroy, *A Treatise on the Law of Water Rights* (St. Paul, Minn., 1893); Wiel, *Water Rights in the Western States*, 405–532; and Kinney, *Law of Irrigation and Water Rights*, 1: 759–818, 863–957. One issue that had not been resolved in 1872 was whether riparian owners could divert water for irrigation. This right was not formally confirmed until the judgment in Lux v. Haggin (1886).

cluding the recruitment of settlers and the plans, provisions, and form of governments of the pueblos. The "Plan of Pitic," drafted in Chihuahua, Mexico, in 1789, provided a model for future settlements in northern Mexico (including Alta California), not just for Pitic in the province of Sonora. The community blueprint— "edict" would be more appropriate—permitted residents of new colonies to pass the title to land on to their heirs but did not extend the same privilege to water. The document recognized irrigation as the "principal means of fertilizing the lands, and the most conducive to the increase of settlement," and ordered the commissioner or *alcalde* who laid out each town to design irrigation canals that would serve as much land as possible. The Plan of Pitic also required the town council to appoint an officer to supervise each main ditch, to set the hours of irrigation, and to determine the amount of water each farmer needed and received. The whole community was responsible for maintaining the irrigation system, each member contributing money or labor according to the number of *suertes* he irrigated. Significantly, the plan provided for the common use of water *outside* each pueblo as well as within it.[13]

Mexico built on the legal principles inherited from Spain. Water was regarded as common property. Occasionally, individuals received exclusive grants, but these did not resemble the water rights later permitted under prior appropriation. The government imposed special conditions or limitations as it saw fit, there was no chronological priority, and older grants were not necessarily superior to later ones. Only water that rose on a private estate and stayed there was beyond the government's jurisdiction. Since peopling and subduing the land was the first priority, water used for farming took precedence over that used for mining. The waters of all nonnavigable streams were dedicated to the use of the com-

13. The Plan of Pitic echoed the "Regulations for the Government of California," issued by the governor of California, Don Felipe De Neve, on June 1, 1779, and approved by the king of Spain on October 24, 1781. See Francis F. Guest, "Municipal Government in Spanish California," *California Historical Society Quarterly* 46 (December 1967): 307; Kinney, *Law of Irrigation and Water Rights*, 1: 994–998; "Plan of Pitic," Addenda no. 7 in John W. Dwinelle, *The Colonial History, City of San Francisco* (San Francisco, 1867), 11–17; John A. Rockwell, *A Compilation of Spanish and Mexican Law* (New York, 1851), 445–450; and Leonidas Hamilton, *Hamilton's Mexican Law* (San Francisco, 1882), 110–127.

munities through which they passed, and public officials had to approve each diversion. Usually, water was apportioned according to the number of acres irrigated. Grants could be increased or decreased month by month or year by year according to the supply, or as old land was abandoned or fallowed and new land opened to cultivation. In times of scarcity, each irrigator, no matter how long he had irrigated his land or where the land was located, was cut back proportionately.[14]

The transcendent right of the community, as opposed to individual water users, can be seen in the regulation of water use at Los Angeles. The pueblo of Los Angeles contained nearly 18,000 acres, approximately 1,500 people, and 103 "town farms" in 1848. Most of these were planted to grapevines, but farmers also raised oranges, peaches, and apricots. The absence of accessible markets for Los Angeles crops kept the irrigated acreage to less than 1,000 acres in 1848. However, irrigation expanded rapidly enough during the 1840s to prompt city officials to issue ordinances regulating the construction, maintenance, and operation of community canals. For example, in 1841 the town council created the office of watermaster to insure that "fairness be observed in the use of water, which will not be wasted." Aside from regulating the amount of water each farmer received and checking to see that the headgates on distribution ditches did not leak, the new official was responsible for supervising the cleaning and repair of the main ditch (*zanja madre*). On order of the watermaster, each irrigator had to provide one or two Indians, depending on the amount of land under ditch, to do the work. The official could also arrest anyone who washed clothes in the canals, polluted them with garbage, or wasted water. The alcaldes approved several similar laws during the 1840s, especially in dry years.[15]

14. Walter Prescott Webb, *The Great Plains* (Boston, 1931), 441; Lux v. Haggin, Argument for Respondent on Rehearing, April 21, 1885, Lux v. Haggin File, California State Archives, Sacramento, pp. 18–19, and 27–28.

15. Hubert Howe Bancroft, *California Pastoral, 1769–1848* (San Francisco, 1888), 355–356; Vincent Ostrom, *Water and Politics: A Study of Water Policies and Administration in the Development of Los Angeles* (Los Angeles, 1953), 27–40; Iris H. Wilson, *William Wolfskill, 1798–1866: Frontier Trapper to California Ranchero* (Glendale, Calif., 1965), 147; and J. Gregg Layne, *Annals of Los Angeles* (San Francisco, 1935), 49, 55.

When California became a state, this system of community control persisted. In 1850, the legislature passed an act incorporating the city of Los Angeles. "The Corporation created by this Act," the lawmakers declared, "shall succeed to all the rights, claims, and powers of the Pueblo de Los Angeles in regard to property."[16] Then, in May, 1852, the legislature ordered that "the Common Council [of Los Angeles] shall have power, and it is hereby made their duty, to pass ordinances providing for the proper distribution of water for irrigating city lands."[17] This mandate was reaffirmed by the legislature in April 1854.[18] The population of Los Angeles doubled between the mid-1840s and 1853. By 1854, the expansion of irrigation required the creation of a special department in the city government to regulate the canal network.[19]

The community water system of Los Angeles was built on what came to be called "pueblo rights." When California Governor Don Felipe De Neve issued regulations for the organization and settlement of new pueblos in 1779, he promised colonists "the common privilege of the water."[20] Beginning in 1797, when the friars at San Fernando Mission dammed the Los Angeles River and diverted part of the flow to mission lands, the city fought to maintain a monopoly over the entire stream. After statehood, both riparian owners and would-be appropriators challenged that claim. Some argued that their rights had ripened through prescription because the city had not contested diversions within the required five years, and virtually all the city's rivals wanted to restrict the city's claim to the water needed within the city limits. (The city often sold surplus water to farmers outside the city.) In 1873, the legislature granted Los Angeles "absolute ownership . . . to all of the water flowing in the River of Los Angeles" from the river's source to the point it left the city limits. Nevertheless, despite the city's dramatic population growth in the late 1870s and 1880s, sev-

16. *Cal. Stats.*, 1850, 155.

17. As reprinted in William H. Hall, *Irrigation in [Southern] California* (Sacramento, 1888), 559.

18. *Cal. Stats.*, 1854, 205.

19. Lewis Publishing Company, *An Illustrated History of Los Angeles County, California* (Chicago, 1889), 262.

20. Rockwell, *Compilation of Spanish and Mexican Law*, 447.

eral supreme court tests of the pueblo rights proved inconclusive.[21] Consequently, in August 1884, the city bought the water right of its chief rival, C. J. Griffith, owner of the Los Feliz Rancho, for $50,000.[22] The city's rights were not completely defined until 1899, when the supreme court ruled that pueblo rights antedated and took precedence over all riparian rights on the river, and that such rights included subterranean as well as surface water.[23] In effect, pueblo rights constituted a third system of water law nearly as indefinite as riparian rights in the nineteenth century. These were elastic rights that expanded with the city's boundaries and needs, even when the size of the community exceeded the pueblo's original four square leagues of land. Clearly, former pueblos such as San Jose and San Diego could claim such rights. But since Spanish authorities expected every mission and presidio to become a pueblo eventually, pueblo rights conceivably extended to *every* Spanish community in Alta California.

A statute approved in May 1854 reflected the substantial contribution of Mexico and Los Angeles to the evolution of California water law.[24] It permitted a majority of voters in any township within the agricultural counties of San Diego, San Bernardino, Santa Barbara, Napa, Los Angeles, Solano, Contra Costa, Colusa, and Tulare to request an election to select a board of three water commissioners and an overseer, or watermaster. In all incorporated cities within these counties, the mayor and city council could assume the commission's responsibilities on their own initiative. The act provided that "the duties of the Commissioners shall be to examine and direct such water courses, and apportion the water thereof among

21. *Cal. Stats.*, 1874, 633; City of Los Angeles v. Baldwin, 53 Cal. 469 (1879); Feliz v. City of Los Angeles, 58 Cal. 73 (1881); Elms v. City of Los Angeles, 58 Cal. 80 (1881).

22. Richard J. Hinton, "Irrigation in the United States," 49th Cong., 2d sess., 1887, S. Misc. Doc. 15 (serial 2450), 96.

23. City of Los Angeles v. Pomeroy, 124 Cal. 597 (1899); Abraham Hoffman, *Vision or Villainy: Origins of the Owens Valley–Los Angeles Water Controversy* (College Station, Tex., 1981), 11, 39–40.

24. San Francisco's *Alta California* of May 14, 1854, commented: "The bill applies chiefly to the southern counties, and refers to vineyards." The *Los Angeles Star* of February 18, 1854, announced that the bill had been introduced by a lawmaker from that city and claimed that its purpose was "to regulate streams so that they may prove of greatest good to the greatest number."

the inhabitants of their district, determine the time of using the same, and upon petition of a majority of the persons liable to work upon ditches, lay out and construct ditches." The overseers could demand up to twelve days' work a year from irrigators, and the commissioners could levy taxes to pay for maintenance and construction in proportion to the amount of water used by each farmer. To facilitate the construction of new canals, the commissioners were given the right to condemn land, at a fair price, and the law provided a simple arbitration process in case of disagreement over the price. It contained two important qualifications. It did not apply in counties where mining was the dominant industry, or to "mining interests" in general; mining took precedence over irrigation where the two uses of water came into conflict. Second, section 14 specified: "No person or persons shall divert the waters of any river, creek or stream from its natural channel to the detriment of any other person or persons located below them on any such stream."[25] This, apparently, acknowledged the primacy of riparian rights.

The 1854 law—expanded in 1857 to include Santa Cruz and San Luis Obispo counties—did not apply to unincorporated townships in which a majority of voters did not want a water commission and overseer. But the voters had a strong incentive to favor community irrigation systems. Common irrigation works prevented or reduced conflicts among individual farmers. And since few private companies invested in irrigation canals during the 1850s and 1860s, the law offered the best hope of extensive agricultural development. From a legal perspective, the most important feature of the law was that it superseded prior appropriation, or at least prevented its application, in the designated farming counties.

By the middle 1860s, most of the counties affected by the 1854 law had persuaded the legislature to adopt specific legislation defining the powers and responsibilities of their individual water commissions. Los Angeles, San Bernardino, Kern, Tulare, Fresno, and Siskiyou counties had mandatory commissions. In San Bernardino County, which had the most extensive system of public control outside Los Angeles, settlers who took up land along a stream already fully utilized for irrigation were criminally liable, as were farmers who lied to the overseer about the number of acres

25. *Cal. Stats.*, 1854, 76.

under cultivation to secure more water.[26] Yet the law's prohibition against any diversion "to the detriment" of water users downstream prevented its application in many watersheds. The legislature tried to remove this obstacle in 1862 by giving the various water commissions power to condemn *any* water rights that interfered with a public system.[27] But apparently many lawmakers questioned the state's constitutional power to condemn private property, even for public purposes. The 1862 law was never used and the privilege of condemning water rights does not appear in any of the water commission laws enacted in 1864 and 1866.[28]

Not only did the water commissions face the challenge of downstream riparian and, perhaps, pueblo rights, but by the middle 1860s they also encountered serious competition from fledgling private water companies. Without recognizing the full implications of its act, in 1862 the legislature, largely at the instigation of two irrigation companies in Yolo County, passed a law giving ditch and canal companies the right to condemn private land for canal rights-of-way. This was a privilege already enjoyed by railroad companies. One section of the act permitted companies to use "waters not previously appropriated." It specifically excluded eleven mining counties but did not mention the agricultural counties.[29] This permitted private water companies to move into river basins whose surplus water (at least in theory) had been reserved for public use under the 1854 law and its amendments. For example, as private water companies and land developers moved into Tulare County (which included most of Kern County until 1866), public control waned. The county had been one of those named in the 1854 legislation, but the legislature did not adopt a statute specifically providing for water commissions in the county until the first wave of settlers entered the Tulare Basin ten years later. The 1864 law

26. *Cal. Stats.*, 1859, 217; 1864, 87.

27. *Cal. Stats.*, 1862, 235.

28. *Cal. Stats.*, 1864, 87 and 375; 1866, 609 and 777.

29. *Cal. Stats.*, 1862, 540. On this law see the *Report of the California State Board of Agriculture for the Years 1864–1865*, Journals of the California Senate and Assembly, (hereafter *JCSA*), 16th sess. (Sacramento, 1866), Appendix, 2: 19–20; Paul Wallace Gates, *California Ranchos and Farms, 1846–1862* (Madison, Wisc., 1967), 81; Gerald D. Nash, *State Government and Economic Development* (Berkeley, 1964), 78; and *Sacramento Daily Union*, January 30, 1866.

was remarkably similar to the general statute of 1854. But in 1866 the law was extensively modified. No longer were Tulare County's water commissioners permitted to lay out ditches and apportion water in proportion to the amount of land each farmer irrigated, and they also lost the right to levy labor and property taxes. (The money had been important not just to build canals but also to pay salaries.) Under the new law, the commissioners could still appoint watermasters, but only individuals recommended to them *by the private water companies*. Moreover, both the commissioners and the watermasters were now paid by those same companies. The law allowed the commissioners to reject any proposed future diversions, insuring that the first companies would be able to monopolize much of the county's water.[30] In *Lux* v. *Haggin*, the California Supreme Court wisely observed that the 1866 law "seems to have been studiously prepared in the interest of the [water] companies then existing."[31]

In 1868, "an Act concerning water ditches and water privileges for agricultural and manufacturing purposes in the County of Tulare" overturned the last vestige of public control. It provided a procedure by which private interests could submit applications to the water commissioners for permission to dig irrigation ditches. Section 5 provided that "nothing herein contained shall be so construed as to affect the right and privileges of those who, by prior appropriation, have secured the right to the use of water from the several rivers and streams in Tulare County." The sanctity of prior appropriation in Fresno County was recognized in 1876, soon after the first large irrigation colonies opened.[32]

The system of community control survived into the 1870s in San Bernardino County and longer in the City of Los Angeles. Los Angeles' pueblo rights made the city relatively immune to the power of private water companies. The construction and maintenance of ditches, and the salaries of watermasters, were financed by water sales. In 1877, about 4,500 acres of land were irrigated within the city limits, as were another 4,000 to 5,000 acres outside

30. *Cal. Stats.*, 1866, 313, 314.
31. Lux v. Haggin, 69 Cal. 365 (1886).
32. *Cal. Stats.*, 1868, 113; 1876, 547.

the town. But a major drought in 1876–1877 coincided with the arrival of the first rail line into the city, and the population boom that followed made land more valuable for housing than for farming. By the end of the 1880s, the largest community ditch had been closed; the last was abandoned in 1904.[33]

The 1854 law won few advocates compared with those who lined up behind riparian rights or prior appropriation in the 1870s. However, California's surveyor-general in the early 1880s, James B. Shanklin, made an effort to sell the idea of public control. By the 1880s, much of the state's running water had already been appropriated and water conflicts had become increasingly common. Shanklin argued that the legislature had never intended prior appropriation to apply in the agricultural counties; the doctrine had been designed exclusively to serve the needs of miners. The 1862 water company act only permitted claims to water "not previously appropriated." Shanklin interpreted "appropriated" in a nonlegal sense: "assigned or reserved to a particular use." The mere act of creating the water commissions constituted an "appropriation" of water to meet their needs, in theory the entire water supply of each agricultural county named in the law. Had the legislature intended otherwise, Shanklin reasoned, the 1862 law would have contained a specific procedure for filing individual water claims. He also noted that when California codified its laws in 1872, section 19 of the Political Code specified that acquiring rights by prior appropriation could not violate any of the laws "creating or regulating Boards of Water Commissioners and Overseers in the several townships or counties of the State." The surveyor-general claimed that the great virtue of the 1854 law was that it was "sufficiently elastic to meet the increasing wants of the people," as opposed to prior appropriation, which gave a distinct advantage to first claimants and first uses of water.[34]

In 1883 Shanklin, then a state legislator, introduced a bill to limit prior appropriation. It provided that the board of supervisors

33. Hinton, "Irrigation in the United States," 95–97; Ostrom, *Water and Politics*, 40.

34. *Report of the Surveyor-General of California, August 1, 1880 to August 1, 1882*, JCSA, 25th sess. (Sacramento, 1883), Appendix, 1: 30–37.

of each county would also serve as a board of water commissioners charged with distributing the county's water supply. The supervisors would be required to give preference to "ditches built by public funds or acquired by the people of the district." Only after these needs had been satisfied would water be provided to companies and individual farmers outside community systems. Not surprisingly, the bill won little support.[35]

The legislature's retreat from public control—or at least its refusal to integrate public control with prior appropriation, riparian, and pueblo rights—left the responsibility for making water law with the courts. As part of the codification of 1872, the legislature established a formal administrative procedure for acquiring water. Sections 1410–1422 of the 1872 Civil Code required claimants to post a notice at the point of diversion indicating the amount of water claimed, the ditch's size, where the water would be used, and the purpose. Within ten days a copy of the claim had to be filed with an appropriate county recorder; rights dated from the posting of the claim. Work had to begin within sixty days after filing a claim, and the claimant had to "prosecute the work diligently and uninterruptedly to completion." All those who had claimed water in the past, but had not begun to construct diversion works, had twenty days to start on pain of forfeiture. Finally, section 1422 read: "The rights of riparian proprietors are not affected by the provisions of this title." As in the 1854 law, the legislature clearly recognized the supremacy of riparian rights.[36]

The new code was not an attempt to impose greater state control over water users. In 1914, the California Supreme Court explained that the purpose of the 1872 law was "to provide evidence whereby parties claiming under hostile diversions [e.g. contested claims] could establish their respective priorities and corresponding rights to the water and avoid the former difficulties in establishing the precise date of the inception of their respective enterprises."[37] The 1872 code sought to prevent or reduce conflict among water users by precisely defining the date that rights became effective and the conditions under which they lapsed.

35. *Visalia Weekly Delta*, August 11, 1882, January 26, 1883.

36. *California Civil Code*, 1872, Title VIII, Sec. 1410–1422, pp. 268–270.

37. Palmer v. Railroad Commission, 167 Cal. 163 (1914).

The new water code was riddled with weaknesses and omissions. It did not create a comprehensive, centralized record of water rights, and potential water users had no quick way of learning how many claims to a stream existed. Appropriators were not required to announce their claims in a local newspaper, as they were in Utah. Consequently, large water projects could be launched in violation of established rights. To make matters worse, many streams passed through two or more counties, and those water users who had acquired water before 1872 were not required to confirm their rights under the new law. So the records of individual counties were fragmentary at best. Even water users who managed to discover how much water had been claimed could not be sure how much water remained available. Most claims were inflated, the volume of streams varied enormously from month to month and year to year, and no attempt had been made to measure stream flow.

The code limited water rights to "beneficial use." However, the state did not provide administrative machinery to investigate whether appropriators used any or all of the water claimed. Beneficial use was not a fail-safe limitation on water rights in any case. Miners, for whom the 1872 law had been largely drafted, usually put their water claim to full use soon after the completion of their diversion works. By contrast, farmers opened land to irrigation section by section. The courts subsequently ruled that farmers who had well-conceived plans to use the entire amount of water claimed did not have to use it all at once. Beneficial use could be defined by the amount of water originally turned onto the land, by the number of acres capable of irrigation, or by the capacity of irrigation ditches. The courts were not consistent on this point, and judgments in mining and irrigation cases often differed. Similarly, "diligence" in the construction of diversion works was hard to define. Work had to begin within sixty days, but the law placed no time limit on completion. Hence, some water rights survived for years simply because claimants performed inexpensive periodic "work" to sustain their claims, such as the excavation of a few feet of ditch. As in cases concerning beneficial use, the courts usually gave water users the benefit of the doubt.

The 1872 code was not simply ambiguous; it also contained many holes. For example, it said nothing about the transfer or sale of water rights. Could water be owned in the same sense as land? If water could be sold, did the second owner have to use it for the same purpose stated in the original claim? Did the water have to be obtained at the same diversion point? These questions were left to the courts, which invariably treated all water rights as property and made no distinction between mining and agriculture. The code also set no limits on the amount of water that could be claimed. In theory, a single appropriator could demand a stream's entire flow and use far more water than his crops needed. The courts rarely considered waste in water rights suits. Some uses of water— such as flooding gopher holes or washing salts out of alkali-choked soils—were not considered beneficial uses. But the courts upheld many wasteful uses, including flooding land to raise pasture grass; in fact, the main measure of water rights became continuous use rather than the best or most efficient use.

Perhaps the greatest weaknesses of the code related to riparian rights and rights created through "prescription." As mentioned earlier, the courts ruled that water users could establish a claim "outside" the law by using water continuously for five years, as long as riparian owners and prior appropriators did not complain to the courts. These rights were valid even if the water user did not file a claim under the 1872 code. The main advantage in filing a formal claim was that the water right then dated from the posting of a claim rather than from the date of actual diversion. But posting a claim also had its dangers. Many appropriators refused to obey the law because the public record of claims in each county alerted riparian owners to diversions that might have gone unnoticed in the absence of publicity.

The 1872 code reaffirmed the primacy of riparian rights but did nothing to define their limits. During the 1870s and 1880s, when claims against the state's water supply dramatically increased, many fundamental legal questions concerning this class of rights remained unresolved; some were not settled until well into the twentieth century. There was general, though not universal, agreement that riparian rights could not exist on the public domain, but did they pertain to the Spanish and Mexican land grants, especially given the system of public control sanctioned by the legislature in 1854 and later? Could riparian owners use their rights to irrigate?

Could they sell water to those who owned nonriparian land? If they bought nonriparian land adjoining their original riparian tract, did their riparian rights apply to the new parcel? If an owner sub-divided a large riparian estate, which parts inherited the riparian right? Could riparian owners claim the flood flow of a river—that heavy runoff occurring in May and June—or just the "normal" flow? Perhaps the most profound question was whether riparian rights could be condemned for "public purposes." However, this question actually aroused less controversy than many others be-cause on long streams condemnation involved dozens of land-owners and an expensive, lengthy proceeding. The time and expense made the condemnation question a moot point.

Looking back over the years from 1850 to 1872, the legislature can easily be condemned for not exercising more effective lead-ership. After all, the courts did not usurp power; in the case of water law, they merely filled a power vacuum. But the legislature's failures were partly justified—and not just by inexperience, by its limited knowledge of water law in other countries, by the increas-ing power of urban, mining, and irrigation water companies, or even by the frequently acknowledged supremacy of riparian rights. Riparian rights posed the greatest obstacle, but if they were con-sidered an insurmountable barrier, why did the legislature try to create a system of public control in the first place? The most likely reason is that during the 1850s and early 1860s, few lawmakers expected irrigation agriculture to expand rapidly in the future. They sought to protect *limited* agricultural needs, needs that did not conflict with riparian rights until the 1870s. Many legislators un-doubtedly assumed the two systems of water law could coexist. Perhaps they also expected riparian owners to compromise their differences with rival water users.

The uncertain nature of federal water rights also limited the legislature's initiative. State ownership of water suggested state regulation of its use. However, the national government never formally transferred sovereignty over water to the states. In 1866, 1870, and 1877, Congress recognized the right of westerners to use streams flowing over the public lands for mining, agriculture, and other purposes. This gave federal approval to prior appropriation, but these laws left the regulation of water rights to the states and territories.[38] None of these statutes defined federal water rights, and the nature of those rights would become a hot issue, particularly in the twentieth century.[39] Proponents of states' rights argued that Congress had tacitly sanctioned state sovereignty. For example, both Colorado and Wyoming claimed that congressional ratification of their constitutions, which declared unappropriated water the property of the people of those states, indirectly acknowledged state control. However, no one could be sure that the federal government did not retain residual rights. If Congress had simply recognized the status quo in 1866, 1870, and 1877 to provide order in the filing and recording of water claims, then no transfer of sovereignty had occurred. The states could not assume a power simply because Congress had neglected to use it. Whether the federal government had deeded individual or corporate claimants *permanent*, irrevocable rights to water was unclear. But even if it had, title passed from the nation to the water user, not from nation to state to water user. According to this argument, the states simply provided administrative systems to facilitate acquisition of rights. This view of federal sovereignty had many critics, but it served as a warning to those who wanted to create systems of state or community control.

What, then, did California contribute to western water law? Clearly, California was not unique. Arizona and New Mexico shared a similar experience, demonstrating the persistence of Mexican laws in the American Southwest. In New Mexico, where irrigation had been practiced on land adjoining the Rio Grande at least since the seventeenth century, the territorial legislature confirmed the Mexican system of public control in 1851 and 1852. These laws bear a striking similarity to the 1854 water commission law adopted in California. Similarly, when Arizona's first territorial legislature met in 1864, it declared that "the regulations of acequias, which have been worked according to the laws and customs of Sonora and the

38. For the July 26, 1866 and July 9, 1870 laws see *U.S. Revised Statutes*, Sections 2339 and 2340. For the March 3, 1877 law see *U.S. Statutes at Large*, 19: 377.

39. For a discussion of the conflict between the states and federal government over western water, see Donald J. Pisani, "State vs. Nation: Federal Reclamation and the Battle over Water Rights in the Progressive Era," *Pacific Historical Review* 51 (1982): 265–282.

usages of the people of Arizona, shall remain as they were made and used up to this day." As in California, eventually private land and water companies undermined public irrigation systems and molded water laws to suit their needs.[40]

In assessing the significance of California water laws, some mention must be made of the Utah experience. Irrigation was practiced in the Salt Lake Valley beginning in the summer of 1847, but the Mormons rejected private ownership of natural resources. Instead, they believed that the church and its members should serve as stewards over the wealth God had created for man's use. Brigham Young warned: "There shall be no private ownership of the streams that come out of the canyons, nor the timber that grows on the hills. These belong to the people; all the people."[41] Before 1852, when the first territorial legislature met, the bishop in charge of each Mormon congregation arranged canal surveys and organized members of his ward into construction crews. Since irrigation works benefited the whole community, all able-bodied farmers were expected to contribute their labor. When the ditches were ready, the church appointed a member of the ward to serve as watermaster. That official, in close consultation with the church hierarchy, established rules of water use and supervised distribution. Water claims were "attached" to particular tracts of irrigated land, and farmers did not receive permanent grants. The Mormon emphasis on group harmony and cooperation made water conflicts rare, but when they occurred, church-appointed arbitrators settled the disputes quickly and cheaply, without resort to the courts. In 1852, the territorial legislature gave the county courts exclusive control over water rights, but the Mormon communitarian ideals persisted. The courts prevented the waste of water, passed judgment on the construction of proposed dams and canals, approved or rejected requests to change diversion points, appointed watermasters, and

40. New Mexico's laws are reprinted in the *Report of the Surveyor-General of California, August 1, 1880 to August 1, 1882*, 15–17. The Arizona law is quoted in R. H. Forbes, *Irrigation in Arizona*, U.S.D.A., Office of Experiment Stations Bulletin no. 235 (Washington, D.C., 1911), 57. Also see Wells Hutchins, "The Community Acequia: Its Origin and Development," *The Southwestern Historical Quarterly* 31 (January 1928): 261–284.

41. As quoted in Leonard Arrington, *Great Basin Kingdom: An Economic History of the Latter-Day Saints, 1830–1900* (Cambridge, Mass., 1958), 52.

adjudicated rights. As before 1852, the church's pervasive influence insured that most disputes were settled internally.

Until 1880, prior appropriation as it evolved in the mining camps of California was unknown, at least in Utah's most densely settled agricultural communities. In the 1840s and 1850s, the territorial legislature approved some enormous special water grants to prominent Mormons, but the legislature always reserved the right to modify or rescind these grants in the future. Moreover, under the 1852 law, grants were usually expressed as a percentage of stream flow rather than as a fixed quantity of water. This ran counter to the practice in California and Colorado, where the courts settled conflicts by granting specific amounts of water to litigants. Prior appropriation in Utah dates to 1880. In that year the fear of an imminent federal takeover persuaded the territorial legislature to define water rights as private property, in specific quantities, to protect the Mormon agricultural investment.[42]

The role of Utah in the development of the arid West's irrigation institutions is controversial. Nineteenth-century critics of Mormonism usually ignored or downplayed Mormon accomplishments. For example, in 1873 a resident of Colorado's fledgling Greeley irrigation colony remarked: "They [the Mormons] have . . . contributed nothing to the world's store of knowledge on the subject [of irrigation], have written no books, advanced no theories, recorded no new facts, and . . . were able to afford us nothing of any practical value."[43] Perhaps partly by way of compensation, recent students of Mormon agriculture have been much more charitable. One historian has incorrectly suggested that the doctrine of prior appropriation, as it developed in California during the 1850s, contained "essentially the same provisions as were found in Utah's practice," and that California miners may have gotten the inspiration for the customs and procedures adopted in the mining camps from "observations in Utah," through which they sometimes passed

42. Arrington, *Great Basin Kingdom*, 52–54, 241–244; Mead, *Irrigation Institutions*, 220–223; George Thomas, *The Development of Institutions under Irrigation: With Special Reference to Early Utah Conditions* (New York, 1920), 48, 54–58, 88–89.

43. The *Greeley Tribune*, September 17, 1873, as reprinted in David Boyd, *A History: Greeley and the Union Colony of Colorado* (Greeley, Colo., 1890), 90.

on their way to the gold fields.[44] The author of a standard reference book on reclamation, a noted engineer, concludes: "Their laws for appropriation of water and its priority of use have been a pattern to other western states."[45]

As usual, the truth is somewhere between the rancorous statements of critics and the effusive praise of idolaters. Cooperation in the construction and maintenance of irrigation works, devotion to small farms cloistered around individual villages, and the diversity of crops made Mormon agriculture unusual if not unique. The Mormons also added some elements to the doctrine of prior appropriation, such as the publication of water claims in local newspapers. Nevertheless, at least until more thorough historical studies of western water law appear, Utah must relinquish credit for introducing the concept of public ownership and control of water to New Mexico and California, and credit (or blame) for prior appropriation to California.

44. George L. Strebel, "Irrigation as a Factor in Western History, 1847–1890" (Ph.D. diss., University of California, Berkeley, 1965), 296. Strebel also ignores the legacy of Mexican water law. Despite its broad title, his study focuses almost entirely on Utah.

45. Alfred R. Golzé, *Reclamation in the United States* (New York, 1952), 8. Leonard Arrington and Dean May have questioned the Mormon contribution to irrigation agriculture in the nineteenth century, including water law. See "'A Different Mode of Life': Irrigation and Society in Nineteenth-Century Utah," *Agricultural History* 49 (January, 1975): 3–20.

3

Panacea or Curse: Attitudes Toward Irrigation in Nineteenth-Century California

In June 1873, the *Pacific Rural Press*, California's leading agricultural journal in the late nineteenth century, observed that the state's farmers had done little to adapt to the physical environment of the arid West. "What California agriculture needs is not this or that man's opinion formed upon any former experience in the country from which he came," lamented the editor; ". . . we want the result of a California experience, with a California soil and climate. Then and not till then will we have entered upon anything like a system of true agricultural progress." The same refrain echoed through the pages of many journals. In March 1887, a Visalia newspaper noted that the recent settlers' "great objection to the California farm is the necessity for artificial irrigation." New residents generally applauded the state's balmy weather and the fertility of its soils. But while farm life was much more demanding in New England or the upper Midwest, old habits died hard, or not at all: "Back in [the farmer's] old home he has solved the problem of cold winters. . . . He knows the process of dodging blizzards and fighting grasshoppers, but when it comes to moistening the soil from a canal, a reservoir, or a well, he is puzzled and has grave doubts about the possibility of successful results." As late as 1905, William Ellsworth Smythe, one of the nation's foremost publicists of arid land reclamation, remarked that Californians "have always divided on the question as to whether irrigation is necessary."[1] While irrigation expanded dramatically during the 1870s and 1880s, it did not become a dominant feature of California agriculture until the twentieth century. Critics of this innovation abounded in the nineteenth century. Nevertheless, "artificial rain" also had its champions, boosters who supported it not just as a way to raise more abundant and valuable crops but also as a cure-all for many of the state's festering social and economic problems.

California's diverse climate and physical environment set it apart from much of the arid West, as well as from the humid East and Midwest. Although the average annual precipitation per square mile in California is 79 percent of the national average, it is only 44 percent of the average for the East Gulf and South Atlantic states. Moreover, the runoff per square mile of watershed is 51 percent of that in the Ohio River basin and 36 percent of the average for New England. Not only does California receive far less precipitation than the nation's older agricultural regions but its water supply is depleted still further by seepage and evaporation.

The annual runoff averages 71,000,000 acre-feet of water, about 33 percent of the mean annual precipitation, but it is unevenly distributed. Rainfall varies from an average 2.3 inches yearly at Brawley in the Imperial Valley, to 6.36 inches at Bakersfield at the south end of the Central Valley, to 18.02 inches in Sacramento and 38.33 inches at Eureka. While the north coast averages more than 75 days of measurable rainfall per year, the region south of the Tehachapi Mountains averages fewer than 40 days. About 28 percent of the stream flow comes from north coast rivers such as the Klamath, the Mad, and the Trinity; the Sacramento River and its tributaries contribute another 31 percent. Irrigators use 87 percent of the water. But while the northern third of the state produces nearly two-thirds of the entire supply, it contains only about 30 percent of the state's irrigated land. The San Joaquin Valley constitutes the largest block of irrigated land, but its streams yield only about nine percent of the state's water. And although more than 60 percent of the Golden State's population lives south of the Tehachapis, that section contributes only 2 percent of the water.

---

1. *Pacific Rural Press* 5 (June 21, 1873): 392; *Visalia Weekly Delta,* March 10, 1887; William Ellsworth Smythe, *The Conquest of Arid America* (New York, 1905), 131.

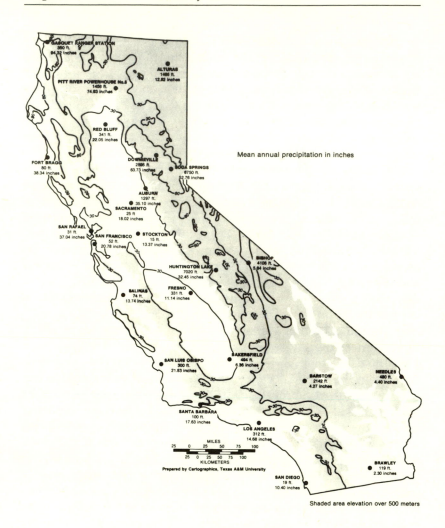

Mean annual precipitation in inches

Shaded area elevation over 500 meters

Mean Annual Precipitation.

Most streams originate on either the Coast Range or the Sierra Nevada, and the latter offers an abundance of good reservoir sites. Irrigation has developed much more rapidly on the east side of the Central Valley—particularly in the San Joaquin basin—than on the west side. Since most precipitation on the Coast Range falls as rain, 75 to 90 percent of its runoff occurs before the end of March in most years. The range's limited storage sites were not developed until after World War II, so most of the water drained into the ocean before the irrigation season began in May and June. Not only does the Sierra Nevada trap much more moisture, but because that range is two to three times higher than its sister range to the West, with colder temperatures, much of the Sierra's precipitation falls as snow. The snowpack insures a gradual runoff; more than 60 percent of the flow occurs after the end of March, reaching a peak in May and June but providing adequate water for irrigation through July. Thus, even without its reservoir system, the Sierra water supply is much more usable by farmers. The climatic influence of the Coast Range also explains why the rich coastal valleys between Marin and Santa Barbara counties receive much more water than the Central Valley. That the western range drains Pacific clouds of much of their moisture is starkly illustrated by comparing rainfall at San Luis Obispo on the coast and Bakersfield 100 miles due east in the Central Valley. San Luis Obispo receives an average 21.5 inches of rainfall annually, but Bakersfield residents can expect only 30 percent of that amount. And, of course, the Sierra Nevada has a similar effect. Nevada's unenviable reputation as the driest state in the West derives in large part from the mountain ranges that extract the moisture from Pacific storms before they reach the edge of the Great Basin.[2]

That California is a state of climatic extremes is evident in its frequent droughts and floods. Drought is an annual event in the arid West, where little rain falls from May through October. But rainfall in California fluctuates more year by year than in most parts of the humid half of the nation. The greatest statewide drought

2. For discussions of California's weather cycles, see William Kahrl et al., *The California Water Atlas* (Sacramento, 1978), 1–14; Sidney T. Harding, *Water in California* (Palo Alto, Calif., 1960), 8–26; and Erwin Cooper, *Aqueduct Empire* (Glendale, Calif., 1968), 19–34.

probably occurred in 1850–1851, when only 7.5 inches of rain fell on San Francisco and only 5 inches on Sacramento, about one-third of normal. The droughts of 1863–1864, 1898–1899, 1923–1924, 1929–1931, and 1976–1977 were also particularly severe. The most serious flood—though the inundations of 1851–1852 and 1955 came close—occurred in 1861–1862 when rain fell steadily during November and December, culminating in a January tropical storm that melted part of the snowpack. The runoff created an inland lake sixty miles wide in the Sacramento Valley, and much of Los Angeles basin was also submerged. Freshwater fish were caught in San Francisco Bay for nearly two months after the flood peaked. However, the state dried out fast. As frequently happens, drought followed flood, and in 1863–1864 the San Joaquin Valley's wheat and barley crops averaged only two bushels per acre.

Drought seldom has the same effect on the various parts of the state. For example, in 1898–1899, San Francisco received 77 percent of its average rainfall, as did Sacramento, while Los Angeles received only 35 percent of normal. In 1899–1900, Sacramento enjoyed 105 percent of its usual rainfall and San Francisco 84 percent, while Los Angeles received only 48 percent of normal. Yet in 1912–1913, Sacramento was at 40 percent of mean while Los Angeles received 80 percent. Similarly, during the drought of 1976–1977, San Diego's rainfall was about average and rainfall in the Imperial Valley well above normal, while many parts of northern California had to get by with little more than 30 percent of their accustomed supply.[3]

Early California explorers and residents quickly noticed the state's "unusual" climate. James Ohio Pattie, the Kentucky-born fur trapper who visited California in the late 1820s, lauded the "exceeding mildness" and healthfulness of the weather. "This is *our* winter as much as yours," a farmer in Sonora wrote the eastern agricultural journal *Country Gentleman* in December 1858, "and yet all my hogs, and all stock everywhere around, are living on grass. Farmers are plowing everywhere. The tilled farms have the ap-

3. For a discussion of droughts that occurred during the years covered by this book, see "Drought in California," *Transactions of the Commonwealth Club* 21 (December 28, 1926): 473–526; and J. M. Guinn, "A History of California Floods and Drought," *Annual Publication of the Historical Society of Southern California* 1 (1890): 33–39.

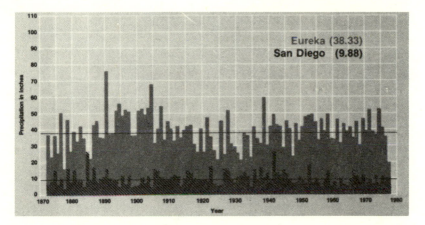

Precipitation Variation, Eureka and San Diego.

pearance of yours next May. . . . My whole farm, except where plowed, is one beautiful emerald green. No winter—no summer—no fall—but all one glorious spring!" Conversely, the climatic variations within California, and the dramatic differences between weather in the arid West and humid East, alarmed some observers. In an editorial on how climate had shaped the "personalities" of various European nations, San Francisco's *Alta California* noted in 1867 that the diverse conditions in California explained the state's early history and boded ill for the future: "The thirst for excitement, the chase for material gratifications and the ruinous habits of wandering and change which characterize our population may all be ascribed, to a certain extent, to the influence of the same climate which endows it with so much vitality and vigor." The article blamed everything from the divisiveness of California politics to the frequency of class conflicts on the weather.[4]

Similarly, early visitors and residents differed sharply over the state's agricultural potential. Auguste Duhaut-Cilly, a captain in the French merchant marine who visited California in 1827–1828, claimed the soil was "highly fertile." That conclusion was echoed

4. James O. Pattie, *The Personal Narrative of James O. Pattie of Kentucky* (Cincinnati, 1831), 216; *Country Gentleman* 13 (February 24, 1859): 122–123; *Alta California*, June 25, 1867.

by Lansford W. Hastings in his famous *Emigrant's Guide to Oregon and California*, published in 1845. Hastings reported that while "the crops of dry seasons are much less abundant than those of the ordinary season, yet . . . the crops even of a dry season are much better here than they are at any time in Oregon, or even in most of the [eastern] States."[5] However, explorer Charles Wilkes, who visited California in the early 1840s, proclaimed California's land "barren and unproductive." "However this may be [adapted] to the character and pursuits of the Mexicans," he scowled, "yet it will be doubted whether our countrymen would look upon it in the light of an agricultural country, or one entitled to be considered as such, and so different from these United States." An 1847 article in *De Bow's Commercial Review* declared that the state's soil was "sterile," presumably because "rains scarcely ever prevail, and for years hardly a shower is known." The author implied that no soil could be productive if deprived of the nutrients carried in rainwater.[6]

The Central Valley in particular seemed a forbidding place to settle, especially when mining promised quick wealth and a speedy return to the comfortable, congenial, familiar environment of home "back East." The fact that trees were sparse in the valley, that its grasses turned green only a few weeks in the year, that soil was volcanic in places, and that the Indians who lived there did not practice agriculture, were all taken as signs of sterility. Moreover, the intense heat of summer alternated with the dense tule fogs of winter and the floods of spring to make the valley seem destined by nature to remain uninhabited. The floods posed great danger to crops, and the mud left behind when the water receded increased the difficulty of planting, harvesting, and transporting crops to market.[7]

Few farmers ventured into the Central Valley until the 1860s.

5. Charles F. Carter, trans., "Duhaut-Cilly's Account of California in the Years 1827–1828," *California Historical Society Quarterly* 8 (December 1929): 320; Lansford W. Hastings, *The Emigrant's Guide to Oregon and California* (Cincinnati, 1845), 84.

6. Charles Wilkes, as quoted in Truman Smith, *On the Physical Character of the Northern States of Mexico* (Washington, D.C., 1848), 23; E. G. Meek, "California," *De Bow's Commercial Review* 3 (June 1847): 547.

7. Kenneth Thompson and Richard A. Eigenheer, "The Agricultural Promise of the Sacramento Valley: Some Early Views," *Journal of the West* 18 (October 1979): 33–41.

Meanwhile, however, boosters began to create a new mythology concerning California agriculture. Many stories printed in the 1850s and 1860s by the *Country Gentleman* and *American Agriculturalist* suggested that fruits and vegetables grew faster and larger in California than in older agricultural regions. Articles reported gigantic turnips that weighed 29 and 36 pounds, 100-pound sugar beets, 200-pound pumpkins, strawberries 7 inches in circumference, pears 21 inches long, and apple trees that grew 12 to 15 feet in a single season.[8] Could any reader doubt that California was a farmer's "gold mine" as well as a storehouse of precious metals?

Nevertheless, such rosy pictures failed to supplant the earlier, somber images of California's agricultural potential. Many transplanted easterners found that life in the new Eden was not as happy as land sharks and promoters made it out to be. In 1865, one disgusted farmer wrote the *Country Gentleman*, warning that successful farming was the exception rather than the rule in California. The flood of 1862 and drought of 1863–1864 had driven many settlers off their lands, and the writer concluded: ". . . examined in its true light, it [California] is no nearer heaven than the good old homesteads of the east." "Old Hurricane," a disgruntled southern California farmer, often wrote the same periodical complaining about droughts, the absence of lumber for fences, the distance from markets, the high price of labor, uncertain land titles, and limited pasture grass. Even the scenery disappointed him: "The scenery is grand, solitary and impressive. Its marked impression on me is desolation—and [it is] dry, sun-burned, dusty and waterless the great portion of the year."[9]

Many settlers considered irrigation as much of an anomaly as the other conditions they encountered. Irrigation had been practiced at Los Angeles and most of the missions for decades. But early visitors usually dismissed it as just another example of the inferiority of Spanish and Mexican institutions, institutions that virtually all American and English travelers held in contempt. James

8. *The Country Gentleman* 3 (April 6, 1854): 216; 9 (June 18, 1857): 399; 14 (December 1, 1859): 350; 17 (June 13, 1861): 387; *American Agriculturalist* 19 (September 1860): 273.

9. *The Country Gentleman* 25 (February 23, 1865): 130–131; 27 (May 3, 1866): 290.

Ohio Pattie declared that the Californios prompted "disgust and hatred." "The priests are omnipotent, and all things are subject to their power." In 1847, another traveler observed: "The greater part of the natives, I think I say without exaggeration . . . think of nothing in the world but gambling, dress, horse-riding, women, and stealing to maintain their vices."[10]

Predictably, European and American observers considered agriculture in Mexican California extremely primitive and inefficient. Simple wooden plows, hewn from branches of trees, lacked mold-boards. They cut only a shallow rut, rather than a furrow, so fields had to be plowed and replowed many times before planting. Harrows were unknown; tree-tops were used to cover the seed after planting. Wheat was separated from chaff by being dumped in a corral and then trampled by horses or mules. Then Indians winnowed the grain by tossing it in the air with wooden shovels and letting the wind blow away the straw. The Mexicans rarely fenced their land, used carts with solid wooden wheels rather than wagons with spoke wheels, seldom built stables or barns, and did not utilize fertilizers, crop rotation, or fallowing. Nor did they grade their irrigated fields. They simply flooded them, leaving some parts high and dry with others submerged in stagnant pools of water. When British merchant Alexander Forbes published the first English history of Upper and Lower California in 1839, he described mission agriculture as "most rude and backward." He treated irrigation with thinly veiled contempt, noting that "I have never seen even by irrigation any thing which could promise a very superior return per acre to a heavy crop in England." Subsequent critics reached similar conclusions. For example, in 1851 John J. Werth described irrigation as a "great bugbear" that had "very limited application."[11]

10. Pattie, *The Personal Narrative*, 216; William R. Gardner, "Letters from California," *Philadelphia North American*, April 26, 1847.

11. Alexander Forbes, *A History of Upper and Lower California* (London, 1839), 257, 246–247; John J. Werth, *A Dissertation on the Resources and Policy of California* (Benecia, Calif. 1851), 77–78. For early attitudes toward irrigation also see Edwin Bryant, *What I Saw in California* (New York, 1848), 307, 449; T. Butler King, *Report of Hon. T. Butler King on California* (Washington, D.C., 1850), 32, 34; and Walter M. Fisher, *The Californians* (San Francisco, 1876), 47–48. On agriculture in Mexican California see Zephyrin Englehardt, *The Missions and Missionaries of California*, 4 vols. (San Francisco, 1908), 2: 258–259; Milo M. Quaife, ed., *Narrative of the Adventures of*

Contempt for Spanish and Mexican institutions fitted well with the booster mentality of the first decades after statehood. Not only did immigrants to California find irrigation "alien" and the legacy of an inferior culture; it was also considered an added expense that put the state at a competitive disadvantage in the contest for new residents. Land speculators and other promoters recognized that the humid Midwest offered plenty of cheaper, more accessible land, so they minimized the need for canals and ditches. Admittedly, some land in the coastal valleys and Sacramento Valley could produce grain without irrigation, at least most years. But this land was either already in private lands or far removed from avenues of transportation. The first volume of the *Overland Monthly*, which began publication in 1868, contained a story entitled "Farming Facts for California Immigrants." Irrigation received only brief mention, and then to deny its need. "Artificial irrigation is not practised but in very few and exceptional cases in this country." This was true, but not for the reasons suggested by the author. He claimed that California's matchless soil was so fertile that irrigation was unnecessary, noting that in the foothill counties, where mining ditch companies offered farmers water at cheap rates, most continued to trust nature. In 1869, one of California's best-known publicists, Bentham Fabian, published his *Agricultural Lands of California*. He completely ignored irrigation and painted a picture of the San Joaquin Valley that bordered on fantasy. For example, in describing Tulare County (a desert where rainfall averaged only eight or nine inches a year), he promised "lands for all, a fertile soil, a delightful climate, and everything the heart of man can desire. Timber and water are abundant, and the woods provide for thousands of swine. Every description of grain, fruit and vegetables can be raised in profusion." In the following year, the *Sacramento Daily Union* chided

---

Zenas Leonard (Chicago, 1934), 167–168; and T. J. Farnham, *Life, Adventures, and Travels in California: To Which are Added the Conquest of California, Travels in Oregon, and History of the Gold Regions* (New York, 1849), 342–344. For a very useful summary of attitudes toward Mexican agriculture see Imre E. Quastler, "American Images of California Agriculture, 1800–1890" (Ph.D. diss., University of Kansas, 1971). American and European observers failed to recognize the obstacles faced by farmers in Spanish and Mexican California. These included poor transportation, a lack of markets, and Spanish and Mexican laws designed to fix prices and prevent competition with merchants and farmers at home. There was little incentive to innovate.

the New York *World* for suggesting that agriculture in California could not flourish without irrigation. The *Union* proudly noted that no more than 2 or 3 percent of the 4,000,000 acres then under cultivation in California were irrigated. "The New York *World* is, therefore, quite wrong in assuming that farming here must be attended with the expense of irrigation," the editor grumbled, "and that this circumstance ought to turn that class of people from the Pacific coast. . . . We have in this state not less than twenty million acres of good arable land that can be cropped every year in grain without irrigation." Finally, Stephen Powers, a newspaper reporter who walked from North Carolina to San Francisco in 1868 and spent most of the next six years tramping through California on foot, warned that the dominance of the wheat industry would persist until "the old Spanish belief, of the absolute necessity of irrigation for all green crops, can be exploded."

> [If] it can be shown that a farmer can produce almost any-where on good land the variety of little crops, which all farm-ers have in the East, then California will have a future—it will have a population. A great number of little farms will absorb this vagabond element continually drifting down out of the mines; whereas, if the land remains in vast ranches, these men will always continue hirelings and tramps. In this view of the matter, it becomes of momentous importance to demonstrate that common land will produce green crops without irrigation.[12]

Nineteenth-century critics of irrigation fell into two camps: those who claimed it was unnecessary and those who considered it positively dangerous. The former group included farmers who maintained that deep plowing in summer, or following the first

12. "Farming Facts for California Immigrants," *Overland Monthly* 1 (August 1868): 176–183; Bentham Fabian, *The Agricultural Lands of California* (San Francisco, 1869), 23; *Sacramento Daily Union*, May 28, 1870; Stephen Powers, *Afoot and Alone: A Walk from Sea to Sea* (Hartford, Conn., 1872), 310. *The California Farmer*, the most popular agricultural journal in California during the 1850s and 1860s, virtually ignored irrigation. Whether the editor, James Warren, did not appreciate the value of irrigation or simply did not want to frighten away prospective immigrants, is unclear. Many California periodicals of that period were widely circulated in the East, and editors generally tried to portray the state in a favorable light.

rains of fall, trapped sufficient moisture to obviate the need for irrigation. Ironically, in rejecting the innovation of irrigation, many California farmers contributed to the development of dry farming, a farming technique that subsequently became very popular on the High Plains. Some farmers also recommended fallowing as a way to "build up" the soil's moisture content, in effect using the land itself as a storage reservoir. The farmer who left part of his land unplanted one year, they claimed, could expect a bumper crop the following year, even during dry years.[13] Others argued there were better ways to get water onto the land than irrigation. Until the devastating drought of the 1880s, many farmers in the arid West believed that "rain follows the plow." In 1883, a booster of southern California's Ontario Colony suggested that the fear of droughts and the cost of irrigation should not trouble prospective settlers because "since the country has been cultivated the rainfall has perceptibly increased and there is a fair prospect that the dry years will almost entirely disappear."[14] Tree planting offered a second alternative. Like the myth that cultivating the land increased rain-fall, it rested on the assumption that the climate of the humid East was "normal," and that arid climates were in some way abnormal. When the first major western irrigation convention met in Denver in 1873, the delegates concluded that tree planting was of the "ut-most importance" and should be given a fair chance before farmers turned to more expensive alternatives of digging canals or tapping artesian wells. California's Central Valley seemed a perfect place to experiment, and in 1869 the *Tulare County Times* proposed that the state use proceeds from the sales of swamp land to pay for tree

13. E. J. Wickson, *Irrigation in Fruit Growing*, U.S.D.A., Farmers' Bulletin no. 116 (Washington, D.C., 1900), 4; Samuel Bowles, *Our New West* (Hartford, Conn., 1869), 441; *The Fresno Expositor*, June 15, 1870; August 17, 1881.

14. As quoted in R. Louis Gentilcore, "Ontario, California and the Agricul-tural Boom of the 1880s," *Agricultural History* 34 (April 1960): 85. For similar state-ments see William H. Bishop, "Southern California," *Harper's Magazine* 65 (November 1882): 869, and the statement by L. N. Holt in Richard J. Hinton's "Irrigation in the United States: Progress Report for 1890," 51st Cong., 2d sess., 1891 S. Ex. Doc. 53, (serial 2818), 23–24. For a superb survey of this important myth, see Henry Nash Smith, "Rain Follows the Plow: The Notion of Increased Rainfall for the Great Plains, 1844–1860," *Huntington Library Quarterly* 10 (February 1947): 169–193. For the view that California's climate was gradually becoming like that of the humid half of the nation, see the *Alta California*, July 25, 1854, October 9, 1869.

planting to make the great valley more attractive to settlers. Sacramento's *Union* confidently predicted in 1871: "Once cover any considerable portion of our plains with forest trees, and marked effects will be seen in the climate and productive power of contiguous lands." In 1883 the U.S. Commissioner of Agriculture conceded that the fact that forests were often on mountains accounted in part for the greater rainfall in their vicinity; still he insisted that more rain fell on forested lowlands than on barren ground. He also considered the forests natural reservoirs. The "exhalation" of moisture from the trees promoted "a richer growth of the grasses and grains of the husbandman." Thus the farmer won in two ways.[15]

In an age when bold engineering schemes captivated the public imagination, plenty of would-be rainmakers stepped forward. Their fanciful ideas included building fires simultaneously over large areas of land and setting off widely scattered gunpowder explosions.[16] One of the most intriguing schemes was designed to eliminate the deserts of southern California by diverting the Colorado River into the Colorado Desert (now the Imperial Valley). The promoters, including Oliver Wozencraft and William S. Chapman, knew that the river had once emptied into the desert. They also knew that Indian cultures had flourished on land adjoining a huge inland lake at the end of the stream. By a leap of faith, they assumed that the desert had once been a garden. By flooding the valley they promised to increase rainfall south of the Tehachapis, perhaps even into the San Joaquin Valley. This scheme, like the tree planting crusade, revealed a profound lack of understanding of the reasons weather varied from region to region. Nevertheless, in January 1873, the *Kern County Weekly Courier* concluded that the theory was "based on principles, the correctness of which are well established." Oliver Wozencraft, the father of the scheme, maintained that "there can be no question but that when it was a sheet of water . . . the surrounding country had a greater rainfall than

15. The *Stockton Daily Independent*, July 11, 1873; *Weekly Colusa Sun*, July 3, 1875; *Forest and Stream*, 1 (October 23, 1873): 168; *Sacramento Daily Union*, Oct. 4, 1871; *Tulare County Times*, May 1, 29, 1869, and May 4, 1871; *San Francisco Examiner*, April 1, 1871; *Report of the Commissioner of Agriculture for the Year 1883* (Washington, D.C., 1883), 453.

16. *Weekly Colusa Sun*, January 13, 1877.

now."[17] The plan is discussed in greater detail in the following chapter.

Those who saw irrigation as a positive danger claimed that it damaged the land and crops. James D. Schuyler, assistant state engineer, surveyed agriculture in Yolo County west of Sacramento in 1879 and discovered a "general prejudice against the irrigation of grain. Those who had tried it found that the soil was too stiff, and when irrigated in the spring the land became baked and soured."[18] Following flood irrigation, especially in the spring or summer, heavy clay soils often baked to the consistency of bricks, rendering the ground impervious to the plow. In addition, since most farmers failed to provide for drainage, irrigation often raised the water table and left behind patches of alkali that stunted crop growth. The damage to crops from irrigation had first been observed in the 1850s and 1860s by nurserymen. Most conceded that the controlled application of water promoted plant growth, but some charged that the foliage of irrigated trees was not as uniform and contained more wood in proportion to fruit. They also claimed that the wood was more "pulpy." The roots of irrigated trees, critics charged, sprouted closer to the surface and never became as efficient in transmitting nutrients. Unirrigated trees were hardier and more resistant to frosts and disease. Moreover, the fruit from irrigated orchards was commonly considered inferior in taste, if not in quantity or size.[19] These attitudes of nurserymen reinforced the views of farmers as irrigation expanded in the 1870s and later. As early as 1861, Agoston Haraszthy, recognized as the father of the modern California wine industry, wrote that "the experience of

17. *Alta California* August 13, 1873; *Kern County Weekly Courier*, January 4, 1873; *Bakersfield Californian*, December 14, 21, 1876; May 19, 1883; *San Francisco Chronicle*, January 13, 1887.

18. James Schuyler's comments are reprinted in De Pue and Company, *Illustrated Atlas and History of Yolo County, California* (San Francisco, 1879), 82. Also see *Fresno Expositor*, September 6, 1876; and Wallace W. Elliott, *History of Tulare County* (San Francisco, 1883), 107.

19. *California Culturalist*, 1 (December 1858): 312–313; 2 (October 1860): 181–182. *Transactions of the Agricultural Society of the State of California for 1859*, Journals of the California Senate and Assembly (hereafter *JCSA*), 11th sess. (Sacramento, 1860), Appendix, 1: 322; and "Western Agricultural Improvements," *Overland Monthly* 4 (February 1870): 150.

France and all wine growing countries in Europe proves that irrigated vines produce weak wines, void of acidity or astringency, possessing an aguish or watery taste, and without any flavor." He also claimed that wine produced from irrigated vineyards did not keep well.[20] Nor were these views restricted to viticulturalists. A grain farmer in Anaheim wrote the *Anaheim Gazette* in 1871 complaining that irrigation produced a hardpan, "an excessive growth of stalk," and a "small product of grain." Once begun it had to be continued because "plants irrigated have only surface roots, as the subsoil becomes too dense to be penetrated."[21]

The views of farmers and nurserymen probably attracted little attention among the public at large, but the argument that irrigation contributed to or even caused disease won wider recognition. Malaria was second only to tuberculosis among serious diseases in nineteenth-century California, and it struck soldiers stationed at posts in the swampy Sacramento Valley particularly hard. Until the turn of the century, when the germ theory of disease won acceptance, doctors and scientists, as well as the general public, believed that a wide range of diseases resulted from the decomposition of plant and organic matter. Standing water promoted decay, but high temperatures, too much sunshine, and even the wind were also believed to contribute to the spread of dangerous "miasmata." Appropriately, the literal definition of malaria was "bad air." Many doctors believed that trees purified the atmosphere and that their virtual absence in the Central Valley and large parts of southern California helped explain the persistence of malaria in those sections of the state.[22] Irrigation, of course, seemed to contribute to the problem, and not just in the Central Valley. The *Pacific Rural Press* noted that "there is no doubt but that the irrigation of lands late in the season in this State is almost sure to be followed

20. *Sacramento Daily Union,* November 11, 1861; Vincent P. Carosso, *The California Wine Industry: A Study of the Formative Years* (Berkeley, 1951), 42–43.

21. As reprinted in the *Alta California,* February 11, 1871.

22. Kenneth Thompson, "Insalubrious California: Perception and Reality," *Annals of the Association of American Geographers* 59 (March 1969): 50–64; idem, "Irrigation as a Menace to Health in California: A Nineteenth Century View," *The Geographical Review* 59 (April 1969): 195–214. For a sampling of newspaper explanations of the causes of malaria, see the *Havilah Weekly Courier* (Havilah, Calif.), September 21, 1869; *Kern County Weekly Courier,* July 27, 1872, June 6, 1874; *Weekly Colusa Sun,* May 31, 1884; and *Visalia Weekly Delta,* January 1, 1885; May 13, 1886.

by a general prevalence of chills and fever and other bilious diseases in the vicinity of such lands. The prevalence of diseases of this character in the portions of the foothills of the Sierras [sic] is now generally attributed, and no doubt truly, to the presence of water in the mining ditches, reservoirs, hydraulic tailings, etc." The *Press* suggested that farmers confine irrigation to the winter months.[23]

The presumed relationship between irrigation and disease worried the State Board of Health during the 1880s, as irrigation expanded rapidly in southern California. The board first recognized the danger in its report for 1873, and in the late 1870s or early 1880s formed an Irrigation and Tree Planting Committee to study the problem. The group devoted most of its attention to the Los Angeles basin. Dr. J. P. Widney concluded that irrigated land near the coast developed "with irrigation, a very active form of malaria." Inland communities near the mountains, such as San Gabriel, Pomona, and Riverside, seemed to have little problem. According to Dr. Widney, the porous soil and heavier native vegetation in these areas protected against malaria. Like Widney, Dr. H. S. Orme emphasized the value of drainage and noted that many diseases similar to malaria were caused by poor sanitation rather than irrigation. Nevertheless, irrigation was extremely dangerous in some places because hot weather spurred "into activity many forms of organic germs, including minute algae confervoids, diatoms, bacteria etc. The germ spores of these organisms require both heat and moisture for their full development. Until then, they remain in a passive condition for weeks, months, and even years; but in the presence of heat and moisture, they develop and become prolific with the most wonderful rapidity." In short, the dormant "germs" posed a threat only under certain conditions. Alluvial land rich in humus offered particularly good breeding ground. When the summer temperature exceeded sixty degrees, according to Dr. Orme, water touched off a process of "poisonous fermentation." He maintained that planting eucalyptus trees in irrigated areas subject to malaria offered the best protection from disease.[24]

23. *Pacific Rural Press* 1 (June 24, 1871): 388.

24. *Seventh Report of the State Board of Health of California for July 1, 1880 to December 1, 1881* (Sacramento, 1882), 104–106; *Eighth Biennial Report . . . for the Years 1882 and 1883* (Sacramento, 1884), 51–59; *Ninth Biennial Report . . . from June 30, 1884 to June 30, 1886* (Sacramento, 1886), 132–133; *Tenth Biennial Report . . . from June 30, 1886 to June 30, 1888* (Sacramento, 1888), 224–227.

Most arguments against irrigation focused on cost or health hazards, but it faced other obstacles as well. Those who favored irrigation were often (and with considerable justification) charged with selfish motives. The *Stockton Daily Independent* observed in 1873: "All at once there is a wonderful mania prevailing in relation to the necessity of watering every poor man's land, and it is worthy of observation that it is not poor men or small farmers who are manifesting any great anxiety upon the subject." A correspondent of the Colusa's *Sun* charged that the irrigation crusade had been "gotten up by speculators—by designing men, or perhaps, from those who have large amounts of poor land in the valley and wish to get up some excitement in order to sell their land." The mere *possibility* that an irrigation project would be undertaken in a particular section caused land prices to soar.[25] During the 1870s and 1880s, some critics feared that irrigation would destroy the family farm. The small farmer could not afford to build his own canals, nor could he compete for markets with large irrigated farms worked by Chinese or other cheap labor. Irrigation would encourage private land and water companies to monopolize those resources in anticipation of huge future profits. Where could the "poor man" get cheap land when the best tracts had been taken up by speculators?

Of course, the proponents of irrigation found arguments of their own. The best was that irrigation paid. The farmer, no less than the land speculator, yearned for quick returns. Irrigation assured a crop even in dry years and could spare the prudent farmer from bankruptcy. Irrigated fruit and vegetables promised the greatest revenue. Land at Riverside, which in the 1880s cost about $500 an acre with a secure water right, yielded an annual profit of $1,000 to $3,000 per acre. Orange trees usually took at least seven years to bear, but within twelve years they could return $10 per tree each year. In 1871, a writer in the *Overland Monthly* claimed that three or four years after planting, a citrus orchard would return "not only a competency, but an independent fortune." In 1874, a writer in the same publication crowed: "As a rule, the first crop [of oranges] is sufficient to pay all current expenses. The second crop will give a fair profit, while the third crop . . . is enough to pay back all the principal invested. . . . This is no fancy picture—the

25. *Stockton Daily Independent*, July 11, 1873; *Weekly Colusa Sun*, July 3, 1875.

dream of an imaginative mind. The Los Angeles and San Gabriel Valleys, in Los Angeles County, afford ample proof of the truthfulness of these assertions." The author claimed that an investment of $25,000 would return $45,000 plus 10 percent interest on the principal in five years and an annual income of $70,000 a year thereafter. These southern Calfornia "gold mines" promised the investor a much better return than mines in the northern part of the state.[26]

Proponents of irrigation insisted that it dramatically increased production, whatever the crop. Estimates varied widely, particularly since some land was completely unproductive without irrigation. In 1867, the U.S. Commissioner of Agriculture suggested that irrigation increased production by one-fourth to one-third—besides insuring a crop. But by the 1880s and 1890s estimates became far more optimistic. For example, Richard J. Hinton estimated in 1890 that irrigation doubled the production of grain over the yield of an equivalent farm in the humid part of the nation, and increased the production of root crops and garden vegetables from five to ten times.[27] Boosters also pointed out that at least in southern California, irrigation allowed crops to be raised all year long, making farmers independent of the seasons. The increased productivity was usually attributed to the fertilizing properties of water, the assumption that water carried an abundance of nutrients and stimulated the effects of fertilizers already in the soil. In California, this argument was often directed toward wheat farmers. Wheat quickly drained the soil of nitrogen, resulting in dramatic declines in production in as few as five or ten years after virgin soils were opened to cultivation. In 1861, one farmer maintained that there was "little

26. Remi Nadeau, *City-Makers* (Garden City, N.Y., 1948), 18; Merlin Stonehouse, *John Wesley North and the Reform Frontier* (Minneapolis, 1965), 231; Benjamin C. Truman, *Semi-Tropical California: Its Climate, Healthfulness, Productivness, and Scenery* (San Francisco, 1874), 43; Josephine Clifford, "Tropical California," *Overland Monthly* 7 (October 1871): 11; T. Evans, "Orange Culture in California," ibid., 12 (March 1874): 241–242.

27. *Report of the Commissioner of Agriculture for the Year 1867* (Washington, D.C., 1868), 195; Hinton, "Irrigation in the United States: Progress Report for 1890," 15. Also see *Visalia Weekly Delta*, November 17, 1882; *Sacramento Daily Union*, Jan. 30, 1866, Aug. 20. 1889; Titus Fey Cronise, *The Natural Wealth of California* (San Francisco, 1868), 381; *Report of the California State Board of Agriculture for the Years 1864–1865, JCSA*, 16th sess. (Sacramento, 1866), Appendix, 2: 21–25.

doubt that the chemical and mineral properties of the soils which enter into the substance or fruit of a plant, can be more readily taken from the soil when it is saturated with water."[28] In 1885, the *Chicago Tribune* commented "that irrigated land does not seem to wear out—the water acting as a perpetual manure and renewing its fertility as fast as exhausted by cropping. . . . The Saints in Utah find their irrigated land unimpoverished after thirty or thirty-five years of constant cropping." Six years later, a writer in the *Rural Californian* remarked that "when used warm" water became "a great stimulant."[29]

Irrigation was also expected to encourage crop diversification. Farmers who raised wheat seldom grew anything else. Consequently, during the 1850s and 1860s, the state imported much of its food as the prosperity of California agriculture came to depend on the whims and caprice of the Liverpool wheat market, halfway around the globe. Tight credit, labor shortages, and shipping monopolies were only three of the reasons wheat farming was such a risky business. In the midst of the great drought of 1872, one newspaper editorialized: "Diversify the crops, then begins the date of a new development in the agricultural resources of California; new wealth will spring up, more varied interests will be directed here, and better prepared will be the State to compete with the States of the East. Corn, wheat, oats, barley, vegetables and fruits should be raised, so as to make every farmer independent of all markets; in fact, make his own farm a market place for such products."[30]

Many critics of wheat culture also expected irrigation to stimulate immigration into the state by encouraging the subdivision of large wheat farms as land prices and taxes increased. In this way, more land would become available to family farmers. In an address before the state agricultural society in September 1874, Morris Estee explained: "Once irrigate the country, and the lands in large tracts under one ownership will, as a rule, be confined to remote or

---

28. *Sacramento Daily Union*, November 11, 1861.

29. *Chicago Tribune*, as reprinted in *Visalia Weekly Delta*, January 15, 1885; *Rural Californian* 14 (July 1891): 395–396.

30. The *Stockton Daily Independent*, February 9, 1872. Also see the *Independent* for April 16, 1873.

mountainous districts, while gardens and orchards will be found on every one hundred acres of land in all the valleys of the State, population will increase, wealth will become more evenly distributed, villages will appear every few miles, [and] a thousand pleasant homes will dot the State where now there are but scores."[31]

The subdivision of large estates would pave the way for a new kind of "irrigation civilization." By the 1890s, such "philosophers" of irrigation as William Ellsworth Smythe promised that the new innovation would transform rural life by demanding cooperation and discouraging the individualism so characteristic of nineteenth-century farming. The cooperation required to build canals and distribute water would carry over to the formation of marketing cooperatives, cooperative dairies and canneries, and a wide range of civic institutions ranging from churches and schools to performing arts groups and literary guilds. The small farms and dense settlement pattern required by irrigation would also dispel the isolation and dreariness of rural life experienced by farm families in the Midwest.

Above all, irrigation would strengthen the family and the middle class, eliminating wealth and poverty along with the servile class of farm laborers so noticeable in California by the 1870s. On small farms, the owner worked side by side with his family and hired hands. And since irrigated crops were often cultivated year around, requiring continuous labor, the worker could avoid becoming a tramp, part of the despised, rootless class whose members drifted from farm to farm and constituted a permanent class of dispossessed.[32]

Some boosters believed that irrigation was part of God's plan to make California a new promised land. Perhaps the state's arid environment had been designed to test man's ingenuity and spur

---

31. *Transactions of the California State Agricultural Society* (hereafter *TCSAS*), *for the Year 1874, JCSA,* 21st sess. (Sacramento, 1876), Appendix, 1: 201. Also see *Stockton Daily Independent,* February 17, 1879; *Weekly Colusa Sun,* July 30, 1887.

32. The best summary of William E. Smythe's thinking is his *Conquest of Arid America* (New York, 1900), based on writings published during the 1890s. His ideas were not new. Many had appeared in the columns of California newspapers and magazines as early as the first years of the 1870s. For the development of irrigation as a tool of social reform on the national level, see Donald J. Pisani, "Reclamation and Social Engineering in the Progressive Era," *Agricultural History* 57 (January 1983): 46–63.

him to use resources more efficiently. Was it just coincidence that so many arid states needed irrigation, but so few had the water to make irrigation possible on a large scale? California contained an abundance of fertile, flat land near the vast natural reservoir of the Sierra Nevada. Those mountains delivered water to the eastern half of the San Joaquin Valley through a series of evenly spaced, roughly parallel rivers. And in the driest part of the valley, Tulare Lake served as another natural reservoir, capable of serving all the dry land in the west half of the valley. Moreover, the gentle slope of the Sacramento and San Joaquin valleys invited the construction of gravity-fed canals. Perhaps the absence of summer floods and rains at harvest time were also part of God's plan. Potentially, California offered the grandest irrigation system in human history. The Creator left man responsible for carrying His design into practice.[33]

Proponents of irrigation cited many other potential benefits. It promised to beautify the state by transforming parched, dusty fields into lush gardens. It raised the prospect that forests could be grown in the Central Valley to supplement the less accessible— and some thought rapidly disappearing—timber supply of the Sierra Nevada. It might even undermine the railroad's transportation monopoly in California by providing canals that could double as avenues of inland transportation. Since irrigation offered so much to so many, in the 1870s and 1880s it became a very significant reform issue.

Irrigation might well have made more headway in the nineteenth century had it rested on a sounder scientific foundation. Irrigators rarely applied water to the land carefully, and the quality of their crops suffered accordingly. Flood irrigation consisted of soaking small sections of land. These were surrounded by dirt embankments or levees, which were opened one by one until an entire tract of land had been watered. Compared to furrow irrigation, flood irrigation was cheap because it required little labor or equipment. But it also resulted in a tremendous waste of water,

33. For examples of the religious impulse in irrigation propaganda, see the *Bulletin*, May 14, 1864; *TCSAS*, 1864, in *JCSA*, 16th sess. (Sacramento, 1866), Appendix, 3: 72; *TCSAS*, *1874*, *JCSA*, 21st sess. (Sacramento, 1874), Appendix, 1: 190; *Fresno Expositor*, November 13, 1872; *Stockton Daily Independent*, April 21, 1877 (reprint from *Tulare County Times*); the *Sacramento Daily Union*, December 29, 1877.

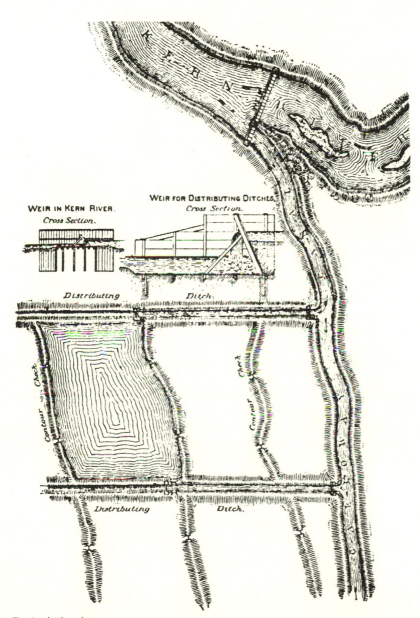

Typical Flood Irrigation System, 1880s. *Report of the Special Committee of the United States Senate on the Irrigation and Reclamation of Arid Lands* (Washington, D.C., 1890).

leaving stagnant pools in low areas.[34] Not until the twentieth century would soil scientists and hydraulic engineers conduct systematic studies of the "duty" of water, the optimum quantity of water needed to raise a particular crop in a particular kind of soil in a particular climate.

As one of California's most perceptive students of agriculture observed in 1923, early misconceptions about irrigation "influenced settlement for some time and delayed development of those vast areas of interior plains and mesas from which the greatest volumes of distinctively Californiana products are now secured."[35] By 1910 or 1920, many myths had been exploded. California farmers recognized that some plants with shallow roots, including virtually all summer vegetables, *required* irrigation to mature well even where rain fell in abundance. And where rainfall was sufficient to produce regular harvests, irrigation improved both the quality and quantity of crops, particularly fruits. Admittedly, too much water produced poor plants. This was true in the humid East as well as in the arid West. Heavy summer rains damaged the quality of fruit as much as did excessive irrigation.

The task of assessing the impact of the critics of irrigation on the institution's development is no easier than trying to measure the impact of inadequate water laws. California certainly was not unique. Everywhere in the West, irrigation was the stepchild of land speculation. And everywhere in the West, stockmen recognized that irrigation and intensive agriculture directly threatened the future of the open range. Yet California was different. Far more irrigation schemes were launched in California, not just because of the availability of investment capital but also because of the wide

---

34. California water laws encouraged waste. "Beneficial use" was not the same as "economical use," and farmers used as much water as possible so that they could, in effect, reserve a supply for future use on additional acreage. The courts rarely questioned the volume of water diverted as long as it was used for a legitimate purpose.

There were four basic methods of irrigation used in California in 1900: flooding (by far the most common); furrow irrigation; the use of ditches to saturate the subsoil and raise the water table to the level of plant roots; and subsurface pipes. See Hinton, "Irrigation in the United States, 1891," 52d Cong., 1st sess., 1893, S. Ex. Doc. 41, (serial 2899) 307; and *Twelfth Census of the United States Taken in the Year 1900: Agriculture, pt. 2, Crops and Irrigation* (Washington, D.C., 1902), 810–814.

35. E. J. Wickson, *Rural California* (New York, 1923), 312.

range of potential projects. And the multitude of social and economic problems faced by California in the last three decades of the nineteenth century, along with the diverse climate of the state, forced irrigation boosters to spend much more time justifying their cause. In the 1890s and early years of the twentieth century, when such advocates of federal arid land reclamation as George Maxwell, William Ellsworth Smythe, and Francis G. Newlands attempted to sell irrigation to the nation, they would rely heavily on the California experience for inspiration. Although Utah was the first state to practice irrigation on an extensive scale, California was the first to appreciate its value as a tool of social reform.

4

Schemers and Dreamers: California's First Irrigation Projects

Long before the Spanish arrived in North America, Indians irrigated as much as 250,000 acres, using 1,000 miles of canals and laterals, in Arizona's Salt River Valley. They also tapped the Gila, Verde, Rio Grande, and Colorado rivers, among others. Kansas, Colorado, and Utah show evidence of pre-Columbian water systems, but in the Southwest irrigation reached such a level of sophistication that gravity-fed canals, sometimes carved through tough volcanic outcroppings and lined with clay to prevent seepage, wound for miles along hillsides and through canyons. Mexicans later improved and extended some of these aqueducts and also built their own. Although oases of irrigated fields could be found outside Mexico, such as the wheat fields cultivated by the Hudson Bay Company at Fort Walla Walla in the 1830s, in the 1840s the largest tracts of irrigated land were the thousands of acres in the Rio Grande Valley and Utah and the hundreds of acres served by Los Angeles' community water system. Outside Los Angeles, irrigation was practiced on a limited scale in other southern California settlements, particularly near present-day Colton on the Santa Ana River, and on land adjoining the missions.[1]

1. George Thomas, *Early Irrigation in the Western States* (Salt Lake City, 1948); Roy Huffman, *Irrigation Development and Public Water Policy* (New York, 1953); and C. S. Kinney, *A Treatise on the Law of Irrigation and Water Rights and the Arid Region Doctrine of Appropriation of Waters* (San Francisco, 1912), 1: 117–125.

The first two decades after statehood witnessed dramatic changes in California's economy, including the rise of the wheat industry and the virtual disappearance of the baronial ranchos of southern California. Irrigation remained limited because transportation was still too primitive, the population too scattered and migratory, land titles too unsettled, and the mining industry too covetous of investment capital. As late as 1870, no more than 60,000 to 100,000 acres were irrigated in California. Nevertheless, these decades saw the first private irrigation ventures, wild and impractical schemes consistent with the speculative, isolated economy of a frontier state. Irrigation quickly became the stepchild of land speculation. Water companies subdivided large blocks of land adjoining the canals that they, or associated land companies, owned. Promoters attempted to secure both land and water monopolies, but they usually constructed shoddy, wasteful canal systems and provided poor service to farmers.

California comprises slightly more than 100,000,000 acres of land. In 1850, only 62,324 were "improved," and the value of the state's farms was the lowest in the nation, only $3,874,041. Aside from mining, cattle raising was the state's largest industry, but it lagged far behind that same industry in Texas, New York, Pennsylvania, the Carolinas, and Georgia, both in the size of herds and the quality of animals. California led the nation only in viticulture—a dubious honor when respectable California hotels refused to serve the state's wines—and Ohio, of all states, was close on its heels.[2] Substantial agricultural growth occurred during the 1850s as the mining camps provided lucrative markets for farmers. Acres cropped increased from 110,748 in 1852 to 461,772 in 1855, and during the decade the number of farms increased from 872 to 18,716. However, almost all were within one hundred miles of San Francisco.[3]

Southern California was particularly isolated in the 1850s. The Butterfield Overland Mail Company's line from Saint Louis to San Francisco via El Paso, Tucson, Yuma, and Los Angeles, did not open until 1858, and the transcontinental telegraph line did not reach the city until 1861. In 1850, fewer than 3,000 of the 3,250,000

2. *Report of the [U.S.] Commissioner of Patents, for the Year 1851* (Washington, D.C., 1852), 494–499, 502–505.

3. Gilbert C. Fite, *The Farmer's Frontier, 1865–1900* (New York, 1966), 157, 164.

acres in Los Angeles County were under cultivation, and the crops they produced returned less than $9,000. Roughly 1,000,000 acres in the county were devoted entirely to feeding 100,000 head of cattle. The city's population increased from 1,610 in 1850 to 4,399 in 1860. This stimulated wheat production, and the demand for grapes, wine, and brandy in the mining camps led to a sevenfold increase in the number of grapevines in the county, most of which were irrigated.[4]

The early development of irrigation in southern California owes much to the Mormon settlement at San Bernardino and the German settlement at Anaheim. In 1851, elders of the Mormon church purchased the San Bernardino Rancho from the Lugo family and sent out a party of settlers. The new residents fenced and planted 2,000 acres of grain, built houses, set up a lumber mill, put out a 40-acre vineyard, and erected a fortress for protection against the Indians. This settlement, the first of a string of towns along the "Mormon corridor" connecting the ports of southern California with Salt Lake City, was expected to serve Mormon converts from the Hawaiian Islands and Asia headed for Utah. But it also provided a source of revenue for the church at a time when the demand for flour in Los Angeles permitted the Saints to sell their grain for as much as $5 a bushel. However, conflict soon broke out between Mormons and non-Mormons, particularly because the Saints insisted that the San Bernardino Rancho contained 80,000 acres, a claim challenged by squatters on the contested land. Following the pattern established in Utah, the settlers, who numbered 1,000 by late 1853, provided labor to construct irrigation ditches in proportion to the number of acres each farmed, though land was never held or worked in common. In 1856, they irrigated 3,000 acres of grain and "city-farms" as small as 1 acre. The Santa Ana River served as the Mormon colony's main source of water.[5]

4. Robert Glass Cleland, *The Cattle on a Thousand Hills: Southern California, 1850–1880* (San Marino, Calif., 1951), 153, 138, 142, 43; Joseph Netz, "The Great Los Angeles Real Estate Boom of 1887," *Annual Publication of the Historical Society of Southern California* 10 (1915–1916): 54–55.

5. George and Helen Beattie, *Heritage of the Valley: San Bernardino's First Century* (Pasadena, Calif., 1939), 170–310; L. A. Ingersoll, *Ingersoll's Century Annals of San Bernardino County, 1769 to 1904* (Los Angeles, 1904), 135–136; Flora B. Houston, "The Mormons in California, 1846–1847" (M.A. thesis, University of California,

How the Mormons doled out the water is still something of a mystery. San Bernardino took full advantage of the 1854 law creating water commissions (discussed in chapter 2). For example, in May 1856, the three commissioners—all Mormon—granted a request from the largely non-Mormon settlers at Timber and City Creek along the Santa Ana to construct canals to serve their communities, even though the two ditches had the capacity to carry more than the entire summer flow of the river. Several months later a Mormon bishop constructed a ditch five miles upstream from the Timber and City Creek diversions to serve old mission lands. Apparently, he had not requested the commission's prior approval because when residents along the two canals asked the commissioners to block the diversion, their request was granted even though the bishop's ditch would have served more land. One history of the county suggests that the Mormons simply affirmed the doctrine of prior appropriation. However, since the exact boundaries of the San Bernardino rancho had not been determined, and since conflict between Mormons and "gentiles" threatened the existence of the colony, the decision may simply have been an expedient to avoid a "water war" rather than a recognition of a legal principle that was not honored in nearby Los Angeles or in Utah. Significantly, in the aftermath of this contest, the Mormons did not abandon the concept of community control of water; they simply extended the old zanja that served lands near present-day Redlands.[6]

In November 1857, following the famous Mountain Meadows Massacre, Brigham Young recalled the San Bernardino Mormons to Utah. Most returned, and the remaining land in the Rancho San Bernardino was sold to non-Mormons in early 1858. Nevertheless,

---

Berkeley, 1929), 69–96. H. F. Raup, in "San Bernardino, California: Settlement and Growth of a Pass-Site City," in University of California Publications in Geography 8 [1940–1962], (Berkeley and Los Angeles, 1962), claims that the Mormons dry-farmed all their land, except for gardens, and that irrigation played little part in the settlement's early history (pp. 11, 30). Raup assumes this because wheat was not usually irrigated in California, and San Bernardino was an infeasible place to practice irrigation. Its proximity to Cajon Pass subjected it to dessicating Santa Ana winds, cold air from nearby mountain canyons, and occasional freezes in the winter. However, virtually all other sources indicate that the Mormons practiced irrigation on an extensive scale, at least for the 1850s.

6. Beattie, *Heritage of the Valley*, 241–242, 246.

irrigation continued and, as noted in chapter 2, the system of community control was expanded in the late 1850s and early 1860s.

San Bernardino had been founded as a way station for Mormons on their way to Utah, not as an experiment in desert agriculture. But in the same year the Mormons departed for Utah, fifty Germans, members of the Los Angeles Vineyard Company, formed the state's first irrigation colony at Anaheim. It was laid out thirty miles southeast of Los Angeles near the Santa Ana River on a 1,165-acre tract purchased for $2 an acre; the most prominent native vegetation was cacti. The land was divided into fifty 20-acre tracts. Ten additional tracts, at the center of the community, were reserved for schools and other public purposes. The colonists included an assortment of carpenters, watchmakers, blacksmiths, teachers, shoemakers, merchants, and even musicians. Within a year, 8 acres of vines had been planted on each farm. By 1861, the Anaheim colony counted 300 residents.[7]

Viticulture could not succeed without irrigation, and leaders of the colony constructed a five-mile ditch from the Santa Ana River before the first settlers arrived. In 1860, the Anaheim Water Company took over the canals and water rights. The company was not designed to make money. It issued fifty shares of stock, one to each landowner, and provided that the stock could not be transferred independent of title to the land. Thus, the Anaheim colonists laid the foundation for the mutual water companies that became dominant in southern California (though not north of the Tehachapis) by the end of the 1880s. Such companies rested on several principles: that irrigation water should be sold at cost; that company policies should be determined by the water users themselves; that the amount of water received and the individual farmer's influence over company policies should depend on the amount of land irrigated; and that water rights should be appurtenant to, or attached to, the land watered. Los Angeles' strong system of public control undoubtedly contributed to these principles. Anaheim was part of

---

7. On the Anaheim Colony see Vincent P. Carosso, *The California Wine Industry: A Study of the Formative Years* (Berkeley, 1951), 60–73; Mildred Yorba MacArthur, *Anaheim: "The Mother Colony"* (Los Angeles, 1959), 26–27; Juan Jose Warner, *An Historical Sketch of Los Angeles County, California* (Los Angeles, 1876), 59; Richard D. Bateman, "Anaheim Was an Oasis in a Wilderness," *Journal of the West* 4 (January 1965): 1–20; and Dorothea Paule, "The German Settlement at Anaheim" (M.A. thesis, University of Southern California, 1952).

Los Angeles County until the formation of Orange County in 1889, and the German settlers used the Spanish words *zanja* and *zanjero* instead of ditch and watermaster. As the colony grew and the first tracts of land were broken up or consolidated, the original policies became impractical. In 1880, the company's stock was divided into 3,000 shares, each representing an acre. Only landowners could buy the stock, but by that time it could be transferred from one parcel to another. Moreover, when company assessments for the maintenance and repair of ditches were not paid, the stock could be advertised and sold at public auction to any interested buyer, landowner or not.[8]

The pattern of public control, seen from the beginning at Los Angeles, San Bernardino, and Anaheim, was not evident in other parts of the state. More typical was the Capay Valley in Yolo County, an extremely fertile block of land twenty miles long and one to three miles wide, sheltered by hills on all sides and fed by Cache Creek, which drained out of Clear Lake. Except for Jerome Davis, who irrigated orchards and vineyards on his 8,000-acre ranch twelve miles west of Sacramento at present-day Davis in the early 1850s, James Moore was the first irrigator in the county. He built his first ditch in 1856, eight miles north of modern Woodland. Since he owned 1,000 acres of riparian land and established the earliest claim to Cache Creek (Yolo County was not among those covered by the 1854 water commission law), Moore exercised an absolute monopoly over water use in the Capay Valley. The Moore Ditch cost $10,000 to dig, but by the late 1870s its owner had spent $50,000 defending his water rights against subsequent appropriators. Still, irrigation was a profitable business for Moore. In the dry years of 1877 and 1878, he collected $12,000 from water sales. In the early 1870s, Moore's network of ditches served about 15,000 acres.[9]

The Moore irrigation system paled in comparison with some of the other schemes hatched in the 1850s and 1860s. The largest involved the southern San Joaquin Valley. The San Joaquin was little more than a wasteland in the first two decades after statehood. Cattle shared the native grasses with herds of antelope, elk, and

8. Paule, "The German Settlement at Anaheim," 60.

9. Rosemary McDonald More, "The Influence of Water Rights Litigation upon Irrigation Farming in Yolo County, California" (M.A. thesis, University of California, Berkeley, 1960), 7–8; De Pue and Company, *Illustrated Atlas and History of Yolo County, California* (San Francisco, 1879), 80–81.

wild horses. The dreary landscape was broken only by fields of tules and by the strings of willow trees which lined the streams that wound out of the Sierra. It was a sunbaked, brick-hard land in the summer, punctuated only by an occasional stream and dangerous marshes, sloughs, and bogs. From Stockton at the head of the valley to the Tehachapi Mountains 250 miles to the south, Visalia was the only town of any size, and its population numbered no more than a few hundred.

In September 1850, the federal government promised to deed to a handful of states the swamp and overflowed land within their borders on condition that the money received from land sales be used to aid in draining the swamps. The act was passed particularly to assist states along the lower Mississippi river where flooding was a year-round problem. In the arid West, most overflowed land was under water no longer than a few weeks a year. The law provided that only those lands designated as "swamp and unfit for cultivation" on federal land office maps would be turned over to the states.[10] In California, however, federal surveys of public land moved painfully slow. Repeatedly, California's officials appealed to Washington for faster service, but to no avail; the state did not receive patents to the land until 1866, by which time most of it was already in private ownership. The delay prompted the California legislature to take the law into its own hands. In April 1855, it requested the county surveyors-general to designate the flood land within their counties. The new law allowed individuals to buy "swamp" land at $1 an acre if they could secure two affidavits swearing that the land was flooded. To prevent speculation, especially in "urban" real estate, the legislature prohibited the sale of land within ten miles of San Francisco, five miles of Sacramento, Stockton, and Oakland, and one mile of the Sacramento River. Still, the law was an open invitation to fraud, and a company of San Francisco land speculators milked it for all it was worth.[11]

10. Roy Robbins, *Our Landed Heritage: The Public Domain, 1776–1970* (Lincoln, Neb., 1976), 154–156.

11. The swampland bill of 1855 is reprinted in full in the *Sacramento Daily Union*, April 18, 1855. Also see Governor John Bigler's address to the California Legislature reprinted in the *Union* of January 10, 1856; *Report of the Surveyor-General for 1858*, Journals of the California Senate and Assembly (hereafter *JCSA*), 9th sess. (Sacramento, 1859), Appendix, vol. 1; *Report of the Surveyor-General for 1860*, *JCSA*, 11th sess. (Sacramento, 1860), Appendix, 1: 12–13.

On April 11, 1857, the legislature promised W. F. Montgomery and Associates (the Tulare Land and Canal Company) all the odd-numbered sections of swamp and overflow land within the public domain between Tulare Lake and the Kings River Slough on the San Joaquin River, along with generous chunks of swampland bordering Tulare, Buena Vista, and Kern lakes, in exchange for the construction of a thirty-four-mile canal linking Tulare Lake and the San Joaquin River. Construction had to begin within one year and finish within five on pain of forfeiture of the franchise. The Company received the right to charge tolls from boats passing through the canal, though control over the waterway would pass to the state after twenty years.[12]

The canal was expected to accomplish three purposes. The company claimed that Tulare Lake was at least thirty-five feet higher than the San Joaquin River, so the canal would drain a large part of that lake as well as Buena Vista and Kern lakes, which fed into the larger lake from the south end of the San Joaquin Valley. The project's promoters predicted that as much as 500,000 to 700,000 acres of reclaimed alluvial soil would become available to new settlers as the waters receded. In addition, the canal would provide a transportation artery linking the southern San Joaquin Valley with Stockton and the San Francisco Bay, via the San Joaquin River. Crops, cattle, and hogs would find easy access to the markets of northern California, and the cost of transporting goods into the southern valley would be dramatically reduced. No longer would most of the San Joaquin Valley be an isolated desert. Finally, though the canal's main purposes were drainage and transportation, it would also provide water to irrigate thousands of parched acres along its route. Early in the fall of 1857, the Stockton *Argus,* one of the scheme's most ardent supporters, confidently predicted: "The opening of this canal will direct the trade of that whole valley to Stockton."[13]

The Tulare Land and Canal Company quickly began selling the swampland to settlers for $1 an acre. In December 1857, a company spokesman wrote to the *Argus* recounting a trip to Tulare Lake during which he claimed to have been lost in the tules for a

---

12. *Cal. Stats.,* 1857, 192.

13. As reprinted in the *Bulletin,* October 6, 1857.

week. He reported that work had begun on a drainage aqueduct that would rival the Erie Canal in width and depth. Settlers, he noted, had begun to flock into the lower valley:

> Already is this work drawing into that valley a large immi-gration, principally from the Coast Range, driven from their former homes by Spanish grants, and drawn there by the fertile soil and prospects of the speedy completion of the canal for irrigating and commercial purposes. The lands thus opened to market are sold by our State Government, to which they belong at the same low price as is demanded for unreclaimed lands under the law of 1855. Large locations are being made there. The Company making the canal receives pay for its labor in lands upon its border, and, of course, their price of lands is governed by that of the State.

The company's agent reported that many prospective settlers were helping to excavate the canal in exchange for land. By spring, he predicted that over 200,000 acres would be reclaimed.[14]

Nevertheless, public criticism of the project surfaced in the opening months of 1858. The *San Jose Tribune* labeled the scheme a "barefaced land stealing operation" and "an intense humbug and swindle." It charged that the company had done no substantial work on the canal and owed a blacksmith $150 for the plow de-signed to "dig" the drainage ditch. The newspaper questioned the company's plan to dig a furrow, then use the flow of water to carve out the canal. Such an aqueduct could not float boats of 80-ton burden, as the 1857 law required. The editorial also charged that the company had taken possession of vast tracts of dry land which did not require reclamation. The *Tribune* called for a legislative inquiry. The *Stockton Democrat* seconded the request and described W. F. Montgomery and Associates as "a party of seedy, hungry, broken down schemers." It pointed out that since the swampland grant had not been surveyed and formally designated by the state, the company, in effect, exercised a virtual monopoly over the best land in the southern San Joaquin Valley. Both newspapers, along

14. As reprinted in the *Alta California*, December 20, 1857, and in the San Francisco *Daily Evening News*, December 14, 1857.

with the powerful *Sacramento Daily Union*, urged the state to rescind the grant.[15]

The California Senate's Swamp and Overflowed Land Committee investigated the project in January 1858 and issued a report on February 2. The company claimed to have invested $40,000 to $50,000 in the project, but the committee discovered that the only work consisted of a furrow several miles long. Theoretically, by beginning work on the canal within one year of passage of the 1857 law, the company had lived up to its part of the bargain. But the legislature, convinced that the company would never complete an acceptable canal, annulled the land grant in the middle of April. Nevertheless, the company continued to sell swampland and sued the state for breach of contract. A year later the Sacramento *Union* editorialized: "The fact appears to be that the Company [has] disposed of large tracts of its land to speculators in San Francisco and elsewhere, and these parties are in nowise disposed to abandon the field without a struggle."[16]

The "struggle" reached all the way to the California Supreme Court, which upheld the company in 1860, ordering the grant restored. The State had a strong incentive to reaffirm the franchise because many legislators wanted to "borrow" money from the swampland fund to pay general state expenses. Since the state retained the even-numbered sections of swampland in the Tulare and Kern valleys, any migration into that region would result in large sales of state as well as company land, swelling state revenue.[17]

The success of W. F. Montgomery and his band proved short-

15. The *San Jose Tribune* editorial was reprinted in the *Sacramento Daily Union*, January 13, 1858; the *Stockton Democrat*'s statement is from the *Union* of January 28, 1858.

16. *Sacramento Daily Union*, February 3, February 9, April 15, 1858. The quote is from the *Union* of April 7, 1859.

17. *Sacramento Daily Union*, December 15, 1860, April 3, 1862, November 28, 1870. Also see the reports of the state surveyor-general and board of swampland commissioners for the early 1860s in the appendixes to the journals of the legislature, and Robert Kelley, "Taming the Sacramento: Hamiltonianism in Action," *Pacific Historical Review* 3 (February 1965): 21–49; Kenneth Thompson, "Historic Flooding in the Sacramento Valley," *Pacific Historical Review* 29 (November 1960): 349–360; Richard H. Peterson, "The Failure to Reclaim: California State Swamp Land Policy and the Sacramento Valley, 1850–1866," *Southern California Quarterly* 56 (Spring 1974): 45–60.

lived. On April 10, 1862, the legislature restored the grant and gave the company three more years to complete the canal. But the Tulare Land and Canal Company, its capital depleted by the legal battle, could not raise the money needed to "continue" excavating the canal. Late in 1862 it sold out to Thomas Baker and Henry S. Brown for $10,000.[18] Baker was a successful lawyer and president of the first Iowa Senate before he came to California in 1849. In the Golden State, he took up civil engineering, helped found Visalia, went to the legislature in 1855, and in 1858 became receiver of the United States Land Office at Visalia. He served a second term in the leg- islature in 1861–1862 as senator from Fresno and Tulare counties.[19] Baker knew his way around Sacramento, and when the legislature met in 1863, following the massive statewide flood of 1861–1862, he persuaded it to drop the requirement that the canal be navigable.[20]

Subsequently, Tulare Lake was discoverd to be *lower* than the San Joaquin River. Without an adequate lock system, the canal might have flooded more land than it reclaimed. As a result, Baker turned his attention to swampland near Buena Vista and Kern lakes at the south end of the valley. In September 1863, he hired thirty Indians from the nearby Tejon Reservation and spent $30,000 ex- cavating a drainage ditch. However, nature did more to "reclaim" the Kern Valley than Baker. The great drought of 1864 temporarily dried up most of the region's swampland, and on November 11, 1867, Governor Frederick Low and the State Land Office approved patents to 89,120 acres. Ironically, only a few months later heavy rains flooded the land again, and the state attorney-general filed suit to nullify the grant. After a long court battle, the Twelfth District Court in San Francisco sustained his appeal in 1878. By that time Thomas Baker and his heirs (he died of typhoid pneu- monia in 1872) had sold virtually all the land for $1 to $1.50 an acre, and the court allowed claimants who had spent at least $1

18. Wallace Morgan, *History of Kern County California* (Los Angeles, 1914), 55; Herbert G. Comfort, *Where Rolls the Kern* (Moorpark, Calif., 1934), 107–110; Norman Berg, *A History of Kern County Land Company* (Bakersfield, Calif., 1971), 1–3; Lilbourne A. Winchell, *History of Fresno County and the San Joaquin Valley* (Fresno, Calif., 1933), 103–104. The law of 1862 (*Cal. Stats.*, 1862, 190) authorized the governor and state surveyor-general to issue patents to the land, but only after they were satisfied that it had been reclaimed.

19. Thelma B. Miller, *History of Kern County, California* (Chicago, 1929), 5–8.

20. Morgan, *History of Kern County, California*, 55; *Cal. Stats.*, 1863, 494.

an acre on reclamation to keep it. Ultimately, most of the land was acquired by James Ben-Ali Haggin and W. B. Carr and became part of the Kern County Land Company's massive holdings.[21]

The Montgomery-Baker swampland reclamation scheme was no more ambitious than the vague plan offered to the legislature in 1859 by Oliver Wozencraft. A Louisiana-born doctor, he helped draft California's first constitution in 1850, negotiated reservation treaties with California Indians, and became one of San Francisco's civic leaders. He lobbied for a wagon road from Fort Kearny, Nebraska, to San Francisco via South Pass and California's Honey Lake and became chairman of the California Emigrant Road Committee which clamored for better transportation and mail service.[22]

Wozencraft entered California through the Colorado Desert in 1849 and recognized at least part of its agricultural potential. In the early 1850s, he lived in San Bernardino, and that town's civic leaders hoped that a road through the desert would put their community on the map—located as it was near a natural pass through the mountains separating the Mojave and Colorado deserts from the Los Angeles basin. In 1853, Congress authorized a series of surveys to map a railroad route to the Pacific Coast. Lieutenant R. S. Williamson of the U.S. Topographic Engineers headed the southern reconnaissance, assisted by William P. Blake, a young graduate of the Yale Scientific School, who served as the expedition's geologist. Apparently, Blake first recognized that the Salton Sink was lower than the Colorado River. Through him, Wozencraft met the surveyor of San Diego County (which at that time stretched from the Pacific Ocean to the Colorado River), Ebenezer Hadley, who discovered that the Colorado could be diverted into the desert if a dry arroyo near Yuma (the Alamo overflow) was dredged out. Hadley and Wozencraft recognized that the alluvial soils of the desert would prove extremely fertile if they could be irrigated.[23]

21. S. T. Harding, *Water in California* (Palo Alto, Calif., 1960), 145; *Sacramento Daily Union*, March 28, 1863; *Bakersfield Californian*, February 14, 1878.

22. W. Turrentine Jackson, *Wagon Roads West: A Study of Federal Road Surveys and Construction in the Trans-Mississippi West, 1846–1869* (New Haven, Conn., 1952), 175–176, 201, 205; Barbara A. Metcalf, "Oliver M. Wozencraft in California, 1849–1887" (M.A. thesis, University of Southern California, 1963).

23. Metcalf, "Oliver M. Wozencraft in California," 81–99; George Kennan, *E. H. Harriman* (Boston, 1922), 2: 97; Paul L. Kleinsorge, *The Boulder Canyon Project* (Stanford, Calif., 1941), 20–21.

The scheme hatched in 1853 and 1854 initially attracted little attention. Regular steamboat service began on the Colorado River in 1852 to serve mining camp supply towns in Arizona,[24] but there was no overland transportation in the region until September 1858, when the Butterfield Overland Mail Company opened its ox-bow route through Arkansas, Oklahoma, Texas, New Mexico, and Arizona. The most dangerous and uncomfortable part of the trip was from Yuma on the California-Arizona border to Los Angeles. To stay close to water and avoid the intense desert heat, the stages traveled a circuitous 180 miles through northern Mexico before swinging north along the Laguna, Vallecito, and Santa Rosa mountains, thence through San Gorgonio Pass into the Los Angeles basin.

A road straight through the Colorado Desert would save time and eliminate the encroachment on Mexican soil. But there was no water for eighty miles west of Fort Yuma. On April 12, 1859, the California legislature asked Congress to deed the entire Colorado Desert to the state, explaining that the "country herein described is known to be a *desert waste*, devoid of water, and vegetation, owing to which it presents a great barrier to travel, and transportation, on the most approved route of land communication between the Atlantic and Pacific." The lawmakers maintained that a series of canals would provide water for travelers and would "cause the desert to yield to the wants of man her latent, reserved, and hidden stores." Since the Colorado basin had flooded in the past, the legislature even suggested that a land grant would be in keeping with the federal swampland law of 1850.[25] On April 15 the lawmakers declared that once the state had received title to the land, it would transfer ownership to Oliver Wozencraft and his associates, though not until a special commission consisting of state and county officials had inspected the finished canals.[26]

---

24. Edwin T. Force, "The Use of the Colorado River in the United States, 1850–1933" (Ph.D. diss., University of California, Berkeley, 1936), 24.

25. *Cal. Stats.*, 1859, 392.

26. *Cal. Stats.*, 1859, 238. Local groups had already tried to build roads across the desert. The *Alta California* reported on June 28, 1858, that enterprising San Bernardino citizens were planning to dig wells at twenty mile intervals across the Colorado Desert. But shifting sand made building a road to Yuma nearly impossible.

The outbreak of the Civil War delayed consideration of Wozencraft's scheme until May 1862. By that time the commissioner of
public lands had raised many objections to the project, which was
embodied in House Bill 417. The land office complained that no
detailed construction plans, or even estimates of construction costs,
had been submitted to Congress; nor had the federal government
investigated the project's feasibility. The commissioner described
the desert land as "third rate," but warned that it might include
valuable mineral deposits. For this reason the House Committee
on Public Lands reduced the size of the grant from the 6,500,000
acres requested by California to 3,000,000 acres. However, the committee rejected the commissioner's call for a grant of alternate sections on grounds that such a grant would prevent Wozencraft from
securing clear rights-of-way for canals. Two congressional reports
had been prepared containing eyewitness descriptions of the desert
and southern California's transportation problems. They emphasized that the land would remain totally worthless without reclamation. Nevertheless, the bill was tabled.[27]

Congress's lack of interest was not surprising. Many northern
congressmen had opposed the ox-bow route, and the outbreak of
war offered a perfect excuse to shift the Butterfield Overland from
the Southwest to the old Oregon Trail—which was done in 1861.
Moreover, most northern Californians had opposed the southern
route because they feared Los Angeles would be built up at the
expense of San Francisco and Sacramento. When President Lincoln
signed the Pacific Railroad Act on July 1, 1862, guaranteeing that
the first "transcontinental" would also follow the central route, the
dream of a wagon road through the sandy wilderness became
anachronistic. The shift to the central route eliminated the need
for way stations in the desert and cut off the stream of stage passengers into southern California, some of whom might have helped
settle the reclaimed desert.[28]

---

27. *Congressional Globe*, 27th Cong., 2d sess., 1861–1862, pt. 3: 2379–2381; S.
Rep. 276, 36th Cong., 1st sess. (serial 1040); H. Rep. 87, 37th Cong., 2d sess. (serial
1145).

28. The standard history of the Butterfield Overland is Roscoe P. Conkling
and Margaret B. Conkling, *The Butterfield Overland Mail, 1857–1869* (Glendale, Calif.,
1947).

Though Wozencraft never published the details of his vague project, it enjoyed amazing vitality. In 1875, the War Department surveyed 326 miles of the Colorado River and concluded that no diversion could be made "at any point along the present channel of the river within the territory of the United States."[29] Its report conceded that such profitable crops as cotton, coffee, sugar, tea, and flax could be grown in the rich desert soil but denied that canals or a desert lake would alter the climate; the amount of moisture reaching the atmosphere through evaporation would be too small to have an appreciable effect on humidity or rainfall. The scheme stayed alive largely because the Southern Pacific's tracks reached Yuma in 1877, a mining boom followed in Arizona, and the desert became even more accessible in 1883 when the Santa Fe crossed the Colorado at Needles. Congress considered the project again in 1877 and 1887, but public hostility toward land grants and monopolies of all kinds, coupled with the criticisms raised by the War Department, doomed the project. J. Ross Browne, a California journalist who promoted many wild schemes of his own, echoed the skepticism of many when he commented: "I can see no great obstacle to success except the porous nature of the sand. By removing the sand from the desert, success would be assured at once."[30] The Imperial Valley was not opened to large-scale irrigation agriculture until the early twentieth century, and the All-American Canal, which roughly follows the diversion route proposed by Wozencraft, was not completed until 1943.

The third large irrigation scheme of the 1850s and 1860s involved the Sacramento Valley. It grew out of the drought of 1864. In the wake of that disaster, Governor Frederick F. Low acknowledged "the necessity of providing a general system of irrigation for our noble expanse of valley land," and the California State Board

29. George M. Wheeler, "Geographical Surveys West of the One-Hundredth Meridian in California, Nevada, Utah, Colorado, Wyoming, New Mexico, Arizona, and Montana," 44th Cong., 2d sess., 1876, H. Exec. Doc. 1, pt. 2, (serial 1745), 290.

30. J. Ross Browne's comments are reprinted in Kennan, *E. H. Harriman*, 2: 100. For Wozencraft's scheme in Congress see *Fresno Expositor*, February 21, 1877; *San Francisco Chronicle*, Jan. 13, 1887. The project reappeared at least twice more in the California legislature. See A.B. 385 (Rockwell), February 14, 1870, *Assembly Bills, 1869–1870*; S.B. 481 (Satterwhite), March 2, 1876, *Senate Bills, 1875–1876*, California State Law Library, Sacramento.

of Agriculture asked Congress to deed all arid land to the state to aid in reclamation. "Why should not these vast plains, lying back from our great rivers, almost valueless without such improvements, be as justly and properly the subject of redemption by Government land aid, as the lesser extent of tule or swamplands bordering immediately on their banks?" the board asked. "If the policy is good, and it certainly is, in the one case, then why not in the other?"[31]

Will S. Green, the father of the Sacramento Valley project, was destined to become a prominent leader in the irrigation crusade of the 1880s and 1890s. Born in Kentucky, Green emigrated to California in 1849 at the age of sixteen. For a few months he piloted the first steam ferry across the Carquinez Straits at the north end of San Francisco Bay, and he also won a government contract to deliver mail in Sonoma and Napa counties. He moved to Colusa County in July 1850 and helped to establish the city of Colusa on the Sacramento River. There he honed his skills as a self-taught engineer, and in 1863 founded the *Colusa Sun*, a weekly, and by the 1890s a daily, newspaper which he owned and edited until his death in 1905. He served as Colusa County's surveyor from 1857 to 1867 and also spent one term in the California legislature during the late 1860s. Though Green owned thousands of acres of land between the Sacramento River and Butte Creek, his attempts at farming were unsuccessful because of droughts and heavy flooding.[32]

Not surprisingly, the *Sun* ran frequent editorials on the value and necessity of irrigation. "Farming must be made a certainty, or else we had as well quit it," Green commented in April 1864. "We cannot compete with other States and other countries if we must

31. Governor Frederick Low's annual address to the legislature was printed in the *Sacramento Daily Union*, December 9, 1865; *Report of the California State Board of Agriculture for the Years 1864–1865, JCSA*, 16th sess. (Sacramento, 1866), Appendix, 2: 22.

32. The best survey of Will S. Green's life is John P. Ryan's "Notes on Will S. Green, Father of the Glenn-Colusa Irrigation District," a typescript manuscript, Bancroft Library. Also see Julian Dana, *The Sacramento: River of Gold* (New York, 1939), 192–193; McGowan, *History of the Sacramento Valley,* 1: 236. Henry George, in *Our Land and Land Policy, National and State* (San Francisco, 1871), 63, listed Green as one of thirteen individuals who had received 20,000 acres or more from the state.

lose an entire crop every few years." And in December of the same year he noted: "We of the Sacramento Valley have been particularly unfortunate . . . having had as many as four [crop] failures in thirteen years, and four short crops." Green drummed home an important truth. While the Sacramento Valley usually received 20 or more inches of rainfall per year—far more than in the San Joaquin Valley or land south of the Tehachapis—the rain rarely fell when it was needed most. The valley was wetter, but hardly immune to the ravages of drought.[33]

In the fall of 1864, County-Surveyor Green and a business associate, C. D. Semple, called several public meetings in Colusa to discuss a scheme to reclaim 200,000 acres. They wanted to dig two canals to divert water from the Sacramento River near the mouth of Stony Creek, roughly forty miles upriver from Colusa. The first canal would follow the high west bank of the river for an undisclosed distance; the other would skirt the Coast Range until it emptied into Putah Creek, about one hundred miles south in Yolo County. Green suggested that the state could afford to spend $100,000 on the project, and the counties of Yolo, Colusa, and Solano could also subsidize the work by buying stock in his canal company; each county could expect a rapid increase in tax revenue once irrigation became available. But the most novel feature of the plan involved the obligation of the farmers served by the canals. Green proposed that each landowner mortgage his land to the Colusa, Yolo and Solano Canal Company in exchange for shares of stock. The greater the amount of land mortgaged, the greater the investment—and, in turn, the more the control over company policies. The company would issue interest-bearing bonds to pay for construction, holding the land as collateral. Water sales would provide a sinking fund to pay off the bonds. Green repeatedly emphasized that the drought of 1864 had cost far more in crop losses than the anticipated $800,000 cost of the main canal. But his primary inducement was in keeping with the speculative nature of farming in nineteenth-century California. Noting that most land along the river sold for less than $6 an acre, with land farther removed from water worth only half that amount, Green promised: "These lands, after the canal is completed, will be worth twenty dollars per acre, so that the money lender would have an immense

33. *Weekly Colusa Sun*, April 16, December 10, 1864.

margin in addition to the canal itself, which will probably be the most profitable as well as the most secure stock in the State of California."[34]

The Colusa, Yolo and Solano Canal Company won little public support. Few Sacramento Valley farmers favored irrigation and fewer still were willing to pay for it. Throughout the arid West during the last third of the nineteenth century, support for irrigation grew out of immediate water shortages, not from a desire for comprehensive water resource planning or scientific farming; most farmers were not willing to commit themselves to agriculture as a long-term investment. Thus when heavy rains fell in the Sacramento Valley in November 1864, ending the drought, the promise of a bumper harvest in 1865 effaced the bitter memory of stunted crops. When Green appealed for public subscriptions of $5 to $10 per landowner to pay the cost of surveying the main canal, his request fell on deaf ears. Though he paid the survey costs out of his own pocket, the lack of public support forced him to scale down the project. The canal surveyed was only 7.5 miles long. Nevertheless, Green and his boosters maintained it would irrigate 80,000 acres of Colusa County's best farmland.[35]

Having received little encouragement from Sacramento Valley wheat barons, Green turned to Sacramento and Washington. His suggestion that Congress grant valley land directly to the counties was impractical, given the swampland frauds of the 1850s and early 1860s,[36] but Green had a bill ready when the California legislature convened in 1866. He now proposed a 120-mile canal from the Colusa-Tehama county line to Cache Creek Slough in Solano County. The canal would be 100 feet wide and 5 to 6 feet deep, capable of carrying barges laden with wheat, barley, and other bulky crops, and able to irrigate "at least" 600,000 acres of land, most of which could not be farmed without irrigation. Green estimated that the canal would increase the annual tax revenue of Colusa, Yolo, and Solano counties by $5,000,000 to $6,000,000. It would also serve as an "overflow channel" when the Sacramento River reached flood

34. Ibid., October 22, November 12, 1864. The quote is from the latter issue.

35. *Weekly Colusa Sun*, November 26, 1864; April 1, 15, 1865; *TCSAS, 1864*, *JCSA*, 16th sess. (Sacramento, 1866), Appendix, 3: 37; *Report of the Surveyor-General for 1865*, *JCSA*, 16th sess. (Sacramento, 1866), Appendix, 108.

36. *Weekly Colusa Sun*, September 9, 23, 1865.

stage, aiding in swampland reclamation and protecting the flood-prone towns along the river. The bill asked for $8,000 to survey the canal, and the Senate's Committee on Agriculture assured the governor and legislature that foreign investors stood ready to build it. Moreover, the survey appropriation would not have to be paid until the surveyor-general and other state officials decided that the plan was feasible. With these assurances, the bill passed. The appropriation was the state's first direct economic encouragement to irrigation.[37]

William H. Bryan, an engineer appointed by the governor, conducted the survey. He concluded: "I can speak with confidence of the adequacy of the plans proposed for the object in view, provided the execution of them is placed in the hands of persons experienced in the building and management of canals, and are properly supervised during their construction." He estimated that the canal would cost $11,381,068 and irrigate 750,000 acres. If, as Bryan thought, the irrigated land would support an average of one person per acre, the Sacramento Valley's population might increase manyfold. He opposed beginning the canal at the Colusa-Tehama line because most arable land in that part of the valley was too far above the river. But he claimed $8,000 was insufficient to survey an alternate route. The only pessimistic note in the report was Bryan's conclusion that a navigable canal was infeasible. He considered irrigation and transportation incompatible because shipping interests required a channel with little current while farmers needed a fairly strong flow.[38]

The attempt to combine irrigation and transportation helped

37. *Report of the Senate Committee on Assembly Bill no. 321, JCSA,* 16th sess. (Sacramento, 1866), Appendix, 3. Also see *Report of Committee on Internal Improvements of the Assembly on Assembly Bill No. 321* in the same volume.

38. *Report of the Engineer of the Sacramento Valley Irrigation and Navigation Canal, JCSA,* 17th sess. (Sacramento, 1868), Appendix, 2. The quote is from p. 31. Also see William Bryan's letter to I. N. Hoag, secretary of the California State Board of Agriculture in the *Biennial Report of the Board of Agriculture of the State Agricultural Society for the Years 1866 and 1867, JCSA,* 17th sess. (Sacramento, 1868), Appendix, 3: 47. Bryan had apparently travelled in Europe and the Middle East and observed irrigation there. Perhaps this explains why he won the appointment to survey the canal. In any case, in April, 1866, he published a series of articles on irrigation in Italy and India, the first systematic attention paid by a California engineer to irrigation outside the United States. See the *Sacramento Daily Union,* April 5, 13, 20, 27, 1866.

kill Green's scheme. Dual-purpose canals cost much more to build, and critics of the plan balked at the potential transportation monopoly that the Colusa, Yolo and Solano Canal Company would enjoy (the Southern Pacific did not complete its line through the Sacramento Valley until the late 1880s). Finding a reliable source of income was the company's biggest problem. The valley's population was thin and scattered, and in wet years farmers could not be expected to buy water to irrigate. The project would have to rely heavily on canal tolls and land sales.

Ironically, while the fear of transportation monopoly soured many northern Californians on the project, it was the multiple-purpose nature of the canal—flood control and transportation along with irrigation—that proponents of the plan hoped would persuade Congress to approve a land grant. In the late 1860s, Green served on a state commission to draft a flood control plan for the Sacramento Valley. The group endorsed his canal project as a flood control measure, but nothing came of the recommendation. Time worked against Green. By the beginning of the 1870s, much of the valley's public land had been taken up by wheat farmers. As noted in chapter 1, from 1866 to 1872 the acreage planted to wheat in California more than tripled. Theoretically, this increased the potential need for irrigation; but it also broke up the virgin tracts of government land coveted by the company. Rival land and ditch companies soon entered the field. For example, by the early 1870s the Clear Lake Water Works Company began to irrigate land adjoining Cache Creek in the Capay Valley. It hoped to win control over Clear Lake, at the head of the stream, and anticipated that its planned reservoir would irrigate as many as 400,000 acres in the Sacramento Valley. Green's company had plenty of competition, at least on paper.[39]

39. *Report of the Commissioners Appointed to Examine into the Practicability of Making a New Outlet for the Flood Waters of the Sacramento Valley, JCSA,* 18th sess. (Sacramento, 1870), Appendix, 3; 43d Cong., 1 sess., H. Ex. Doc. 290, (serial 1615), 29. A modified scheme to build a dual-purpose canal from Red Bluff to Benecia was proposed in 1873 by John K. Luttrell, a candidate for the state legislature. He suggested that the state issue bonds at 7 percent interest and loan the money to Tehama, Colusa, Yolo, and Solano counties to finance construction. The counties would repay the state using revenue from water sales. However, the scheme was unrealistic because the California constitution forbade the state from loaning its credit to any person, corporation, or county, and the cost of the project exceeded the constitutional debt ceiling. See *Weekly Colusa Sun,* December 6, 1873.

The Montgomery-Baker, Wozencraft, and Green projects had much in common. Each was the creation of an individual entrepreneur who, even with the aid of associates, lacked the financial resources to accomplish his objectives. In all three instances, the promoters faced a dilemma. To interest potential investors, they needed either free land or some other government subsidy. The public domain was particularly attractive because it offered large blocks of uninhabited land. Most good farmland near the coast was either tied up in litigation or far too expensive for large-scale development. Moreover, if the canal builders confined their activities largely or exclusively to the public domain, they did not have to worry about securing rights of way. Still, by depending so heavily on land subsidies, the promoters opened themselves to the charge of being land grabbers with no serious interest in agricultural development. The speculative nature of the ventures, and the absence of reliable engineering data to support them, encouraged these charges. Since the projects were located in remote parts of the state not served by adequate transportation—which helps explain the popularity of irrigation canals that could double as arteries of transportation—each was doomed to failure if only because of its distance from potential markets.

Despite the failure of the three largest irrigation projects proposed during the 1850s and 1860s, much was happening in California to influence the expansion of irrigation in subsequent decades. The mining industry's contributions to California agriculture are easy to overlook. The same ingenuity that drove men to move mountains in the quest for precious metals had, by 1867, constructed over 300 ditch systems covering 6,000 miles in mountainous counties from Siskiyou to Tulare. By the 1880s, the U.S. Army Engineers estimated that over $100,000,000 had been invested in these artificial channels.[40] Many of California's early engineers learned their trade in the mining camps, and the skills they acquired—for example, techniques to construct flumes, pipelines,

40. *The California Water Atlas* (Sacramento, 1978), 16. The *Report of the Commissioner of the General Land Office, 1868* (Washington, D.C., 1869), 74, claimed there were 617 irrigation ditches in the state in 1866 irrigating 37,813 acres and pegged the value of these ditches at $16,000,000. The acreage figure is fairly accurate, but the other numbers were probably inflated by counting mining canals that also provided water for irrigation.

and pumps—were later applied to the design and construction of irrigation works and municipal water systems. Similarly, foundries and ships devised to fabricate mining tools were easily adapted to the construction of agricultural and industrial machinery.

Ironically, the two most tangible legacies of the mining industry—its reservoirs and canals—proved of less value. By the 1870s, miners stored over 150,000 feet of water. In Nevada County, the center of hydraulic mining, North Bloomfield's system alone impounded over 23,000 acre-feet, most of it in Bowman Lake. However, by the mid-1880s the mining industry was moribund, and the crusade to "store the floods" did not blossom until the late 1890s and the early years of the twentieth century. By that time, the brush, log, and earth-filled dams popular in the 1870s had given way to larger, more substantial concrete structures. Mining dams were often used by irrigation districts and hydroelectric power companies, but the methods and materials used to build them were outmoded by the twentieth century. In 1880, the state engineer reported that about 9,000 acres of land in the Sierra foothills were irrigated from mining ditches, most near Auburn and Placerville. This constituted no more than one or two percent of the irrigated land in California in that year. Despite the valiant efforts of northern California boosters, irrigation in the foothills lagged far behind the rate of growth in southern California and the Central Valley.[41]

Though irrigation expanded little during the 1850s and 1860s, several barriers to future expansion were overcome. The decline of the mining industry freed capital for investment in irrigation companies during the 1870s and 1880s, and the destruction of southern California's livestock industry opened huge tracts suitable for irrigation, ushering in the horticultural revolution.

High cattle prices made the ranchos of southern California very profitable in the late 1840s and early 1850s. But by the end of the decade, declining demand and competition from cattlemen in other states had reduced the number of cattle driven north each year by about two-thirds; in the summer of 1861, stockmen slaughtered 15,000 animals for their hides, tallow, and jerky rather than take the low prices offered in the north. Rancho owners also faced high land taxes, interest rates of 4 or 5 percent a month, and

41. Harding, *Water in California*, 65–66.

expensive litigation over land titles. The livestock industry was clearly in deep trouble even before the drought of 1864.[42]

Nevertheless, the effects of that drought were devastating. An officer in the First Cavalry, California Volunteers, visited Los Angeles in the spring of 1864, and the *Sacramento Daily Union* reported that "in a walk of five or six miles from the City of Los Angeles, he counted on the bottoms of creeks and small streams, as many as eight hundred dead cattle which had perished from starvation. The entire grazing area is as clear of vegetation as a desert."[43] The air was so thick with the stench of decaying carcasses that thousands of emaciated animals were driven off cliffs into the Pacific Ocean. A mounted, armed guard patrolled the fence surrounding Anaheim's two square miles of irrigated vineyards to prevent desperate, starving cattle from breaking into the oasis of vegetation. The assessed value of Los Angeles County property fell from $3,650,330 in 1860 to $1,622,370 by the end of 1864. In the drought's aftermath, rancho land sold for as little as 25¢ to 50¢ an acre.

Since the land had little value itself, when the cattle died the rancho owners were left with nothing but staggering debts. In 1864, much of Abel Stearns's estate was sold at auction for $4,000 in back taxes; at the time his land was assessed at 10¢ an acre. Subsequently, a land syndicate acquired the 200,000-acre estate, subdivided it into 40-acre tracts, and began selling it for $2 to $10 an acre in 1868. Extensive advertising lured would-be farmers from the East as well as from northern California, and the promoters pocketed $2,000,000 in profit. The pattern was repeated throughout the Los Angeles basin, with the promise of irrigated vines or trees being an important inducement.[44]

Southern California, like the San Joaquin Valley, still awaited

---

42. The standard work is still Cleland's *The Cattle on a Thousand Hills*.

43. *Sacramento Daily Union*, March 30, 1864. For other contemporary reports on the drought see *Country Gentleman* 23 (April 14, 1864): 244; 23 (May 19, 1864): 321; *California Farmer* 21 (March 11, 1864): 41.

44. J. M. Guinn, "The Passing of the Cattle Barons of California," *Annual Publications of the Historical Society of Southern California, 1909–1910*, 8: 59–60; Netz, "The Great Los Angeles Real Estate Boom of 1887," 54.

an adequate transportation system before it could begin sustained agricultural development. In the absence of a state-owned or state-regulated irrigation system, irrigation was left entirely to private enterprise. The 1870s brought the most ambitious canal scheme launched in the nineteenth-century arid West. They also brought the first sustained criticism of the monopolization of California's land and water. For the first time, irrigation became a public issue central to the state's economic future.

5

Irrigation in the 1870s: The Origins of Corporate Reclamation in the Arid West

The decline of the mining industry, the adoption of no-fence laws, and the expansion of rail transportation into the San Joaquin Valley and southern California contributed to a dramatic increase in irrigation during the 1870s. Irrigated land nearly tripled during the decade, even as the state suffered through its first protracted economic depression. The transition from a frontier economy characterized by individual entrepreneurs and loose confederations of investors, to an economic system dominated by large corporations, accelerated during the "terrible seventies." The "growth pains" manifested themselves in many ways, particularly in public hostility toward big business. The fear of land and transportation monopolies had been building since the 1850s, but the danger posed by water monopolies was new. Proponents of corporate water development pointed to the irrigation colonies adjoining Fresno as proof of how private enterprise had encouraged diversified, small farms; but critics maintained that farmers could never be independent and self-sufficient when they depended on greedy capitalists for their water.

The number of miners in California fell from 83,000 in 1860 to 36,000 in 1870 as the production of precious metals declined from an average $34,000,000 a year from 1860–1864 to $18,000,000 a year from 1865–1869. By contrast, the number of farmers increased from 20,000 to 48,000 during the decade.[1] Equally important, in 1870 the Southern Pacific Company chose Lathrop, south of Stockton, to begin its railroad through the San Joaquin Valley. The line reached Modesto in the same year, Fresno in 1872, Goshen, Tulare and Delano in 1873, and Bakersfield in 1874. Modesto and Turlock were but two of the new communities created by the railroad. In the 1870s, the S.P. also built a spur-line from Coalinga on the west side of the southern valley east to Goshen, where it intersected with the north-south rails. This railroad improved transportation between the southern San Joaquin Valley and the Santa Clara and Salinas valleys, though there was no rail service across the Coast Range.[2]

Wheat farmers had already expanded into the San Joaquin Valley, using the San Joaquin River to carry their grain to San Francisco Bay. Between 1866 and 1869, acres planted to wheat in the valley increased from 89,563 to 380,547. Most of the increase occurred in Stanislaus County, which became the state's largest wheat-producing county by the end of the 1860s.[3] San Joaquin Valley stockmen already faced serious problems from overstocking the open range and from declining livestock prices. Before barbed wire came into widespread use in the early 1880s, fencing a quarter-section of land cost more than $2,000. Not surprisingly, farmers demanded that stock owners pay for damages to crops caused by

1. *Ninth Census of the United States, 1870* (Washington, D.C., 1872), 3: 820; *Eighth Census of the United States, 1860* (Washington, D.C., 1864), 662; Rodman Paul, *California Gold: The Beginnings of Mining in the Far West* (Cambridge, Mass., 1947), 345.

2. Alice L. Carothers, "The History of the Southern Pacific Railroad in the San Joaquin Valley" (M.A. thesis, University of Southern California, 1934), 38–44; Marion N. Jewell, "Agricultural Development in Tulare County, 1870–1900" (M.A. thesis, University of Southern California, 1950), 6–7. Construction on the Southern Pacific's rail line from San Francisco Bay (Antioch) along the west side of the San Joaquin Valley did not begin until 1888, about the same time the Southern Pacific began building north through the Sacramento Valley.

3. *Transactions of the State Agricultural Society* (hereafter *TCSAS*), *1866–1867* (Sacramento, 1868), 548; *TCSAS, 1870–1871* (Sacramento, 1872), 172. Wheat production increased from 8,805,411 bushels harvested from 361,351 acres in 1861 to 20,000,000 bushels from 1,098,901 acres in the dry year of 1869. See *Report of the Commissioner of Agriculture for the Year 1862* (Washington, D.C., 1863), 577; *Report . . . 1869* (Washington, D.C., 1870), 31.

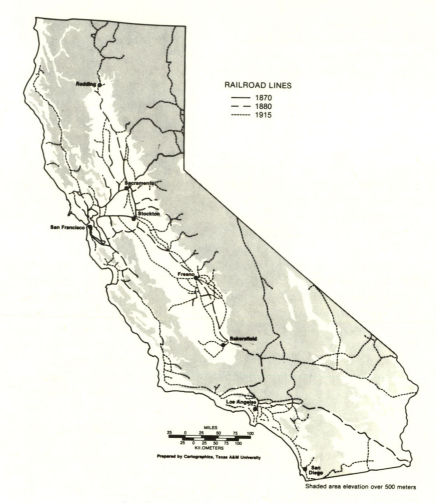

Railroads of California, 1870, 1880, 1915.

roving herds of cattle and sheep. The state's first no-fence law, passed in 1866, pertained exclusively to the Sacramento Valley. But in 1874, the legislature extended the law to cover the San Joaquin Valley's most important agricultural counties. The statute was not enforced until 1878 because it was difficult to establish ownership of stock and costly to bring suit. However, when farmers began to shoot animals that entered their fields, the stockmen took the law more seriously.[4] One more agricultural barrier had toppled.

Another obstacle was more difficult to surmount. The state's first protracted drought occurred in 1868 and lingered on until 1872 or 1873. The west side of the San Joaquin Valley suffered most. There the wheat crop was a total loss in 1870 and 1871, and by the fall of the latter year the parched valley had turned to dust. In October, San Francisco's *Bulletin* reported that "not less than 2,500 people in this valley are so destitute that they cannot procure seed for sowing their fields." Herdsmen drove cattle and sheep into Nevada in a desperate search for forage; embittered farmers abandoned their homesteads; banks foreclosed on mortgages; and the state's newspapers reported an abundant crop of irrigation schemes designed to render California immune from nature's capricious cycles.[5] In June 1870, the *Sacramento Daily Union* described irrigation projects as the "coming California epidemic." Most ventures launched in the 1870s were speculative and ephemeral, but by 1880, irrigation companies had spent $400,000 on irrigation south of the Kings River alone.[6]

The most ambitious and controversial reclamation project originally belonged to San Francisco capitalist John Bensley. During his dazzling business career, Bensley promoted a variety of ven-

4. *California Statutes,* 1874, 50; Jewell, "Agricultural Development in Tulare County," 14–15; William D. Lawrence, "Henry Miller and the San Joaquin Valley" (M.A. thesis, University of California, Berkeley, 1933), 85; Carothers, "History of the Southern Pacific Railroad in the San Joaquin Valley," 24–28.

5. *Bulletin,* October 6, 1871. Also see the *Bulletin* of November 3, 1871; the *Alta California,* March 31, 1871; *Weekly Colusa Sun,* July 8, 1871; *California Mail Bag,* 1 (December 1871): 3.

6. *Sacramento Daily Union,* June 9, 1870; *Pacific Rural Press* 1 (April 15, 1871): 232; 2 (November 18, 1871): 309; *Weekly Colusa Sun,* December 2, 1871; Wallace Smith, *The Garden of the Sun* (Los Angeles, 1939), 449; and *Fresno Expositor,* December 7, 1870; April 5, June 21, September 20, 1871; October 29, 1873; May 13, December 2, 1874; May 5, 1875; February 28, May 16, 1877.

tures ranging from lead, iron, and coal mines to the California Steam Navigation Company. But he became particularly interested in water resource development. He organized San Francisco's first water company in 1857, and it provided most of the city's water until 1865. On a trip to Chile, Bensley marveled at irrigated fields of grain and alfalfa, which prompted him to claim a large part of the San Joaquin River and organize the San Joaquin and Kings River Canal Company on March 7, 1866.[7]

At the time, rail transportation into the San Joaquin Valley was six or seven years in the future. The region's backwardness can be seen in a traveler's droll description of Fresno—destined to be a bustling city—in 1866: "This city (God save the mark!) is situated on a slough of the San Joaquin, and some wags will gravely tell you it is the head of navigation, and destined at a not distant period to become a 'great place.' The 'city' comprises two houses— one a hotel, where there is also a store, and the other I think [is] now uninhabited." Fresno County's population in 1866 numbered only 3,000 people, and Millerton, the county seat, had fewer than 200 residents. The county's taxable property was worth only $1,000,000, with 4,500 acres under cultivation; pasture grass constituted the main crop. Tulare and Kern counties to the south were equally undeveloped and remote from civilization.[8]

Not surprisingly, Bensley's scheme to build a huge irrigation canal through the San Joaquin Valley was considered visionary by potential investors. His company began a canal in 1868 but quickly ran out of money. Nevertheless, the drought, and construction of the Southern Pacific's rail line into the valley, finally kindled interest in the project. Early in 1871, Bensley won the financial support of William C. Ralston, a founder, and after 1872, president, of the Bank of California. Ralston's bank was the largest financial institution in the arid West, and the financier cut a wide swath in both business and politics. He had worked closely with Bensley in the California Steam Navigation Company and various San Francisco water projects.[9]

7. George H. Morrisson, typescript biography of John Bensley, Hubert Howe Bancroft Collection, Bancroft Library, University of California, Berkeley.

8. *Alta California*, January 9, 1866.

9. David Lavender, *Nothing Seemed Impossible: William C. Ralston and Early San Francisco* (Palo Alto, Calif., 1975), 353–355.

The San Joaquin and Kings River Canal Company reorganized in May 1871, with Ralston in command. The crew of prominent investors included William S. Chapman; Isaac Friedlander, the "Grain King," who monopolized the shipment of California's wheat crop to England; Nicholas Luning and A. J. Pope, directors of the Bank of California; and Lloyd Tevis, president of Wells Fargo Company. The new canal company also had two "silent partners." The cattle barons Henry Miller and Charles Lux owned so much land in the San Joaquin Valley that a canal could not be built without passing through their vast estate. Hence, the two men received company stock in exchange for lukewarm support. Ralston and company hoped one day to irrigate the entire Central Valley. As a first step they wanted to build a 230-mile aqueduct skirting the Coast Range from Buena Vista Lake, at the south end of the valley, to Antioch, on the upper arm of San Francisco Bay. The promoters also planned eight east-west canals to connect the main line with the San Joaquin River and its tributaries. These canals would serve as a transportation network—for barges carrying grain, farm equipment, and lumber—as well as provide water for irrigation. The San Joaquin *Republican* commented that the project was so vast as to appear impractical, "but when we read further and learn the names of incorporators . . . we are reassured, and we lay down the paper with the idea dawning upon us that those men are capable of performing anything they undertake."[10]

The promoters needed more than money to make their venture successful. The vast project rivaled even the Erie Canal and demanded an engineer with a national or international reputation and plenty of experience. The San Joaquin and Kings River Canal Company found such a man in Robert Maitland Brereton. Brereton was born in England but spent the early years of his career in India. In the late 1850s, he began designing the Indian Peninsula Railway, which connected Bombay with Calcutta and Madras. His talents

---

10. The *Republican's* editorial was reprinted in the *Sacramento Bee* May 23, 1871. On the organization and objectives of the company, see the *Pacific Rural Press*, May 13, 20, 27, 1871; *Kern County Weekly Courier*, June 1, 1872. Bensley gave up his primary interest in the canal company in May, but the company's board of trustees did not hold its first formal meeting until September 9, 1871. At that time the San Joaquin and King's River Canal Company was renamed the San Joaquin and Kings River Canal and Irrigation Company.

gained immediate recognition in England. During more than a decade's stay in India, Brereton carefully studied the massive British irrigation system then under construction.

After the Bombay-Calcutta-Madras line opened in 1870, the Indian government asked Brereton to visit the United States and report on American techniques of railroad construction. By this time, his fame and social position had won him introductions to Cornelius Vanderbilt, William B. Astor, Cyrus Field, James B. Eads, Jay Cooke—and William Ralston. In July 1871, at a stopover in Victoria, British Columbia, Brereton received an urgent telegram from Ralston asking him to come to California and prepare comprehensive engineering plans for the project. The banker used a high salary and gifts of company stock to persuade the Englishman to assume the permanent position of chief engineer. Many years later Brereton lamented abandoning a career that paid $12,000 a year and promised a comfortable retirement and a knighthood, for such a speculative scheme. He also turned down an offer to become Japan's Chief Engineer of Railroads in 1872.[11]

Brereton reached California in July 1871 and quickly completed his initial survey work in the San Joaquin Valley. He recommended reducing the canal's length from 230 to 160 miles by beginning the aqueduct at Tulare Lake, rather than at Buena Vista Lake at the south end of the valley. The shorter ditch would cost much less, about $2,600,000, but would still serve more than 600,000 acres at an average cost of $4.33 per acre. If farmers paid $1.25 per acre per crop for water, then once the land adjoining the main canal had been settled, the company could expect a yearly revenue of at least $800,000 from water sales alone. Brereton estimated the cost of reclaiming all the San Joaquin Valley's arable land at $14,350,000, including $7,600,000 for irrigation works and $6,690,000 for levees and other swampland reclamation. He predicted that ultimately the company would salvage nearly 4,000,000 acres from swamps and desert. Nevertheless, he cautioned against building too much too fast. He recommended that the canal network be constructed in sections, as population increases warranted, predicting that the entire job would probably take fifty years. The main canal would

11. Robert M. Brereton, *Reminiscences of an Old English Civil Engineer* (Portland, Oreg., 1908), 23–24; Brereton to George Davidson, September 3, 1910 and April 14, 1911, George Davidson Collection, Bancroft Library.

provide cheap transportation at a time when opposition to the railroad's monopoly in California had reached fever pitch, and canal tolls promised the company a steady source of revenue until the anticipated population boom increased the demand for irrigation. "With water, rich soils and heat combined," Brereton maintained, "the productiveness of this country will be so great that the present wandering and never-settled population, who are in the San Joaquin Valley this year and next year in Oregon . . . will give place to settlers who will delight in making California their permanent abode. By carrying out *gradually* a sensible and practical system of irrigation and land reclamation, you insure California the possibility of its becoming the most populous and richest state in the Union."[12]

For all his optimism, Brereton recognized that his company faced many real and potential obstacles. The first involved its uneasy partnership with Miller and Lux. Ralston and his allies had agreed to build the first 50 miles of canal through the baronial estate of the cattle kings, from Firebaugh's Ferry and the great bend of the San Joaquin River northeast to Los Banos, strategically located at the end of a 75-mile long wagon road which linked it to Gilroy on the other side of the Coast Range. The Southern Pacific had just completed a railroad connecting Gilroy to the communities surrounding San Francisco Bay. The line would facilitate construction by carrying workers, heavy equipment, and supplies as well as potential settlers. This region seemed likely to experience a population boom. Moreover, Ralston and company had strong financial incentives to build the first stretch of canal through the Miller and Lux ranch. The two stockmen had promised to provide rights-of-way in exchange for company stock. In other parts of the valley, securing rights-of-way would have been much more complicated and costly, if only because the company would have had to negotiate with many more landowners. Moreover, Miller and Lux promised to pay $20,000 toward the cost of construction and allowed the canal company to use their workers, horses, and equipment. Finally, they agreed to buy sufficient water to irrigate 16,667

12. Brereton filed two reports with the San Joaquin and Kings River Canal and Irrigation Company, one dated August 19, 1871, and the other October 6, 1871. Both are reprinted in his *Reminiscences of Irrigation-Enterprise in California* (Portland, Oreg., 1903), 54–73. The quote is from p. 72. Also see his statement in Ezra S. Carr, *The Patrons of Husbandry on the Pacific Coast* (San Francisco, 1875), 310–313.

acres of pasture land in 1872, 33,334 acres in 1873, and 50,000 acres in 1874—at $1.25 per acre per crop.

The promised revenue and rights-of-way had been necessary to get the project off the ground. Brereton warned that Miller and Lux might one day use their vast riparian rights to block downstream diversions from the canal, and that they might also refuse to sell their land to prospective settlers. Clearly, large landowners throughout the valley would have to cooperate with the company to render the project successful. In particular, since land adjoining the surveyed canal route had already dramatically appreciated in value, the engineer wondered whether the company would be able to afford rights-of-way when it extended the canal north to Antioch and south to Tulare Lake. Even though several of the San Joaquin and Kings River Canal and Irrigation Company's leading investors owned land in the valley, the company itself did not.

Nevertheless, by the middle of August 1871, 300 teams of horses and 400 men were at work on the canal, excavating an average of about two-thirds of a mile daily. The ditch was thirty-two feet wide at the bottom, forty-two feet wide at the top, and two feet deep. However, four-foot embankments increased the overall depth to six feet. By the end of 1871, the company had completed forty miles of aqueduct.[13]

Much more money was needed to complete the canal, but capital was hard to find in depression-ridden California. In the spring of 1872, Ralston sent Brereton to England on a hunt for investors. The engineer found English capitalists reluctant to pour money into any western scheme. Many had been badly burned in wildcat mining ventures, and they soon learned that the stock Brereton peddled for $7 to $8 a share could be purchased for half that amount in San Francisco. They also discovered that most of the original stock had been exchanged for water rights and rights-of-way; few California investors had been rash enough to buy it. C. J. F. Stuart, head of the Oriental Bank of London, had worked closely with Ralston and the Bank of California in promoting earlier western projects. In August 1872, he warned the California fi-

13. *Stockton Daily Independent*, August 15, October 18, 1871; January 20, 1872; *Pacific Rural Press*, 2 (October 7, 1871): 216; Lawrence, "Henry Miller," 101; E. F. Treadwell, *The Cattle King* (Boston, 1950), 67.

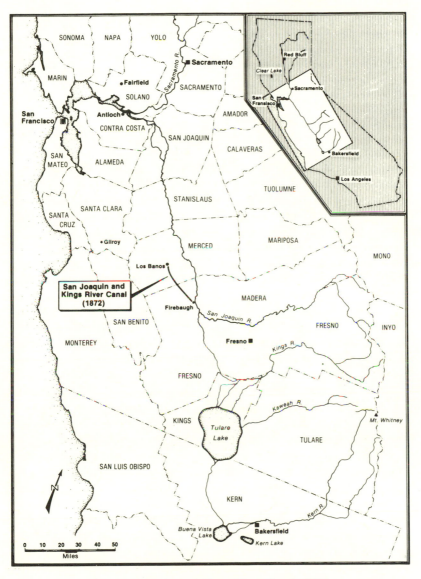

The San Joaquin and Kings River Canal, 1872.

nancier that English investors wanted investments of a "solid moderate character" with stock held in California on the same terms as in London. Yet he conceded that Brereton had come to England "when John Bull was suffering from a too credulous belief in Pyramids of Silver and the like, not to mention the Erie Railway and would not look at anything American—so that the San Joaquin Valley has not had a fair chance."[14]

After Brereton returned empty-handed to the United States, the San Joaquin and Kings River Canal and Irrigation Company launched its first concerted effort to win a federal land grant. Such a grant promised rights-of-way, security to investors, a way to finance construction, and a source of speculative profits. In Washington, Brereton served ably as a lobbyist during the 1872–1873 session, carrying the scheme to President Grant, Speaker of the House James G. Blaine, generals William Tecumseh Sherman, George B. McClellan and W. S. Hancock, the California congressional delegation, and Ralston's old friend, Nevada's powerful Senator William Morris Stewart.

On January 17, 1873, Senator Cornelius Cole introduced the company's bill in the Senate, and Representative Sherman Houghton proposed the same bill in the House on February 10. It granted rights-of-way through Kern, Tulare, Fresno, Merced, Stanislaus, San Joaquin, Contra Costa, and Alameda counties, along with two even-numbered sections of public land per mile of canal and an additional 100 acres per mile to pay for building reservoirs. In all, the land grant totaled 256,000 acres. The company also received the promise of a virtual monopoly over the waters of Buena Vista and Tulare lakes and of the Kern and San Joaquin rivers. The bill subjected the land to state taxation, and the company's transportation tolls and irrigation charges to state regulation. On Cole's request, the legislation went to the Senate Committee on Public Lands. Since Senators Eugene Casserly of California and Stewart of Nevada sat on the committee, the legislation had powerful support. However, widespread public hostility toward railroad land grants and monopoly made the committee reluctant to approve the grant without more evidence concerning the project's feasibility.

14. C. J. F. Stuart to William Ralston, August 15, 1872, Oriental Bank Corporation files, William Ralston Collection, Bancroft Library. Also see Stuart to Ralston, July 27, 1872; Brereton to Ralston, July 17, 1872.

As Casserly reported at the end of February: "Much care has been taken by the committee to mature the bill and to provide the proper guards for the general interests. We desire, however, all the aids possible toward the best possible measure . . . before the next session of Congress."[15]

On February 28, 1873, the day following Casserly's statement in the Senate, Stewart introduced a bill to provide for a federal irrigation survey in California. The legislation created a five-member commission, including two representatives from the Army Corps of Engineers, one from the Coast Geodetic Survey, the chief of the California Geologic Survey, and a "disinterested" engineer to "make a full report to the President on the best system of irrigation for said [Sacramento and San Joaquin] valleys, with all necessary plans, details, engineering, statistical, and otherwise, which report the President shall transmit to Congress at its next session, with such recommendations as he shall think proper."[16] The bill prompted Senator Lyman Trumbull of Illinois to grumble that "if this survey is allowed . . . the time will come when you will be called upon for a very large appropriation to complete the work, if it should be recommended." Nevertheless, the bill passed, and President Grant signed it into law on March 3, 1873. Most Californians failed to see the connection between the survey and Ralston's canal project, but the link was obvious. The first version of Stewart's bill limited the survey to the San Joaquin Valley; only at the suggestion of Senator Casserly did the Nevada Senator add the Sacramento Valley. Moreover, the new commission's head, General B. S. Alexander of the Army Corps of Engineers, quickly offered the "consultant" post to Brereton.[17]

The Alexander Commission first met in May 1873 and did most of its survey work in June, July, and August, spending a scant

15. For the bill itself see the *Sacramento Daily Union*, March 1, 1873. Also see *Stockton Daily Independent*, February 14, 1873; and Cornelius Cole to William Ralston, February 2, 1873, Ralston Collection. The quote is from the *Congressional Globe*, 42d Cong., 3d sess., February 27, 1873, p. 1846. Also see the *San Francisco Chronicle*, February 12, 1873; *San Francisco Examiner*, February 18, 1873.

16. The survey bill (S. 1584) is reprinted in the *Congressional Globe*, 42d Cong., 3d sess., 46: 3, 305.

17. *Congressional Globe*, 42 Cong., 3 sess., 46: 3, 1930. The *Sacramento Bee*, February 25, March 27, 1873; *Stockton Daily Independent*, February 25, March 3, 8, 1873; *San Francisco Chronicle*, March 17, 1873; *Weekly Colusa Sun*, March 1, 1873.

six weeks in the field.[18] The commission's budget was only $5,000, and its members all held full-time jobs; consequently, the reconnaissance was hurried and superficial. Robert Brereton refused to serve formally on the commission for fear his presence would hurt the canal company's chances for a land grant or federal loan. However, he accompanied the group on its trips and provided its members with valuable information, including documents he had gathered concerning irrigation in California, India, Egypt, and Europe. As early as June, the Coast Geodetic Survey's representative, George Davidson, acknowledged that "the surveys & engineering works will require comprehensiveness, time, skill & larger amounts of money." For example, the commission could not afford to run soil or water tests, or to prepare the topographical maps necessary to designate the irrigable lands.[19]

The commission's report, published in 1874, neither endorsed the San Joaquin and Kings River Canal and Irrigation Company's scheme nor recommended a detailed blueprint for a comprehensive irrigation system in the Central Valley. However, it did preach the need for central planning to insure the proper construction of irrigation works, limit waste, and prevent diseases caused by improper drainage. The commissioners predicted that a staggering 8,500,000 acres could be irrigated in the Central Valley, including swampland—12,000,000 acres if the low foothills were included. They estimated the cost of a complete irrigation system at $10 an acre, far more than the existing population could pay and several times the value of land in most parts of the San Joaquin Valley. But the promised return would be great: fifty to eighty bushels of wheat per year and as many as five crops of alfalfa. Even though the system would take decades to complete, "the works should be properly planned and located in the beginning, so that whatever is done to meet the present requirements of a sparse population

18. "Letters Received, 1871–1876" (1873:927), Office of the Chief of Engineers, U.S. Army, Record Group (RG) 77, National Archives, Washington, D.C. This file contains monthly reports outlining the commission's activities and itinerary.

19. George Davidson to Superintendent, U.S. Coast Geodetic Survey, June 23, 1873, Letterbook, vol. 26 (1873), Davidson Collection, Bancroft Library. Also see Davidson's letters to the Superintendent, July 19, August 6, November 24, 1873; idem, "Annual Report for 1873," microfilm, California Academy of Sciences, Bancroft Library; *Weekly Colusa Sun,* May 17, 1873; and *Stockton Daily Independent,* May 16, 1873.

may form a part of those that will be necessary to meet the demands of a population of millions by simply enlarging them."

Though the commission did not provide the detailed construction plans expected by irrigation companies and land speculators, it did suggest the location of major canals. The whole system depended on gravity. Both Red Bluff at the north end of the Central Valley and Bakersfield 450 miles to the south were several hundred feet above sea level. The west side of the valley was much drier than the east side because its few meandering streams flowed from the modest Coast Range rather than from the soaring peaks of the water-rich Sierra Nevada. Thus, the commission proposed that the largest canal extend from the Sacramento River at Red Bluff along the Coast Range to Fairfield, across the northern arm of San Francisco Bay from Antioch. The other half of this canal essentially followed the route from Buena Vista Lake, near Bakersfield, to Antioch favored by the San Joaquin and Kings River Canal and Irrigation Company. Because the main aqueduct would carry a limited water supply, the commissioners decided it could not feed an elaborate network of branch canals. Only in Yolo County, where the aqueduct could be replenished with water from Clear Lake, did the group consider branch lines feasible. On the east side of the valley, a continuous north-south canal was impractical because of the enormous expense of carrying the water over or under so many Sierra streams. Instead, the investigators suggested a broken aqueduct connected to the Sacramento and San Joaquin rivers by a series of east-west canals.[20] The group paid little attention to potential storage reservoir sites because plenty of unclaimed surface water remained if it could be moved where it was needed. It did recommend damming some Sierra streams, but only to raise the water above river banks into irrigation ditches, store water for hydraulic miners, and aid swampland reclamation in the valley.[21]

20. 43d Cong., 1st sess., H. Ex. Doc. 290, (serial 1615), 16, 23–24, 27, 34. The quote is from p. 38. See the map of proposed and existing canals appended to the report.

21. The Alexander Commission's report estimated that reservoirs could help reclaim as much as 1,000,000 acres of additional land in the Central Valley, but the construction of canals to tap normal stream flow took precedence (H. Ex. Doc. 290, pp. 39–40). Not until the late 1880s and 1890s, when much of the region's normal flow had been appropriated, did westerners pay much attention to storing floodwater.

Ironically, the commissioners never questioned the engineering feasibility of moving water over such great distances; like most nineteenth-century westerners, they exhibited complete faith in technology and in California's master builders. They were much more concerned with how to pay for the project and who should do the work. Ultimately, increasing land values were expected to more than offset the cost of reclamation, but in the short run, private companies and the farmers themselves could not pay for comprehensive irrigation works when the Central Valley's population was so thin and scattered. Nor were the state or counties likely to tackle the job. Since irrigation would increase tax revenue along with property values, government could probably afford the work. But many Californians balked at state reclamation on principle, and many more believed that massive public works offered too many opportunities for fraud.

The commission posed an alternative: the state and nation should jointly conduct topographical surveys, gauge stream flow, designate natural irrigation districts, map out a comprehensive network of canals, and then turn the construction work over to private companies. These companies would be required to follow a master plan, with state officials supervising every step in construction and regulating the rates charged for water. To prevent waste, the state would determine the amount of water farmers needed to grow different crops in different soils, though the distribution of water would be left to state-chartered associations of irrigators. To prevent monopoly, the commission recommended that all water rights be attached to the land, so that water could not be bought and sold as a separate commodity. It proposed that no private company be permitted to construct irrigation projects unless that company agreed to transfer ownership of the completed works to the state or irrigators after a stipulated period. The state would hold the option to purchase the company's property earlier, if necessary.

The Alexander Commission concluded that the federal government should promote or regulate irrigation only where substantial tracts of the public domain remained open to settlement, as between Visalia and Bakersfield. It did not propose federal construction of canals, or federal loans or land grants to private companies. The report concluded on a somber and prophetic note, warning that if the state failed to assert control over the acquisition of water rights and the distribution of water, California could look forward to a "fruitful crop of contentions in the future."[22]

The commissioners refused to recommend a land grant to the San Joaquin and Kings River Canal and Irrigation Company because, strangely enough for the time, they looked at irrigation in California as scientists, not as promoters or boosters. They foresaw the danger of monopoly, as California's private canal companies began to snatch up the state's land and water. The commission recognized that the track record of private irrigation companies had been poor. Their irrigation works were poorly designed, flimsy, and wasteful, constructed for speculative profit rather than for permanence. Moreover, until federal and state officials completed a thorough hydrographic survey, and until the state created institutions to supervise the construction of reclamation works, regulate the issuance of water rights, and prevent waste, any grant was premature. In any case, Congress probably would have rejected the grant—it showed no interest in the hydrographic survey. The depression of 1873 dictated economy, and many Congressmen undoubtedly feared that a comprehensive investigation would be followed by new appeals from California for federal construction, or at least for money subsidies.

Meanwhile, the critics of monopoly in California declared open war on Ralston's company. The *Sacramento Bee* warned that the perpetual land and water grants demanded by the company would give it absolute control over the San Joaquin Valley: "This is the worst kind of subsidy, worse a thousand times than any railroad subsidy ever given, for it enables a company to monopolize not business, but the main elements of all life."[23] The *Pacific Rural Press* commented: "the settlers [in the San Joaquin Valley] are unanimous in the opinion that the water should belong to the State or [federal] Government, and if they must pay for it, they do not want

22. The commission's conclusions were reprinted on pp. 77–80 of the report. For a concise summary of the Alexander Commission's report, see the *Report of the Commissioner of Agriculture for the Year 1874* (Washington, D.C., 1875), 352–362. In December 1873, Nevada Congressman C. W. Kendall introduced a bill (H. R. 759) providing for an irrigation survey in Nevada, including a $20,000 appropriation to pay for drilling test wells. The bill stalled in committee. "Letters Received, 1871–1887" (1873: 2414), Office of the Chief of Engineers, U.S. Army, RG 77, National Archives.

23. *Sacramento Bee*, May 23, 1873.

to enrich monopolies by doing so."[24] Stockton's *Independent* charged
that the forty-mile stretch of canal completed in 1871 had been
poorly engineered. Apparently, the canal's grade, one foot per
mile, resulted in a swift current. The canal banks washed out frequently, so the aqueduct had been lined with willow branches.
The company had promised to irrigate several hundred thousand
acres from the ditch, but the swift current restricted irrigation to
about 30,000 acres, virtually all of which belonged to Miller and
Lux. The *Independent* warned that diversions through the canal
would destroy the value of the San Joaquin River for irrigation and
shipping. It favored state or federal appropriations to improve navigability on the stream, where transportation was free, but the canal
company claimed the right to divert water into its ditch year-round,
not just during the growing season.[25]

By the end of 1873, the San Joaquin and Kings River Canal
and Irrigation Company had exhausted virtually every potential
source of financial support. On December 9, the company's trustees, claiming that over $300,000 had already been spent on the
project, wrote Governor Newton Booth offering to sell out to the
state. Booth ignored the proposal.[26] Nevertheless, the company's
prospects brightened in the spring of the following year. On May
18, Robert Brereton persuaded thirty of California's most prominent
capitalists and landowners, including the largest speculators in San
Joaquin Valley land, to accompany him on a tour of a 6,000-acre
model irrigated wheat farm adjoining the canal that the company
had established on land leased from Miller and Lux. This drought-
resistant farm demonstrated that wheat profited as much from
irrigation as did vegetables or fruit. It yielded an average of fifty
bushels an acre.[27] Apparently, this farm, and perhaps the pros-

24. *Pacific Rural Press* 6 (October 18, 1873): 248. Also see the *Sacramento Daily
Union*, July 2, 1873; *Stockton Daily Independent*, July 2, 4, 10, 1873; *Weekly Colusa Sun*,
July 5, 1873; *Fresno Expositor*, June 15, 1870; June 25, November 12, 1873.

25. *Stockton Daily Independent*, February 1, 15, 22, 1873. The judgment of the
Independent concerning the canal was echoed by John S. Hittell in *Commerce and
Industries of the Pacific Coast* (San Francisco, 1882), 404–405.

26. *Stockton Daily Independent*, December 20, 1873; *Weekly Colusa Sun*, December 20, 1873; *Pacific Rural Press* 7 (April 4, 1874): 210.

27. Robert M. Brereton to George Davidson, September 29, 1911, Davidson
Collection, Bancroft Library. Also see Brereton's *Reminiscences of Irrigation-Enterprise
in California*, 7–8 ; *Bulletin*, May 18, 1874; *Evening Post*, May 18, 1874.

perity enjoyed by several irrigation colonies Brereton had laid out for Ralston near Fresno, finally won the active support of speculators in the valley. W. S. Chapman and Isaac Friedlander, as well as Miller and Lux, agreed to give the company 100,000 acres of land along the proposed canal route in exchange for 100,000 shares of company stock, bearing a par value of $25 a share. Brereton then returned to London in late 1874 or early 1875. Using the land as collateral and bait, he won the promise of an English syndicate to complete the canal. Disaster struck when Chapman and Friedlander went bankrupt in 1875, as did the Bank of California; William C. Ralston died in August. Much of Chapman's and Friedlander's best land, and the canal, ended up in the hands of Miller and Lux, while Brereton, who had invested $40,000 in the company, was forced to sell his stock to the cattle barons for $1,000.[28]

One of the nineteenth century's boldest dreams died with Ralston. By 1880, the cattlemen had extended the canal an additional twenty-seven miles and provided water to more than 150,000 acres.[29] But Ralston's plan to reclaim the entire Central Valley through private enterprise would never be revived.[30] The project failed for many reasons, including the enormous cost, the economic depression and antimonopoly sentiment of the 1870s, the return of wet weather in the winter of 1873–1874, the San Joaquin Valley's scattered population, poor engineering, a lack of strong scientific evidence concerning the benefits of irrigation, the overwhelming commitment of most farmers to wheat, the separation of land and water rights, and California's inadequate water laws. The project was far too ambitious for a frontier society.

Nevertheless, the canal project and the Alexander Commission were historically significant for California and the entire arid

28. Brereton, *Reminiscences of an Old English Civil Engineer*, 25, 30; Brereton to George Davidson, September 3, 1910, Davidson Collection, Bancroft Library; Henry Miller dictation, Hubert Howe Bancroft Collection, Bancroft; W. S. Chapman letter to the *Fresno Expositor*, July 30, 1873; Treadwell, *The Cattle King*, 73. Apparently, Miller and Lux did not secure a majority of the canal company's stock until 1877, when John Bensley went bankrupt.

29. S. T. Harding, *Water in California* (Palo Alto, Calif., 1960), 100; Treadwell, *The Cattle King*, 77.

30. After finishing work in the San Joaquin Valley, Ralston hoped to build canals in the Sacramento Valley. See his letter to "My Dear Lawton," May 5, 1873, Ralston Collection, Bancroft Library; J. Ross Browne, "Agricultural Capacity of California: Overflows and Droughts," *Overland Monthly* 20 (April 1873): 313.

West. In the 1860s, California contained only one major canal: the Moore Ditch out of Cache Creek in Yolo County, completed during the drought of 1864. But the drought of the late 1860s and early 1870s produced a canal boom. The ditches constructed thereafter included the Gould Canal at Fresno (1871); the Chowchilla Canal out of the San Joaquin River north of Fresno (1874); the Goose Lake Canal in the Tulare Valley (1874); and the Kern Island Canal at Bakersfield (1872). Ralston's canal was simply the largest of many ambitious projects. Although a much greater increase in irrigated land occurred during the 1880s, the takeoff came in the 1870s. Since California offered the best opportunities for reclamation and land speculation, corporate development during that decade outstripped that in other arid states.[31]

The San Joaquin and Kings River Canal and Irrigation Company, like the large water projects that preceded it in the 1850s and 1860s, focused on reclaiming the land rather than on building desert communities. Yet many critics of the scheme argued that irrigation could succeed only in *settled* communities designed specifically to make the best use of ditches and canals. Random settlement usually resulted in isolated clusters of farms. That isolation deprived settlers of the benefits of community life and contributed to a drab, dreary existence. It also drove up the cost of irrigation because the longer the canals the greater the cost—and the greater the loss of water through evaporation and seepage. Ralston's company had paid little attention to these problems.

The irrigation colony boom of the 1870s and 1880s challenged the individualistic pattern of land settlement so characteristic of the nineteenth century. Aside from the settlements in Utah and at Anaheim, the first colony was established at Greeley, Colorado, fifty miles northeast of Denver, in the spring of 1870. The promoters, *New York Tribune* editor Horace Greeley and his agricultural editor, Nathan Cook Meeker, modeled the settlement after the New England village and Mormon towns in Utah. Civilization, they believed, could only flourish in the arid West if farmers lived in close-knit villages surrounded by their farms. This pattern permitted the establishment of schools, churches, performing arts

31. *Biennial Report of the Surveyor-General of the State of California from August 1, 1877–1879* (Sacramento, 1879), 26–29, 48–49.

groups, and cooperative institutions ranging from libraries to laundries to bakeries. Meeker hoped that the method of dividing up the colony's 12,000 acres would serve as a lesson: "In the success of this colony a model will be presented for settling the remainder of the vast territory of our country."[32]

Unfortunately, the colony's leaders knew little about the cost and construction of canals, or the techniques of irrigation. Initially, Meeker estimated the cost of irrigating the 12,000 acres at $20,000, promising the first batch of settlers that this sum had already been raised through land sales. He assured the colonists that they would pay only for the repair and maintenance of ditches. But the first canal alone cost $27,000 and served less than 2,000 acres; a second ditch, expected to irrigate 5,000 acres, watered only 200. By 1890, the cost of constructing the four main canals stood at $412,000. The colony's early history demonstrated the need to determine the amount of water needed to raise different crops, the amount of water lost to evaporation and seepage, and the best gradient for gravity canals.[33]

Neither the Greeley colony nor a colony established at Riverside, California, in 1870 were designed to make money for the promoters. Both were temperance communities whose locations were chosen, in large part, to permit settlers to separate from a sordid world rather than profit from it. But the colonies established around Fresno were speculative operations which enjoyed many advantages. Fresno was centrally located between the Kings and San Joaquin rivers and usually received an abundant water supply. It was also located on the S.P. line, and the San Joaquin River served as an auxiliary source of transportation. Speculators had acquired land in huge blocks, suitable for large-scale reclamation and subdivision, and the ten-foot-high wild sunflowers that welcomed newcomers testified to the richness of the soil. At Fresno, many different companies and promoters competed for settlers, and the competition forced them to offer inducements not extended to purchasers of irrigated land in other parts of the state.

32. David Boyd, *A History: Greeley and the Union Colony of Colorado* (Greeley, Colo., 1890), 34. Also see Boyd's *Irrigation Near Greeley, Colorado* (Washington, D.C., 1897).

33. Boyd, *Greeley and the Union Colony of Colorado*, 59–60.

In 1868, a syndicate of San Francisco capitalists, led by William
S. Chapman and Isaac Friedlander, used agricultural college land
scrip to acquire 80,000 acres in a nearly solid block around present-
day Fresno. But the syndicate was dissolved in 1873 and the land
distributed among its members. These men, including E. B. Perrin
and Moses J. Church, subsequently vied with one another to attract
different ethnic groups of farmers, particularly Germans and
Scandinavians.[34]

The large landowners learned from the disastrous experience
suffered by the first Fresno colony. The so-called Alabama Colony
was established in 1868 when nine war-weary families from Alabama, Mississippi, and Georgia left their devastated homes and
settled on 30,000 acres puchased from Isaac Friedlander in north
Fresno County. They planted 2,500 acres of wheat in the spring of
1869. Unfortunately, the southerners knew little about desert farming, and no preparations had been made prior to their arrival. They
practiced irrigation but lacked a reliable canal network. To com-
pound their woes, the first settlers arrived at the beginning of
California's longest nineteenth-century drought and suffered repeated damages from roving herds of livestock. The colony was
abandoned in 1874–1875 and most of its residents scattered.[35]

The Central Colony, founded in 1875, profited from the Alabama Colony's mistakes. It was organized by Chapman and Bernard Marks, a former miner and public school principal who became
one of San Francisco's leading agricultural land salesmen. Marks
and Chapman selected twenty-one square miles and began work
on three main irrigation canals in August 1875. At the end of the
year, the colony contained only fifty-four residents, but artful and
extensive advertising paid off; by 1878 every twenty-acre farm had
been sold. The settlement offered many amenities, including twenty-
three miles of tree-lined roads, two acres of vines from Spanish
cuttings laid out on each tract, free cultivation for two years by

34. Arthur Maass and Raymond L. Anderson, . . . and the Desert Shall Rejoice
(Cambridge, Mass., 1978), 158.

35. Virginia E. Thickens, "Pioneer Colonies of Fresno County" (M.A. thesis,
University of California Berkeley, 1939), 8–10. Also see Thickens, "Pioneer Agricultural Colonies of Fresno County," California Historical Quarterly 25 (March, June
1946): 17–38, 169–177; Moses J. Church oral history transcript, microfilm, Bancroft
Library; Fresno Expositor, May 14, 1873.

employees of the Central Canal Company, distribution ditches with headgates, guaranteed water rights, and terms of $1,000 at $100 down and $12.50 per month for six years with no interest. The colony was particularly attractive to city dwellers who lacked farm experience. Expert farmers employed by the company assisted the novice, but there was no need to live on the land until the vines began bearing, if then. The ditch company charged for tending uninhabited farms as well as for operating and maintaining canals. Those who could not afford the expense were encouraged to plant part of their land to alfalfa to produce a modest income to tide them over the first four or five years. At the end of that period, the vineyards—which produced mainly raisin grapes—often returned $100 to $160 an acre. So while Chapman made a good return on his purchase price of $2 an acre, so did the settlers. Nevertheless, they bitterly resented their dependence on his water company.[36]

Another popular enterprise was the Washington Colony organized in March 1878, by Wendell Easton, J. P. Whitney, and A. T. Covell. The investors purchased 7,000 acres of land eight miles south of Fresno, adjacent to the Central Colony. They hired excursion trains, advertised widely, and quickly sold 385 tracts, most 20 acres. The advertising was designed to take full advantage of the protracted depression of the 1870s:

> Business men who wish to retire from their monotonous toil, can here find change and rest, pleasant surroundings and a better income than they have often realized from their life-long drudgery. People in reduced circumstances who are struggling to keep up appearances in the city, with children unemployed, and with young men and women seeking in vain for employment, should at once seek a good location and direct their energies to the cultivation of vines and fruits. In this way every member of the family can find pleasant and profitable employment, and all can enjoy health and the pleasures of a pleasant and happy home. The thousands of young

---

36. Thickens, "Pioneer Colonies of Fresno County," 21–40; Maass and Anderson, . . . and the Desert Shall Rejoice, 165; Paul Vandor, History of Fresno County California (Los Angeles, 1919), 262–265. Maass and Anderson note that drought destroyed most of the first vines set out. The company subsequently offered to plant vines for $15 an acre and tend them for $1 an acre per month.

men who are always seeking clerkships, or "waiting for something to turn up," would do much better to come to the country and go to work, earning money and getting a vineyard and orchard started, that, if well attended to, will make them fortunes for life. Nor should they think that because they have no money their case is hopeless. Every young man of good sense, industry and a will, can easily work his way to a competency if he will determine to be self-reliant and do it.

Inducements included 160 acres for a town site, free town lots to each purchaser of a 20-acre tract, wide streets, and a community dairy. Just as the Central Colony's residents were encouraged to raise alfalfa to get started, promoters of the Washington Colony sold milk cows at low prices to enable new settlers to sell butter and cheese. The businessmen avoided Chapman's mistake and turned the distribution of water over to the farmers themselves. Despite drought and depression, the colony survived and soon boasted literary and dramatic societies as well as its own brass band.[37]

By 1886, twenty-one irrigation colonies covered 45,000 acres adjoining Fresno. They provided homes for 1,500 families, about 7,500 people. By 1890, the value of unimproved land which sold for $2 an acre in 1870 reached $50 to $100, and land under ditch sold for two to three times that amount. The colonies stood in sharp contrast to unplanned agricultural settlements in other parts of the state. Most were served by private water companies, and residents, especially in the Central Colony, often complained of poor service. Nevertheless, water rights permanently attached to each parcel of land; in effect, the purchase of land conferred a permanent right to a certain quantity of water so long as the farmer contributed to the cleaning and repair of canals.[38]

Recently, political scientists Arthur Maass and Raymond Anderson have asked why irrigation around Fresno developed so much more rapidly than in the Mussel Slough region near Hanford

37. Thickens, "Pioneer Colonies of Fresno County," 40–47. The advertisement is reprinted in Maass and Anderson, . . . and the Desert Shall Rejoice, 167.

38. Visalia Weekly Delta, September 24, 1886; Hinton, "Irrigation Progress Report for 1890," 51st Cong., 2d sess., 1891, Sen. Ex. Doc. 53 (serial 2818) 27; Maass and Anderson, . . . and the Desert Shall Rejoice, 166.

(also served by the Kings River but thirty miles to the south). They note that in the Hanford area speculators were unable to monopolize land because the railroad passed through the region before speculation became feasible, in effect putting half the land off limits. As a result, settlement followed traditional patterns and colonies did not develop. Maass and Anderson conclude: "The Fresno and Hanford data lend themselves to a conclusion that concentration of land in the hands of a few individuals, with the prospects of speculative profit, stimulated economic development and formed a desirable state in the transition from barren public land to intensive agriculture."[39]

This raises the old question of whether land speculation and monopoly contributed to economic growth. Unquestionably, many farmers saw them as threats to their welfare. For example, farmers needed Isaac Friedlander's transoceanic shipping monopoly to carry wheat to European markets. Yet wheat that sold for $60 a ton in Liverpool fetched less than half that amount at the San Francisco waterfront, and rail transportation charges to the city reduced profits even further. To make matters worse, the cost of shipping a ton of wheat fluctuated as widely as the value of the grain itself, and for no apparent reason. The price declined from $2.70 per hundredweight in 1871 to $1.28 and $1.60 in 1872, but freight costs to Liverpool increased from $10 to $25 a ton.[40]

Farmers in California were little different from those in the rest of the nation: they saw monopoly everywhere, from the cost of burlap sacks and farm machinery to the cornering of offices by political bosses and clubs. The *Tulare County Times* commented on December 1, 1882, that a "tremendous conflict" was underway pitting the "giant monopolies" against the "yeomanry of the county":

> Continued corporate defiance of the popular will; their inordinate greed and rapacity; their persistent refusal to consider the public welfare; their evident purpose to wring the

---

39. Maass and Anderson, . . . *and the Desert Shall Rejoice*, 213; *Fresno Expositor*, April 23, 1879. Maass and Anderson's stimulating book compares the development and management of irrigation in different parts of Spain with Colorado and Utah, as well as the Kings River region of California. Their analysis does not seem to apply to irrigation in Kern County, where conditions were similar in most respects to Fresno. See the Lux v. Haggin chapter.

40. *Tulare County Times*, September 28, 1872.

last cent out of the industries and commerce of the State, thus
ruthlessly robbing labor and enterprise of their legitimate profits; their assumption of powers transcending those exercised
by the most potent princes and oligarchies of feudal times;
and their corrupting use of money and patronage to effect
their malign objects, all demand the unqualified condemnation of every citizen.

Land speculators were convenient, visible scapegoats in the midst
of an economic depression, especially because they often resorted
to deceptive or even fraudulent means, such as dummy entrymen,
to acquire their land. Not surprisingly, during the 1870s and 1880s,
California newspapers bristled with plans to limit the size of estates
through massive taxes on land held for speculation, the elimination
of taxes on crops, buildings, and other improvements, and confis-
catory inheritance levies.

Speculators created many of their own problems. They wanted
to sell as much land as possible, and in most rural parts of the
state had little incentive to hold undeveloped tracts for long periods. To increase the number of potential buyers, they portrayed
farming as an easy occupation and sold land on time. Since irrigated
land almost always increased in value, they often pandered to
small-fry speculators with no agricultural experience who expected
to profit even without cultivating the land. Yet farmers with inadequate resources or experience were the first to flee at the sign
of drought or hard times, blaming irrigation promoters for their
own greed, shortsightedness, laziness, or incompetence. The *Bakersfield Californian* commented in 1875: "It is well known that a large
number of the settlers on the public lands in the valley are mere
speculators, who build a shanty, never till the ground, go through
the form of pre-emption, and wait for a purchaser."[41] Land speculation came under heaviest fire in the state's large metropolitan
newspapers. In part, this was because rural papers were often
owned, controlled, or heavily influenced by the speculators themselves. But it was also because rural editors had a better understanding of the problems of developing arid land. In February 1876,
the *Fresno Expositor* lashed out at the San Francisco *Evening Post* for
branding William S. Chapman a "land shark." "It is stating the

41. *Bakersfield Californian*, November 11, 1875.

case only moderately," the *Expositor*'s editor wrote, "to say that throughout the length and breadth of the San Joaquin Valley the name of William S. Chapman is honored and is a synonym for fair-dealing and a generosity that is often more lavish than prudent." The journal went on to recount the many irrigation schemes Chapman had sponsored and their value to the valley's economy. And when Isaac Friedlander died in 1878, the same paper described him as "the main stay of our poor farmers, furnishing them money and seed, year after year when they were bankrupt by drought. . . . At Borden, so great was the respect for him, that, on the news of his death, the flags were placed at half mast."[42]

More study is needed of agricultural development in other parts of California before one can conclude that land speculation was a spur to the expansion of irrigation. However, the changes in agriculture during the 1870s were clearly substantial. Though the average farm declined in size by only 20 acres during the decade (482 to 462), the number of farms increased by more than 50 percent, from 23,724 to 35,934. Southern California's population more than doubled during the decade, largely because an S.P. line reached Los Angeles in 1876, but wheat, not citrus, remained the dominant California crop. Production fluctuated year by year, but the 21,504,000 bushels harvested in 1873 paled beside the 41,990,000 bushels produced in 1878—though the acreage planted increased only from 1,592,889 to 2,470,000 acres. Using statistics culled from reports of the state surveyors general, John Ganoe notes that irrigated land nearly tripled during the decade, from 90,344 acres in 1870 to 255,646 acres in 1879; the California State Board of Agriculture estimated the amount of land under irrigation in 1879 as 292,885 acres.[43]

42. *Fresno Expositor*, February 2, 1876; July 17, 1878.

43. *Report of the Commissioner of Agriculture for the Year 1873* (Washington, D.C., 1874), 24; *Report . . . 1878* (Washington, D.C., 1879), 264; Gilbert C. Fite, *The Farmer's Frontier, 1865–1900* (New York, 1966), 168; John T. Ganoe, "The Beginnings of Irrigation in the United States," *Mississippi Valley Historical Review* 25 (June 1938): 65. In 1878, California's leading irrigation counties were Los Angeles and Merced, each of which contained roughly 37,000 acres under ditch. San Bernardino County residents irrigated roughly 20,000 acres and Tulare County's canals served about 18,000 acres. However, in the northern part of the state irrigation was almost a novelty. San Joaquin County contained only 2,000 acres fed by canals, and the Sacramento Valley's most heavily irrigated county, Yolo, contained only 12,250 irrigated areas. Only 5 percent of the state's population lived in the San Joaquin Valley and 10 percent south of the Tehachapis.

Such statistics did not comfort the critics of monopoly. They hoped to turn the development of irrigation over to the state government, local irrigation districts, or some combination of the two. Whatever the success of private irrigation colonies at Fresno in the 1870s, or at Ontario, Etiwanda, and other southern California towns in the 1880s, most farmers saw themselves as businessmen, not community builders. While many farmers profited substantially from private development, they wanted to reduce the cost of reclamation and restore or preserve their independence.

6

Response to Monopoly: Institutional Roots of the Irrigation District, 1868–1885

Public hostility toward monopoly manifested itself in many ways during the 1870s and 1880s. Some reformers argued that California's agricultural promise could not be fulfilled without a state-financed, state-operated network of irrigation canals accompanied by condemnation of all private water rights. However, this was an extreme position because most Californians feared "big government" as much as "big business." All large institutions appeared likely to abuse power and trust. The irrigation district offered an alternative to both state and corporate reclamation. It was an institutional mechanism designed to raise money to build irrigation works, but also a way to promote local control over water. The dreams of the proponents of irrigation districts were not realized in the years from 1868 to 1885. But public debate of the district concept, along with the legislative experimentation conducted in Sacramento, laid the foundation for the Wright District Act adopted in 1887.

Utah adopted the first western irrigation district law in 1865. It provided that when a majority of the residents of any county or part of a county requested their county court to designate a particular tract of land as an irrigation district, local officials would comply if the region contained sufficient water to serve the proposed district's needs. Once the district had been formed, the voters decided whether the taxes needed to build dams and canals should be assessed exclusively against the land benefited, or against all land within the district. They also elected a board of trustees to survey canal routes, determine the amount of irrigable land, and prepare a formal construction plan for submission to the voters. Once the cost of reclamation had been determined, two-thirds of the electorate had to approve the required tax or the district was dissolved. In 1866, the act was amended to permit the inclusion of existing canals, ditches, and dams in new districts.[1]

Immediately following passage of the act, many districts were formed in Salt Lake and Utah counties, especially along the Jordan River. By the 1890s, about 100 districts had been established; but few, if any, raised the money needed to build new irrigation works. The first federal land office did not open until 1869, and for some years thereafter many Utah farmers remained, technically, squatters on the public domain. This prevented the legislature from giving districts the power to sell lands whose owners refused to pay their taxes. In effect, the 1865 law was inoperable. Controversy also erupted over the board of trustees' power to decide which district lands would be taxed or excluded from taxation. The law was repealed in 1897 after the Utah Supreme Court ruled the taxing provisions of the statute discriminatory, and, hence, unconstitutional.[2]

The Utah law had little or no influence in California, where the irrigation district form was copied from the swampland district. In 1861, the legislature passed a law providing for the formation of reclamation districts on the request of one-third of the landowners within a proposed district. The law also created a state board of swampland commissioners and charged it with supervising all private reclamation as well as with planning and construct-

1. George Thomas, *The Development of Institutions Under Irrigation* (New York, 1920), 117–120.

2. Thomas, *Institutions Under Irrigation,* 122, 125–126. On the 1865 Utah irrigation district law also see Elwood Mead, *Irrigation Institutions* (New York, 1903), 234; Ray P. Teele, *Irrigation in the United States* (New York, 1915), 76, 102; and C. S. Kinney, *A Treatise on the Law of Irrigation and Water Rights and the Arid Region Doctrine of Appropriation of Waters* (San Francisco, 1912), 3: 2515, 2518–2519.

ing comprehensive levee and drainage works on the Sacramento and San Joaquin rivers. The board could spend up to $1 an acre from a state fund derived from swampland sales. Any additional money had to come from the landowners themselves, and in 1862 the legislature authorized the districts—if one-third of their residents approved—to levy taxes to pay for drainage and flood protection. By the end of the year, the state contained thirty-eight swampland districts covering 485,000 acres. However, in 1866 the legislature, under strong pressure from land speculators and large wheat farmers, abolished the state board and transferred its functions to the counties, ending the prospect of an integrated, comprehensive state reclamation plan.[3]

In 1868 the legislature dropped the 640-acre limit on the amount of flood land an individual could acquire, touching off a mad scramble for the remaining acreage. The 1868 law, drafted and sponsored by Will Green, specified a procedure for creating local swampland districts. These districts could not be formed unless the owners of 50 percent or more of the acreage within a proposed district petitioned the board of supervisors. Only landowners could vote for the board of trustees, which was responsible for drafting reclamation plans and administering district affairs, and the law required at least half the landowners to approve the district bylaws. The board of supervisors appointed three special commissioners to assess taxes according to the benefits provided each tract of land. All money in the state swampland fund was transferred to the counties where the land had been purchased. The swampland legislation of the 1860s established the principle of using special districts to solve problems not easily addressed by city or county governments. As in Utah, the districts enjoyed the power to tax and condemn land but no authority to issue bonds or float loans. However, the California law did not differentiate between district assessments and any other state or county taxes; the county district

3. Robert Kelley, "Taming the Sacramento: Hamiltonianism in Action," *Pacific Historical Review* 34 (February 1965): 21–49; Kenneth Thompson, "Historic Flooding in the Sacramento Valley," *Pacific Historical Review* 29 (November 1960): 349–360; and Richard H. Peterson, "The Failure to Reclaim: California State Swamp Land Policy and the Sacramento Valley, 1850–1866," *Southern California Quarterly* 56 (Spring 1974): 45–60. Also see S. T. Harding, *Water in California* (Palo Alto, Calif., 1960), 142; W. W. Robinson, *Land in California* (Berkeley, 1948), 191–192; *Sacramento Daily Union*, March 5, 1861; February 25, 1862; November 27, 1869; November 28, 1870.

attorney enforced collection.[4] Only four years elapsed between the passage of Green's law and the first irrigation district act. From 1872 to 1879, with the help of the California Grange, the district idea came into its own.

The irrigation crusade of the 1880s was largely the creation of land, water, and mining companies rather than of farmers. But during the 1870s, the California Grange dominated the movement. The Grange had its birth in informal farmers' clubs that sprang up in 1871. These soon gave way to local Granges, and on July 15, 1873, delegates from 28 chapters met in Napa to form a state organization. The new group engaged in a variety of associational activities, including trade and marketing cooperatives, farmer-owned banks and insurance companies, and stores. It also tried to use the power of numbers to lower transportation rates charged by the railroads and other shippers. By October 1873, when the first annual convention met, 104 subordinate Granges counted 3,168 members, and at the peak of the membership in 1874, the 231 chapters—located mainly in the wheat-growing counties of Napa, Sonoma, Santa Clara, Sacramento, San Joaquin, Santa Cruz, Sutter, El Dorado, and Los Angeles—claimed 14,910 members.[5]

Historians have disagreed about the nature of the organization's membership and leadership in California. Rodman Paul and Gerald Nash, among others, have argued that the Grange consisted of large, speculative wheat farmers, distinguished mainly by being latecomers to the business. They note that most Grangers purchased their land after the Civil War, for high prices and at high interest rates. However, a recent article by Gerald Prescott suggests

4. *California Statutes*, 1868, 507. The principle of creating new institutions to solve problems beyond the reach of local, county, or state governments dated back to the mining camps. Since many of California's wheat farmers had tried their hand at mining before turning to farming, their earlier experience may help explain their willingness to experiment.

5. Clarke A. Chambers, *California Farm Organizations* (Berkeley, 1952), 9. Also see Solon J. Buck, *The Granger Movement: A Study of Agricultural Organization and its Political, Economic and Social Manifestations, 1870–1880* (Cambridge, Mass., 1913); Ezra Carr, *The Patrons of Husbandry on the Pacific Coast* (San Francisco, 1875); Gerald L. Prescott, "Farm Gentry vs. the Grangers: Conflict in Rural America," *California Historical Quarterly* 56 (Fall 1977): 328–345; and Rodman Paul, "The Great California Grain War: The Grangers Challenge the Wheat King," *Pacific Historical Review* 27 (November 1958): 331–350.

that the Grangers may well have constituted an entirely different breed of bonanza farmer. Prescott contrasts Grange leaders with the leadership of California's State Agricultural Society during the 1870s and 1880s, and finds that only four of the agricultural society's forty-five spokesmen also belonged to the Grange. In addition, Grange leaders owned smaller parcels of land, and 93 percent lived on their farms—in sharp contrast to the high percentage of non-resident farmers who led the agricultural society. Perhaps the directors of the older agricultural society, many of whom were also wheat farmers, had closer ties to San Francisco's business elite. In any case, during the 1870s the Grange provided some of the state's most strident critics of monopoly and placed many of its members in the state assembly. It did much more to promote irrigation than had the agricultural society. Just as the Grange stood for cooperation in marketing and purchasing, it favored irrigation as a co-operative endeavor to increase productivity and land values, as well as to guard against the caprices of nature.[6]

The Grangers eagerly joined the newspaper assault on monopoly, and the San Joaquin and Kings River Canal and Irrigation Company became the group's favorite target during the last half of 1873 and 1874. The Congress that met at the beginning of 1873 refused to consider a land grant until the Alexander Commission completed its survey, so the company returned to Congress in June with a new proposition designed to help finance its project. Ralston and his associates had seen the rapid increase in land values adjoining the completed section of canal, and also along the proposed route from the San Joaquin River to Antioch. So they decided to try to force those who owned land through which the canal would pass to help pay its cost. The bill required *all* landowners to pay the company a flat assessment of $1.50 an acre, payable in two annual installments. Those who refused would be required to pay 10 percent interest per year on their debt to the company, and that debt would constitute a lien or mortgage on the land. For five years following adoption of the bill, farmers would also be required to pay the company one-sixteenth of the value of their crops, whether

6. Gerald Nash, *State Government and Economic Development* (Berkeley, 1964), 161; Paul, "The Great California Grain War," 333–334, 344; Prescott, "Farm Gentry vs. the Grangers," 345 (fn. 46).

they irrigated their land or not. After the canal had been completed, they would be obligated to pay an additional levy of $1 per acre, as well as the "usual," or prevailing, rate for the water they used. The bill set no limit on rates. One San Joaquin Valley farmer bitterly assailed the company's request:

> Why, sirs, you would own us, we would be but your serfs, beholden to your mercy for the bread we would put in our children's mouths. You would, with a high hand, backed by legal authority, rob a large community of their homesteads and their birthright, and with the combined wealth of these spoils would make yourselves millionaires. What do you take us for? Fools outright? Slaves from some foreign lands, used to despotism, and ready and willing to bow our necks for the burden you would place upon us?"[7]

Grange chapters and farmers clubs throughout the San Joaquin Valley flooded Congress with petitions urging rejection of the bill and any future land and water grants.[8]

Amid growing public alarm over the power of land and water companies, the Grange held its first convention in Napa in July and ordered its committee on irrigation to prepare a bill for introduction at the next session of the legislature, "having for its objects the utilizing of all the inland waters of the State, and their uniform and equitable division and distribution, under the authority and control of the State, among the actual land owners of the State." The delegates concluded that such a bill should provide for a thorough hydrographic survey by the state, state designation of irrigation districts, and state supervision over the distribution of water. The districts themselves should pay for their irrigation works by issuing bonds. In order to secure reliable water supplies, the Grange urged that the legislature "provide a way for condemning every and all actual, asserted or pretended prior right[s], privilege[s], or franchise[s] to . . . any of the inland waters of this State, whether

---

7. *Sacramento Daily Union*, July 19, 1873. The terms of the bill were discussed in a separate column in the same issue.

8. *Sacramento Daily Union*, July 2, 1873; *Stockton Daily Independent*, July 2, 4, 10, 1873; *Weekly Colusa Sun*, July 5, 1873.

held or claimed by individuals or corporations." The state organization urged the subordinate Granges to apply as much pressure as possible on the legislature to insure "immediate action."[9]

The Grange did not propose a district bill until 1874, but California's first irrigation statute passed the legislature two years earlier with little fanfare or debate. The lawmakers who convened in December 1871 arrived in Sacramento under the shadow of the great drought. Will Green predicted that private companies might spend as much as $25,000,000 to $30,000,000 on canals before the next legislature met but warned that "this can hardly be expected if some encouragement is not given by the state." Senator Thomas J. Keyes of Stanislaus County (then the center of the San Joaquin Valley wheat industry) introduced two bills drafted in the interest of the Tuolumne Water Company. The company wanted to dam the Tuolumne River at La Grange and construct two canals linking the dam to the San Joaquin River. It began work in 1871 but quickly ran out of money. Consequently, it devised a plan by which the county would issue $150,000 in bonds and pay the interest for the first five years. The water company would pay the interest for the second five years and pay off the bonds themselves in ten years. Meanwhile, it would enjoy the right to charge farmers $1.25 an acre per year for irrigation water. The legislature received petitions from 130 proponents and 400 opponents of the scheme. The latter group argued that if the project was sound, money could be found elsewhere, and also that such subsidies set a bad precedent. The law passed anyway, only to receive Governor Newton Booth's pocket veto. The second piece of legislation, the district act, was probably a backup or compromise measure. It simply copied the swamp district law of 1868, including the tax provisions. That it was designed to aid private companies, if not specifically the Tuolumne Water Company, can be seen in several places. Section 24 permitted groups of landowners to create and administer irrigation districts "without the intervention of Trustees or the establishment

9. Carr, *The Patrons of Husbandry*, 147–148. The July convention was held to organize the Grange. The first annual convention was held the following November, at which time conferees simply ratified the statement of principles adopted in July. See the *Pacific Rural Press* 6 (November 15, 1873): 308; *California Granger* 1 (November 8, 1873); Fourth of July address by Grange Master J. W. A. Wright reprinted in the *Stockton Daily Independent*, July 20, 1875.

of by-laws." This was doubtless designed to permit landowners to tax themselves to pay private ditch companies for constructing irrigation works. The law did not permit irrigation districts to condemn established water rights, and one clause specifically excluded Fresno, Kern, Tulare, and Yolo counties, probably to insure that existing canals were not absorbed into districts. No districts were created using the 1872 statute.[10]

The 1872 law was far too limited for the Grangers, and on January 21, 1874, Assemblyman J. W. Venable of Los Angeles introduced a much bolder irrigation bill accompanied by a statement of support signed by 1,700 farmers in southern California. The legislation provided that the governor would designate all the land suited to irrigation in California, determine the size of the state's water supply, and decide the best way to use it. A state board of engineers would prepare construction plans for each district. At least one-third of the landowners within a proposed district had to approve of the state plan before a bond election could be held. Bonds could be issued only if a majority of the landowners approved but, after the election, district residents would exercise little direct influence over administration. The state, not local officials, would levy tax assessments to construct the irrigation works, supervise the distribution of water, and determine the price farmers would pay for the water. Moreover, all proceeds would be paid into the state treasury, for use in retiring the bonds, and the salaries of the board members and the engineering staff would be paid by the state from the general fund. The board could "acquire by purchase all property necessary to carry out and maintain the system of irrigation provided for in this Act." This represented a step back

10. For Will Green's comments see the *Weekly Colusa Sun*, Dec. 2, 1871. Also see the *Stockton Daily Independent*, March 1, 7, 1872. On the Tuolumne Water Company scheme see Benjamin F. Rhodes, Jr., "Thirsty Land: The Modesto Irrigation District, A Case Study of Irrigation Under the Wright Law" (Ph.D. diss., University of California, Berkeley, 1943), 23–28; Thomas E. Malone, "The California Irrigation Crisis of 1886; Origins of the Wright Act" (Ph.D. diss., Stanford University, 1965), 45–46. For the Keyes district law see *Cal. Stats.*, 1872, 945. The *Fresno Expositor* endorsed the Keyes bill on March 20, 1872, claiming that the rail line then under construction through the San Joaquin Valley would enable farmers to raise large enough crops to warrant the expense of irrigation. "It [the Keyes bill] is what is needed to force capitalists to build the much needed irrigating ditches immediately or forever lose the opportunity of doing so," the editor commented.

form the Grange's earlier demand for the condemnation of *all* private water rights in the state, but the bill did give the state exclusive control over all the water used within the districts.[11]

The Venable bill passed the assembly, the stronghold of Granger political power, on February 27, 1874, and was reported on favorably by the Senate Irrigation Committee. However, the full senate rejected the legislation near the close of the session by a vote of thirteen to twenty-six.[12] The bill failed for several reasons. The *Sacramento Daily Union* maintained that "the system proposed by this bill is admirably suited to a paternal Government, like France under Napoleon, but is entirely contrary to the spirit of both our people and our Government. It is centralism gone to seed." Many of the bill's critics argued that the irrigation districts should have been given more power to manage their own affairs. The *Stockton Daily Independent*, which supported the measure, claimed that parochialism killed the legislation; some lawmakers opposed any measure that did not directly benefit their own district, especially if it cost money: "Some of the spiggot economists who disgrace the seats to which they have been elected . . . complained of the expense, and because it was to cost the people of the whole State $40,000 per annum to make the preliminary surveys and do the necessary work of inaugurating this great system of irrigation, they must vote against the appropriation." The *Independent* broadly hinted that land and water companies had worked hard to defeat the bill because its condemnation powers would have sharply restricted corporate control over water. Nevertheless, defeat of the bill probably owed as much to the effects of the Depression of 1873 as to any other cause. The bill raised the possibility that the state would construct extensive irrigation systems in places where the scattered population could not possibly repay the cost of construction in a reasonable period of time. Such an ambitious scheme did not stand much chance during a period of retrenchment. The lost cause was deeply mourned. On hearing of

11. A.B. 172 (Venable), January 21, 1874, *Assembly Bills, 1873–1874*, vol. 1, California State Law Library. Also see the *Pacific Rural Press* 7 (January 31, 1874): 73; *Stockton Daily Independent*, February 3, 1874.

12. *Stockton Daily Independent*, February 27, March 20, 1874; *Pacific Rural Press*, 7 (March 28, 1874).

the senate's decision, Grangers at Ellis Station on the west side of the San Joaquin River flew their flags at half staff.[13]

The legislature of 1875–1876 considered a bill similar to the Venable legislation proposed by Senator Creed Haymond of Sacramento County. However, this legislation made the board elective rather than appointive, required the consent of two-thirds of landowners before a district could be formed, limited the cost of irrigation works to 30 percent of the assessed value of district property, gave the state a lien on district lands, and required all money from water sales to go toward paying the costs of the board and its engineering and administrative staff before any money was used to pay off bonds. Haymond avoided many of the Venable bill's weaknesses, but critics still complained that the legislation would give the state too much power and lead to a bitter struggle between proponents and opponents of irrigation in each district.[14]

Other district bills considered in 1874, 1876, and 1878 gave the state much less power, or bypassed the state entirely in favor of local officials. A bill introduced on January 20, 1876, would have required a state board to designate district boundaries, formulate plans for irrigation systems, and supervise construction. But when the works were finished, locally elected district officials would have been responsible for issuing the bonds, setting tax and water rates, and distributing the water. The proposed legislation contained a novel feature. During the first ten years after the issuance of bonds, when farmers were just opening their land to cultivation, taxes and water sales would be used exclusively to pay interest. Then, in the second decade, taxes would be increased gradually to pay off the principal as well. Presumably, the value of irrigated crops and soaring land prices would more than offset the higher assess-

13. *Sacramento Daily Union*, March 7, 1874; *Stockton Daily Independent*, March 27, 1874; *Weekly Colusa Sun*, February 28, March 8, 1874.

14. S.B. 80 (Haymond), December 20, 1875, *Senate Bills, 1875–1876*, California State Law Library. For editorials opposing the Haymond bill see the *Sacramento Daily Union*, December 22, 1875; *Weekly Colusa Sun*, Jan. 8, 1876; *Stockton Daily Independent*, Jan. 14, 1876; *Bakersfield Californian*, January 13, 1876. Haymond reintroduced his bill at the next session of the legislature with no greater success, even though he served as chairman of the senate committee on irrigation in 1877–1878. See S.B. 13, February 21, 1878, *Senate Bills 1877–1878*, vol. 1. Also see the *Stockton Daily Independent*, Janurary 9, 1878; *Tulare County Times*, December 29, 1877, *Fresno Expositor*, December 27, 1877; January 2, March 20, 1878.

ments. However, the $25,000 appropriation to set up the state office and pay for preliminary surveys doomed the bill, as did the persistent opposition of most stockmen and some land speculators to the higher taxes that usually resulted from irrigation.[15]

A second bill proposed during the 1876 session pertained exclusively to Fresno, Tulare, and Kern counties, where corporate development was well underway. It would have required county boards of supervisors to divide each county's arable land into irrigation districts defined either by governmental subdivisions or natural boundaries, such as river basins. Each district would be supervised by an elected board of water commissioners. That board could condemn existing water systems—though it had to provide established users with sufficient water from the new system to irrigate all their land—and issue bonds not exceeding 10 percent of the assessed value of property within the proposed district. A tax on all property within the district, not just on irrigated farmland, would retire the bonds and pay the board's salaries and expenses. Watermasters appointed by the board would supervise the distribution of water and maintain the ditches. The bill also set penalties for wasting water. But since it contained a $5,000 appropriation to pay for initial surveys, it made no greater headway than the other legislation.[16] By 1878, the *Tulare County Times* concluded that no comprehensive district bill could or should be enacted in California. "What might suit one set of counties similarly located would not fit another set of counties," the editor wrote. "Therefore, our legislature should endeavor to secure the passage of local bills adapted to the needs of individual counties, or collections of counties, whose requirements are of a mutual character."[17]

15. S.B. 233 (O'Connor), Jan. 20, 1876, *Senate Bills, 1875–1876*, California State Law Library. Also see the *Stockton Daily Independent*, January 28, 1876.

16. A.B. 89 (McConnell), Jan. 4, 1876, *Assembly Bills, 1875–1876*, California State Law Library. Also see the *Fresno Expositor*, October 27, 1875; *Kern County Weekly Courier*, September 25, 1875, November 27, 1875; *Bakersfield Californian*, December 16, 1875.

17. *Tulare County Times*, January 5, 1878. For other bills introduced during the 1874, 1876, and 1878 sessions of the legislature see Malone, "The California Irrigation Crisis of 1886." Malone's excellent dissertation is much broader than the title suggests. It provides a comprehensive, thoughtful analysis of the evolution of district legislation from 1872 to 1887.

Most members of the Grange arrived at this conclusion following the defeat of the Venable bill in 1874. During the 1876 legislative session, Grangers in the western San Joaquin Valley strongly supported the so-called West Side Bill which passed the legislature in March and won the governor's approval on April 3, 1876.[18] The bankruptcy of William Ralston, and the subsequent acquisition of the San Joaquin and Kings River Canal and Irrigation Company by Miller and Lux, opened the way for the wheat farmers to build the Antioch to Tulare Lake canal on their own. From 1868 to 1876, the west side produced only two good wheat crops; three harvests were total failures. Consequently, farmers in the region had plenty of incentive to support irrigation. In April and May, 1875, local Grange chapters met several times at Graysonville, west of Modesto, to discuss district legislation. The basic features of the bill signed by the governor in April 1876 had been discussed for at least a year.[19]

The law created an immense district containing several million acres of land. It did not designate precise boundaries, but included all the land between the west shore of Tulare Lake to the south, the foothills of the Diablo Range to the west, the shore of Suisun Bay to the north, and the state survey line which defined the boundary between "high" lands and "swamp and overflow" lands

18. *Cal. Stats.*, 1876, 731. The Grange also proposed a $13,000,000 irrigation and transportation canal down the *east* side of the San Joaquin Valley from the headwaters of the San Joaquin River to Stockton connecting the rivers which flowed out of the Sierra. It wanted to create an irrigation district 130 miles long and 20 miles wide, issue bonds to pay for construction, and retire them through a $1 an acre annual tax on the 1,600,000 acres expected to be reclaimed and tolls of 25¢ per ton of freight. See *Fresno Expositor*, January 21, 1874.

19. *Pacific Rural Press* 9 (April 17, 1875): 250; (May 1, 1875): 283; (May 29, 1875): 360; *Weekly Colusa Sun*, May 1, 29, October 16, 1875; *Sacramento Daily Union*, May 22, 25, 1875; *Fresno Expositor*, May 26, 1875; *Stanislaus County News*, January 14, 1876; Smith, *Garden of the Sun*, 450–451. The *Weekly Colusa Sun* of May 29, 1875, also revealed that a public meeting had been held on May 22, 1875, to draft a district bill for the Sacramento Valley. However, irrigation districts did not enjoy much popularity in that part of the state. Will Green reported that "the advantages of irrigation was admitted by all, but most of them were opposed to the State lending its credit for the purpose of carrying out a system, and also against allowing [district] Trustees or Commissioners to tax land for the purpose." Green outlined his own modest district scheme, based on the formation of joint-stock companies rather than special tax districts, in the *Sun* of September 25, 1875. His proposal was also discussed in the *Pacific Rural Press* 10 (October 2, 1875): 216.

to the east. Soon after signing the act, the governor appointed a board of five commissioners to represent the district until the first general election was held within sixty days. Each commissioner had to live in a different one of the five counties in the district. The board's main job was to survey the canal route, establish precise district boundaries, and prepare construction plans.

The draftsmen of the West Side Act faced a serious dilemma. Originally, they wanted district taxes to apply solely to irrigated cropland, though they expected the value of *all* land within the district to benefit indirectly from irrigation, including town lots. Still, it was important to avoid antagonizing the cattle barons in the district, especially Miller and Lux. The *Sacramento Bee* estimated that of the 500,000 acres then utilized for farming or grazing within the proposed district, about 86 percent belonged to a half dozen men. Only 20,000 acres were held by wheat farmers.[20] However, lawyers advised the Grangers that the state constitution probably prohibited property qualifications for voting. So if range land and town lots were excluded from district tax assessments, control of the district would, in effect, belong to those who paid no taxes. Consequently, the law allowed all "legally qualified electors" who met residence requirements to select district officials and approve or reject a $4,000,000 bond issue. The West Side Act specified that if voters rejected the bond issue or tax, the district would be dissolved. The legislature granted the district all the unappropriated water it needed, and the right to condemn all land and existing canals.[21]

On the same day the West Side Bill was approved, a supplemental act, sponsored by Miller and Lux and other large landowners in the valley, also took effect. This law was designed to buy time for opponents of the district. It postponed the first scheduled district election for one year, to May 1877, and required the governor to appoint three commissioners to conduct a thorough survey of the canal. These officials had to report to the governor by March 1, 1877, and publish their findings in one newspaper in each county within the district for thirty days preceding the May

20. The *Sacramento Bee* article was reprinted in the *Bakersfield Californian,* November 18, 1875.

21. *Stanislaus County News,* December 17, 1875.

election. The supplemental act represented the first in a series of legal maneuvers designed to checkmate the Grange's plans.[22]

The district's opponents must have hoped that since neither piece of legislation appropriated money to pay for the survey—which would have been financed by tax revenue had the bond election been held within sixty days, as the West Side Act required—the project would never be launched. But the governor promised to ask the next legislature to reimburse all farmers who helped pay for the survey if the voters rejected the scheme at the May 1877 election.[23] Raising the money proved a difficult task, but nature provided a convenient reminder of the need for irrigation because the winter of 1876–1877 was the driest in twenty-five years. The drought's worst effects were felt in southern California and the San Joaquin Valley, which received less than 50 percent of average annual precipitation. The Kern River ran dry by late spring, and cowboys used ropes to pull weakened cattle out of the mud. Grass disappeared quickly in the spring, and in some parts of the valley, stockmen chopped down oak trees to permit cattle to feed on the leaves and tender shoots. Desperate herdsmen slaughtered bands of sheep to salvage their pelts, which fetched only 12½¢ each; normally sheep sold for $2 or $3 a head. Many animals were simply left to starve, and the valley was littered with carcasses, much like southern California during the drought of 1864.[24]

At the end of April, the governor appointed the West Side Commission. As chairman, he selected J. R. McDonald, who had moved to Grayson in 1869, opened a general store, and became a large landowner. By 1886, McDonald owned 3,600 acres of farmland and another 4,000 acres of range. He was a leading Granger and a draftsman of the West Side Act. Apparently, San Joaquin Valley farmers had sent him to Sacramento as a lobbyist in 1874. According

---

22. For the supplemental West Side law see *Cal. Stats.*, 1876, 885. Also see the *Bulletin*, April 7, 1876; the *Pacific Rural Press* 11 (Apr. 1, 1876): 209; (April 8, 1876): 233; *Stockton Daily Independent*, April 22, 1876; *Stanislaus County News*, April 7, 1876.

23. *Pacific Rural Press* 11 (June 10, 1876): 369; 12 (July 8, 1876): 36; *Alta California*, November 26, 1876.

24. Ira B. Cross, *A History of the Labor Movement in California* (Berkeley, 1935), 70; William D. Lawrence, "Henry Miller and the San Joaquin Valley" (M.A. thesis, University of California, Berkeley, 1933), 116; *Stockton Daily Independent*, January 3, 1877; *Country Gentleman* 43 (February 14, 1878): 105.

to McDonald, the farmers had already decided to build the Tulare Lake–Antioch canal on their own, though there is no evidence that a formal bill was ready before the early months of 1875.[25] The other two commissioners—Francis Williams, a resident of Antioch, and Henry DeVeuve, a civil engineer who resided in the Bay Area— seem to have left most of the commission's responsibilities in McDonald's hands.[26]

The commission ultimately paid for the canal survey with voluntary contributions ranging from $25 to $1,000. The survey work was entrusted to William Hammond Hall, who had been appointed the district's chief engineer, and General B. S. Alexander of the Army Corps of Engineers, who served as a consultant. (Alexander, of course, had been the ranking member of the federal irrigation commission of 1873–1874.) The two engineers concluded that the 185.5-mile canal was entirely feasible, though they recognized several potential problems. The district would have to condemn the San Joaquin and Kings River Canal and Irrigation Company's aqueduct to insure an adequate water supply. It could expect further opposition from swampland owners around Tulare Lake, who wanted the lake drained, while the district wanted to maintain a consistent level of water in it. Moreover, both Alexander and Hall recommended against building a navigable canal because it would have to be larger than an irrigation conduit, with a stone lining and locks. It would also require a smaller grade and wide embankments for tow paths. The canal's projected $4,305,786 cost could be cut by 75 percent if it were constructed solely for irrigation. Hall suggested that the additional money would be better spent improving the navigability of the San Joaquin River, and Alexander predicted that a rail line down the west side of the valley would follow the inevitable migration of farmers into the region. Alexander also recommended that the state ease the financial burden on district farmers by paying the interest and principal on the bonds, at least until the population had increased sufficiently so that farmers could bear the cost. He predicted that the value of land served by the canal would increase ten times and suggested

25. See the J. R. McDonald autobiographical dictation, June 10, 1886, Hubert Howe Bancroft Collection, Bancroft Library.

26. *Stanislaus County News*, April 28, May 12, June 2, 1876.

that the added tax revenue from the valley would more than offset the cost of state aid. Hall warned that the rivers feeding into Tulare Lake had never been measured and that reservoirs might be needed on these streams to keep the lake at a consistent level. Still, he concluded: "While I do not wish to be understood as saying that Tulare Lake will afford an abundant, unfailing supply of water for the West Side District, I have no hesitancy in asserting that . . . the West Side lands may be irrigated with a degree of regularity wherein the failure of an abundant supply of water . . . will be of rare occurrence." The cost of reclamation would range from $1 to $20 an acre, according to the engineer, and 340,000 acres would be within range of the main canal.[27]

After the commission submitted its report to the governor in March, 1877, criticism of the project intensified. On April 14, a convention called to nominate candidates for West Side District offices adjourned on a sour note as Democrats charged that too many Republicans had been selected. Moreover, Contra Costa County, at the canal's outlet into San Francisco Bay, refused to send a delegate to the convention. Residents of that county grumbled that the canal would serve little of the county's prime farmland and that they would be saddled with a much larger tax burden than other district residents. Contra Costa County contained a much larger population and much more taxable property than the counties to the south. Moreover, even many supporters of the scheme admitted that a dual-purpose canal would be too costly. The Grangers hoped to use the canal to break the Southern Pacific's new transportation monopoly in the valley, but could the canal attract sufficient freight traffic away from the railroad to pay for the added expense and maintenance costs? The money saved by building a canal exclusively for irrigation, some critics predicted, would more than pay for distribution ditches.

Rarely did critics challenge the engineering feasibility of the project, or discuss the multitude of legal questions it posed. Instead, they focused on two points: the canal's cost and its uneven

27. William Hammond Hall's comment is from the *Report of the Board of Commissioners of the West Side Irrigation District* (Sacramento, 1877), 82. Also see the *Sacramento Daily Union,* March 7, 1877; *Pacific Rural Press* 13 (March 17, 1877): 168; *Alta California,* March 29, 1877; *Tulare County Times,* March 17, April 14, 28, 1877; *Fresno Expositor,* April 4, 1877; *Stanislaus County News,* April 13, 1877.

benefits. Theoretically, taxes could continue to be levied even if the district failed to complete the canal, or if breaks in the channel rendered it unusable for extended periods. And while *all* district residents would begin paying irrigation taxes soon after the bonds had been issued, construction would begin at Tulare Lake, so those living at the north end of the valley would wait months or perhaps years to reap any benefits. Then, too, how could residents of the district be sure that additional bond issues would not be required in the future? To attract investors, the West Side law specified that $500,000 in bonds could be sold at 75 percent of par and the remainder at 90 percent. Thus, even if all the bonds sold, they might return as little as $3,525,000, far less than the cost of the project. Could anyone predict how high taxes might rise? On the eve of the election, San Francisco's *Alta California*, a bitter foe of the district, warned: "Much of the land in the district is under mortgage, and one of the first results of an affirmative vote will be that the mortgagees will sell out under foreclosure, probably at a loss, to avoid further risk. Of many wild schemes offered to the public in California as great improvements, this is the worst."[28]

Within the San Joaquin Valley, however, drought-plagued farmers ignored most of the reasonable objections to the law and bitterly assailed the stockmen and speculators whom they considered responsible for the criticism. The *Stockton Daily Independent*, which usually spoke for the Grangers, editorialized:

> The stock-raisers would like very well to have the farming land abandoned as unproductive that they might have the privilege of grazing their herds upon it during the Winter months when some feed exists. That class of stock-raisers, who usually own no land and make no attempt to produce the fodder that their stock consumes, but depend upon what nature gives them, have been the curse of California, and the stock-breeding interests will never be what they should. . . . It is irrigation that is required to raise alfalfa and other grasses for feeding stock.

28. *Pacific Rural Press* 13 (March 31, 1877): 193; (April 21, 1877): 249; (April 28, 1877): 257; *Stockton Daily Independent*, April 21, 1877; *Sacramento Daily Union*, April 24, 30, 1877; *Alta California*, April 27, 30, 1877. The quote is from the *Alta* of April 30.

J. R. McDonald charged that Henry Miller—hardly a landless stock-man—was using his money to defeat the canal project. Most valley farmers opposed the dual-purpose canal, so Hall prepared a sup-plemental report describing alternate canal routes that excluded Contra Costa County from the district. District supporters warned against further delay but promised that defects in the law would be repaired prior to the next session of the legislature.[29]

The May 1877 election represented an overwhelming victory for the district and the Grange. The scant population of the western San Joaquin Valley was reflected in the total vote: 476 for, 224 against. Only Contra Costa County, where the vote ran 3 in favor to 134 opposed, offered formidable opposition. Outside that county, the canal project lost in only one precinct. In Firebaugh's Ferry, where Henry Miller owned most of the land, the district lost by one vote. In Los Banos, a farm community also dominated by the Miller estate, the canal carried by a comfortable two-to-one margin. On June 5 a celebration ball was held at Grayson, a Granger strong-hold, where the vote had been 73 to 8 in favor of the project. The governor, Will Irwin, and the lieutenant governor joined the Grang-ers in toasting the dawn of a new epoch in the history of California. Many who voted for the scheme expected the next legislature to modify the project by eliminating Contra Costa County and the requirement that the canal provide navigation as well as irrigation. But the future looked very bright in the fall of 1877. The *Pacific Rural Press* suggested that the Grange had elected so many mem-bers to the legislature that it held the balance of power, and would use that power to defend the irrigation cause in Sacramento.[30]

When the new legislature met in December, the West Side Irrigation District stood high on its agenda. The need to revise the 1876 statute became imperative shortly after the May bond election when Henry Miller won an injunction suit "based on the uncon-

29. The *Stockton Daily Independent*, April 24, 28, 1877. The quote is from the issue of April 24. J. R. McDonald's charges were contained in a long letter published in the *Stanislaus County News*, April 27, 1877.

30. *Sacramento Daily Union*, May 4, 11, 1877; *Stockton Daily Independent*, May 7, June 9, 1877; *Pacific Rural Press* 13 (May 12, 1877): 296; *Tulare County Times*, May 12, 1877; and *Stanislaus County News*, June 8, 1877. The *Union* of January 12, 1878, reported that fifty Grangers held seats in the legislature, nearly 42 percent of the total.

stitutionality of the act and all proceedings under it were enjoined."[31] Not surprisingly, the 1878 West Side law excluded Miller's land and the aqueduct he controlled. The Assembly Committee on Irrigation and Water Rights argued that Miller already sold water to farmers, providing a useful "public service." Hence, the courts might rule against the district's "higher good" argument in a condemnation suit. In any case, a suit would be lengthy and expensive, and construction on the canal could not begin until the courts had arranged a financial settlement. The committee estimated the value of the Miller and Lux canal at $1,300,000—more than half the $2,000,000 bond issue authorized by the 1878 law. Thus, condemnation would virtually exhaust the district treasury. In this event, the district would be forced to appeal to the state for more bonds or a direct subsidy. The district faced a cruel dilemma, one that must have comforted Henry Miller and the stockmen. Since the 1878 law excluded Contra Costa and Alameda counties, as well as Miller's land and the canal, the district's tax base, now restricted exclusively to sparsely settled land, had virtually disappeared. The $2,000,000 twenty-year bonds that went on sale in December 1878 contained liens on 325,000 acres within the district, valued optimistically at $6 an acre, and $487,000 in personal property. The total value of the collateral was about equal to the face value of the bonds and much less than the total interest debt. Nevertheless, the district promoters promised that irrigation would "more than quadruple" the value of the land, enough of an increase, in the words of the district prospectus, to "convince any reasonable person that the security is more than ample."[32]

San Francisco's *Alta California* accurately predicted on December 16, 1878, that "the best legal and engineering talent of the State will not recommend the investment." The bonds did not sell, though the reason remains something of a mystery. In 1886, J. R. Mc-

---

31. E. F. Treadwell, *The Cattle King* (Boston, 1950), 345.

32. *Cal. Stats.*, 1878, 468. The legislature also repealed both the 1876 West Side laws; see p. 887 of the statutes. Also see the *Stockton Daily Independent*, January 5, March 15, 18, 1878; *Pacific Rural Press* 15 (January 19, 1878): 33; (Mar. 23, 1878): 185; 16 (Dec. 14, 1878): 376; 19 (Jan. 31, 1880): 65; "Report of the Committee on Irrigation and Water Rights re A.B. 73," *Journal of the California Senate, 1877–1878* (Sacramento, 1878), 391–392; and *Prospectus of the West Side Irrigation District, San Joaquin Valley, California* (n.p., n.d.), Bancroft Library.

Donald, who remained chairman of the district commissioners to the end, noted that for "some unknown cause the scheme was dropped . . . and the bill remains intact to the present day." The beginning of a wet cycle in 1878–1879, the apparent defection of an important district leader to the stockmen, the state's refusal to back the bonds, the depression of the 1870s, and the untried nature of irrigation district bonds as an investment, all contributed to the failure. But, most important, Tulare Lake turned out to be an inadequate water source.[33]

The level of the lake dropped rapidly during the dry years of the late 1860s and 1870s. Farmers who took up land adjoining the lake in the middle of the latter decade and used its water for irrigation found that their grain "burned up" when hot weather came. Irrigated garden vegetables looked healthy until they approached maturity, then suddenly withered. In 1876, the University of California formed the state's first agricultural experiment station under the direction of E. W. Hilgard. Although he did not conduct formal analyses of the water until March 1880, Hilgard concluded in 1878 that it was too charged with alkali and other substances destructive to plant life to use for irrigation, particularly since much of the land in the southern San Joaquin Valley already contained high levels of alkali. Apparently, many farmers concluded that irrigation *created* alkali problems rather than simply exacerbated them, as Hilgard argued. In his report to the university board of regents published in 1881, the soil scientist predicted that "a very few years" use of water from Tulare or Kern lakes for irrigation would be "promptly fatal to the productiveness of the lands irrigated. . . . The lake water cannot serve for general irrigation." In a federal report published in the following year he concluded: "With the light now before us, it can hardly be regretted that the old Westside ditch, which was to irrigate the lower country with the corrosive waters of Tulare Lake, was not successful. The lake level is now several feet below that [former] outlet, and the lake keeps receding annually, and its alkali becomes stronger."[34]

33. McDonald dictation, Bancroft Library; Malone, "The California Irrigation Crisis of 1886," 97–111.

34. The first quote is from the University of California, College of Agriculture, *Report of the Professor in Charge to the Board of Regents* (Sacramento, 1881), 22, 24; the second is from E. W. Hilgard, T. C. Jones, and R. W. Furnas, *Report on the Climatic and Agricultural Features and the Agricultural Promise and Needs of the Arid Regions of*

The failure of the West Side project was not the only abortive district scheme. The 1878 legislature approved a second law which pertained exclusively to land between the Stanislaus and Tuolumne rivers in Stanislaus County. As noted earlier, when the county first tried to promote irrigation in 1872, the fear of monopoly turned most of its residents against private ditch and canal companies. Consequently, in 1878 the legislature approved a novel institutional alternative, the joint-stock company, to encourage canal building. The plan grew out of the assumption that irrigation drove up tax revenue as it increased land values. The law provided that any five or more people could incorporate to irrigate land within the "Modesto Irrigation District"—roughly that land presently included in the Turlock and Modesto irrigation districts. The incorporators would issue stock, but only landowners could purchase shares, and then no more than one share for each acre of land. Once Stanislaus County farmers had purchased 50,000 shares, the County Board of Supervisors would issue $25,000 in 6 to 8 percent bonds. Upon the completion of each five-mile stretch of canal, the board would issue an additional $25,000 in bonds, up to a total of $500,000. The law pledged that any increase in county or state tax revenue from the district would go toward paying the interest and principal on the loan, but neither the county nor state assumed financial responsibility for repayment. The statute also forbade raising taxes to pay off the debt, and limited assessments against the stock necessary to maintain the canal to a maximum of $1.50 per acre. Nothing came of the Modesto Irrigation District, in part because the need for irrigation was not keenly felt in Stanislaus County during the rainy years from 1879 through 1881, in part because critics of the Modesto Irrigation Act opposed using all new tax revenue exclusively for irrigation. Nevertheless, the law illustrated the range of irrigation district measures considered during the 1870s to raise money and secure a reliable water supply.[35]

With the exception of the 1872 Keyes Act, all the irrigation

the *Pacific Slope, With Notes on Arizona and New Mexico* (Washington, D.C., 1882), 58. Hilgard first raised the possibility that Tulare Lake could not be used for irrigation in his *Report . . . to the Board of Regents* (Sacramento, 1879), 29. Also see his *Alkali Lands, Irrigation and Drainage in their Mutual Relations* (Sacramento, 1886), 22–26,and the *First Report of the Secretary of Agriculture, 1889* (Washington, D.C., 1889), 497.

35. *Cal. Stats.*, 1878, 820; Rhodes, "Thirsty Land: The Modesto Irrigation District," 33–36.

district legislation enacted during the 1870s pertained to particular parts of the state. However, the new California constitution adopted in 1879 attempted to reduce the number and expense of private bills by prohibiting laws restricted to particular counties, municipalities, or regions. Strong sectional rivalries and animosities, the fear that district bills represented yet another power grab by large land and water companies, the expectation that the state would be forced to subsidize districts in some way, and the sharp decline of the Grange in numbers and political influence in the late 1870s, all made the adoption of new district laws unlikely during the first half of the 1880s. Still there was no shortage of proposed legislation.

In the 1881 legislature, Senator Grove Johnson of Sacramento (Hiram Johnson's father) tried to get around the restriction on special legislation by dividing the state into two massive districts: one devoted to drainage and flood control, the other to irrigation. The bill declared that the "interests of the State are so . . . diversified in character that no general law in relation to the ownership, use, and distribution of water for irrigable purposes can be made applicable to the whole State. Such a law, calculated to foster and protect the interests of one section, might be injurious to another of equal magnitude and importance." Consequently, Johnson suggested the drainage district should encompass the land from Stanislaus County north, and the irrigation district include all of southern California as well as Merced, San Benito, Monterey, Fresno, Inyo, Tulare, Kern, San Luis Obispo, and Ventura counties. The bill also would have prohibited application of the riparian doctrine in the irrigable district. It passed the senate, but riparian interests blocked it in the assembly.[36]

Senator J. P. West, a Workingmen's party representative from Los Angeles, proposed an even more controversial plan in the same year. It was inspired, if not written, by William Hammond Hall, the state engineer. The bill went far beyond the Venable and Hay-

---

36. S.B. 287 (Johnson), January 28, 1881, *Senate Bills*, 1881, vol. 3; California State Law Library. Also see *Journal of the Senate During the Twenty-Fourth Session of the Legislature* (Sacramento, 1881), 146; *Sacramento Daily Union*, February 14, May 23, 1881. Many northern Californians were reluctant to abandon the riparian doctrine because they believed it provided some protection against the debris dumped in streams by hydraulic miners. The doctrine promised downstream riparian owners a flow of water undiminished in *quality* as well as quantity.

mond legislation of 1874 and 1876. It called for a state board of irrigation—consisting of the governor, surveyor-general, attorney-general, and state engineer—with vast powers. The board was to designate irrigation districts on the advice of the state engineer. Once this was done, the governor would appoint a board of directors for each district, and the directors could accept, reject, or modify the state engineer's blueprint for dams and canals. In October of each year, the district officers would order the county tax assessor to impose a tax of one-half of one percent of the value of all property within the district. Once the land had been reclaimed, a second tax not to exceed $3 an acre, payable in six annual installments, would be levied. The act also required the board of irrigation—in effect the state engineer—to approve or reject all future private water claims and irrigation projects. Water not already owned and used was declared to be "and . . . shall remain public waters of the State." A second bill offered by Assemblyman John Bost of Mariposa and Merced counties closely resembled West's bill except that it gave the district board of directors greater authority over the use and distribution of water, including the specific power to prohibit riparian owners from diverting water, other than a "reasonable supply" for stock-watering and household purposes. The residents of each district would exercise no control at all over the district leadership or over the taxes they paid. In most other years the West and Bost bills would have created a storm of controversy; but both the 1880 and 1881 legislatures were preoccupied with the danger of flooding and with the hydraulic debris bill discussed in the following chapter.[37]

By 1883, the irrigation district had become the tool of large land and water companies dedicated to abridging or abolishing the riparian doctrine. In both 1883 and 1885, the most noteworthy district bills were part of legislative packages. The 1883 bill was one of five offered by Assemblyman Wharton of Fresno County. His bill permitted the formation of irrigation districts upon the request of landowners after engineering and hydrographic studies had been performed by county surveyors. Each district would be governed

37. S.B. 309 (West), January 31, 1881, *Senate Bills, 1881*, vol. 3; A.B. 464 (Bost), February 7, 1881, *Assembly Bills, 1881*, vol. 6, the California State Law Library. Also see the *Pacific Rural Press* 21 (February 5, 1881): 89.

by an elected board of water commissioners, and if two-thirds of the voters within a proposed district agreed, bonds could be issued to pay for the works. The bonds would be retired through taxes on the irrigated land. The most radical feature of the bill was a section that gave the board of water commissioners power to condemn *all* water rights within the district. A second Wharton bill, A.B. 365, specifically invalidated the riparian doctrine.[38] In 1885, much of the legislature's attention was focused on the so-called Fresno bills. These included a modest district bill that exalted local control.[39]

By 1885, after thirteen years of legislative experimentation, most lawmakers agreed that any district law should contain certain features. Few Californians supported a centralized state district system after 1881. Those who did usually considered full state control over water and water rights the only way to break the alleged water monopoly held by land and water companies. The reasons for opposition to a state system were many, but three stood out: the state could not afford it, the state could not be trusted to build or administer it, and such a system was undemocratic and dictatorial. The principle of "home rule" dictated that the residents of proposed districts should decide whether they wanted a district, who should run it, and how they would pay for it. The issuance of twenty-year bonds—as opposed to special taxes, levies, or pay-as-you-go water rates—became very attractive. The total interest debt far exceeded the face value of the loan, particularly since bonds would probably have to be discounted to attract buyers; but repayment could be spread over twenty years which would give new farmers "breathing time" to get started and would also allow farmers already living within district boundaries time to sell their surplus land to new residents. There was, by contrast, less agreement

38. A.B. 155 (Wharton), January 13, 1883, *Assembly Bills, 1883*, vol. 2. The other bills pertaining to irrigation districts considered by the 1883 legislature were S.B. 87 (Whitney), *Senate Bills, 1883*, vol. 1; S.B. 257 (Reddy), *Senate Bills* (1883), vol. 2; A.B. 423 (Keeler), *Assembly Bills, 1883*, vol. 3.

39. S.B. 38 (Reddy), January 19, 1885, *Senate Bills, 1885*, vol 1. On the same day an identical bill (A.B. 171, Weaver) was introduced in the Assembly. See *Assembly Bills, 1885*, vol. 2. For more thorough analyses of the irrigation legislation considered during the 1880s, especially water rights issues, see the following two chapters.

on who should vote in district elections and what property should be taxed to retire the bonds. Most lawyers argued that the state and federal constitutions demanded that all electors, not just landowners or irrigators, participate in district elections. And while many friends of irrigation admitted that only those who wanted water should have to pay for it, they recognized that acceptance of this principle would doom most districts to failure. The tax base would be too small and farms too scattered. To avoid lawsuits based on charges of discrimination, all land—though, perhaps, not personal property or some improvements—would be taxed.

Nevertheless, a substantial number of questions remained unresolved. How could the interests of stockmen and irrigators within irrigation districts be reconciled? What size majorities should be required in district elections? Should the majority required to pass bond issues be greater than that needed to, say, elect district officials? Where would the districts get the plans for their irrigation systems—from independent engineers, private companies, or the state? Who should build the works? How much supervision should the state exercise over the construction of dams and canals, the acquisition and regulation of water rights, and the issuance of bonds? Although some of these questions were answered by the Wright Act adopted in 1887, many persisted well into the twentieth century. Ultimately, the irrigation district became a great success, both in California and most of the arid West. But success did not come easily, or quickly.

7

William Hammond Hall and State Administrative Control over Water in the Nineteenth Century

In the nineteenth century, the California state government did little directly to aid or promote irrigation, though it did provide farmers with plenty of subsidies, incentives, and valuable information. For example, in 1859, in an effort to stimulate the wine industry, the legislature excluded the state's vineyards from taxation for four years, and three years later it offered $1,000 cash prizes to the first farmers who produced 1,000 pounds of such "exotic" products as tobacco, hemp, flax, and molasses. The state also printed the proceedings of local agricultural societies and of the state society as well. That the state did not go further is not a reflection of an inflexible commitment to the doctrine of laissez-faire.[1] The public fear of monopoly and corruption, which extended to institutions of government as well as to business, was a much more formidable philosophical barrier; the state's limited revenue and constitutional

1. For an excellent overview of the state's efforts to stimulate agricultural development, see Gerald Nash, *State Government and Economic Development: A History of Administrative Policies in California, 1849–1933* (Berkeley, 1964), 63–80.

restriction on debts were equally important. Though unwilling to approve a comprehensive state irrigation system, or even a master blueprint to guide future water development, the lawmakers tried to assist irrigators in several ways. Not only did they pass laws to regulate the rates and service provided by canal companies, they also authorized the first comprehensive survey of water resources ever undertaken in the United States. William Hammond Hall, California's first state engineer, directed those surveys. His career foundered in the storm of controversy over hydraulic mining in the early 1880s, though he remained in office until 1889. The state wasted his considerable talents and vision, but his quest to modernize antiquated water laws touched off a significant western reform movement which heralded the Progressive Movement's dedication to the dictates of science and efficiency.

California's early governors paid scant attention to the state's water problems, but a few state officials recognized the need for planning even during the 1850s. An 1850 statute ordered the new state surveyor-general to provide the legislature with "plans and suggestions" for improving river navigation as well as for "the draining of marshes, prevention of overflows and the irrigation of arable lands by means of reservoirs, canals, artesian wells or otherwise." The surveyors-general frequently lacked the skills, usually lacked the interest, and always lacked the money, to carry out that part of their mission. However, in his 1856 report, John A. Brewster noted that irrigation, flood control, and swampland reclamation were closely related. Reservoirs could be used to store the annual spring floods for irrigation and mining; California abounded with suitable reservoir sites. Similarly, dykes and bypass channels could be used in the Central Valley both to prevent flooding and to reclaim floodland. Brewster concluded: "A system of reclamation similar to the one proposed, or in fact any other should not be left to individuals or counties, but be general for the whole State where required and under the care of a State officer. Now is the proper time for a determination of the State policy in regard to this matter, and when a proper system is once adopted, all direct legislation thereupon should be in accordance with it." In his reports for 1854 and 1855, the state geologist also acknowledged the need for irrigation, but when the surveyor-general offered to prepare a comprehensive water plan, he received no encouragment from either the governor or the legislature.[2]

Nevertheless, the essential features of the multiple-use concept of water planning won some attention during the 1860s. For example, following the 1864 drought, state officials ranging from the governor to the board of agriculture urged Congress to donate the irrigable public lands to the state. John Bidwell, a congressman from northern California, strongly supported such legislation and maintained that "canals for irrigation should be made upon a system, so as to harmonize with the reclamation of swamp and tule lands, and equalize the distribution of water for the benefit of all." Such a system implied coordinated state supervision, if not direct state construction and operation, of waterworks. Bidwell knew that piecemeal water planning, such as individual or swamp district construction of ever higher levees built in the hope that flood breaks would occur downstream where embankments were lower, was short-sighted and counterproductive.[3]

Few Californians could argue with the need for a water plan, but the preparation and implementation of that plan threatened a wholesale raid on the state treasury. In 1854, two successful artesian wells drilled near San Jose suggested that subterranean sources might provide abundant water for farmers as well as miners. Subsequently, test wells were sunk in the Los Angeles basin as well as in the Santa Clara Valley. In 1856, the legislature considered an "Act to Encourage Agricultural and Mining Interests of the State" which would have created a state commission to investigate the potential of artesian wells. A special committee, to which the assembly referred the legislation, concluded that "with water for the purposes of irrigation, the plains referred to [the Central Valley and coastal plains of southern California] are capable of sustaining a vast number of inhabitants. . . . This done by the State at large, opens the field for private enterprise, and there can be no doubt

2. *Annual Report of the State Surveyor General, 1856,* Journals of the California Senate and Assembly (hereafter *JCSA*), 8th sess. (Sacramento, 1857), appendix, 24–26. The quote is from p. 26. Also see the *Third Biennial Report of the Department of Engineering of the State of California, December 1, 1910 to November 30, 1912,* in *JCSA*, 40th sess. (Sacramento, 1913), appendix, vol. 4.

3. The quote is from the *Sacramento Daily Union,* September 1, 1865. Also see the *Alta California,* October 19, 1865, and the *Union,* September 12, 1867.

that monied power thereafter will seek the wants of the people."
However, both the committee and the legislature as a whole considered the $20,000 appropriation to pay for drilling test holes ex-
cessive, and the bill made no headway.[4]

Nevertheless, the legislature did much to encourage private
water companies. In 1858, it helped set the nineteenth-century
pattern of corporate control over municipal water when it passed
an act providing for the incorporation of water companies. San
Francisco's rapid growth demanded more water, and the fledgling
Spring Valley Company promised to provide it—in exchange for
an exclusive market. Since the company needed to secure rights-
of-way and reservoir sites in San Mateo County south of the city,
the legislature modified a law passed on April 22, 1853, which
permitted railroad companies to condemn land. The new law allowed municipal water companies to condemn land and water
rights as well. It required companies to provide service at "reasonable rates" set by special administrative boards of four members, two appointed by the town, city, or county, and the other
two by the company itself. However, sharp disagreements among
members of the San Francisco commission rendered that body ineffective. For a variety of reasons, most communities failed to establish regulatory agencies, and the state's urban water companies
set their own rates.[5]

As mentioned in chapter 2, in 1862 the legislature extended
the condemnation privilege to companies formed "for the transportation of passengers and freights, or for the purpose of irrigation
or water power, or for the conveyance of water for mining or
manufacturing purposes." Such companies could claim all the unappropriated water they needed. Since this restriction did not apply
to municipal water companies formed under the 1858 law, the

---

4. Paul Wallace Gates, *California Ranchos and Farms, 1846–1862* (Madison,
Wisc., 1967), 79–80; *Sacramento Daily Union*, April 5, 1856; A.B. 84 (Leihy), February
12, 1856. The California State Law Library's collection of bills begins with the 1868
session. Bills introduced in earlier sessions may be found in the California State
Archives. Also see *Majority Report of Special Committee to whom was Referred Assembly
Bill No. 84, JCSA*, 7th sess. (Sacramento, 1856), Appendix.

5. *Cal. Stats.*, 1858, 218. On the effects of this law see *Debates and Proceedings
of the Constitutional Convention of the State of California Convened at the City of Sacramento,
Saturday, September 28, 1878* (Sacramento, 1880), 3: 1371.

legislature implied that the needs of towns and cities transcended all others, save perhaps mining. Water rates charged by ditch and canal companies were "subject to regulation" by county boards of supervisors, but in no case could the boards set combined rates lower than one and one-half percent per month of the capital actually invested in the enterprise. Not surprisingly, given the potential conflict between different groups of water users, the law did not apply to the mining counties of Nevada, Placer, Amador, Sierra, Klamath, Del Norte, Trinity, Butte, Calaveras, and Tuolumne.[6]

Both the 1858 and 1862 laws failed to regulate water rates. However, the large number of private water companies formed during the 1870s and 1880s, along with the growth of California's towns and cities, kept the issue alive. In 1876, a conflict erupted in Riverside County between farmers and the Riverside Canal Company. As in many other parts of the West, the ditch company was an offshoot of a speculative land company; the same investors ran both. Initially, the water company provided settlers with cheap water as an inducement to buy land and to reduce the expense of setting up a farm. It also sold surplus water to some irrigators whose land adjoined the main ditch but had not been purchased from the land company. In the middle 1870s, the water company, maintaining that it had been operating at a loss, substantially increased rates. Apparently, it also began to discriminate among water users, reducing or cutting off the water provided to some farmers. Settlers bitterly complained that the water company's directors had already reaped huge profits as investors in the land company. Though each farmer was supposed to receive water according to the amount of stock he owned in the water company, irrigators charged that the company had issued far more stock than the water supply warranted. Consequently, the 1876 legislature passed a law requiring irrigation companies to provide water on equal terms to all users, though the companies retained the exclusive power to set rates. Unfortunately, the legislature could not anticipate the drought of 1877–1878, and the 1876 law did not provide an effective way to reduce service in dry years. The drought, and an avalanche of complaints from customers, persuaded the

6. *California Statutes*, 1862, 541. Also see *Cal. Stats.*, 1870, 660, and *Cal. Stats.*, 1872, 732.

directors of many water companies to sell out to their customers, as the Riverside Canal Company and Riverside Land and Irrigating Company did in the mid-1880s.[7]

The constitutional convention that met in 1878 presented an opportunity to expand state control over water companies and water rights.[8] However, it encountered several major obstacles. Most Grange and Workingmen party delegates had plenty of energy and bold ideas but little political or legal experience. Consequently, they could not be sure their proposals were practical or consistent with the federal constitution. For example, any proposal for radical changes in California water laws raised the specter of inchoate federal water rights. Few questioned the nation's sovereignty over navigable waters, but what about nonnavigable streams that originated on or flowed through the public domain? What about nonnavigable streams tributary to the Sacramento and San Joaquin rivers, both of which were navigable for part of their course? What about diversions from navigable streams that had no apparent effect on shipping? No one knew how Congress or the United States Supreme Court would react to the new constitution, and the lawyers at the convention preached caution and prudence. Radical changes in water laws threatened to create more problems than they solved. Moreover, the convention mirrored the state's geographical diversity, parochialism, sectionalism, and wide range of

7. William Hammond Hall, *Irrigation in [Southern] California* (Sacramento, 1888), 228–235; *Bakersfield Californian*, August 3, 1876. At least two unsuccessful bills considered during the early 1880s, and probably more, would have required ditch companies to distribute water pro rata during dry years, according to the amount of land irrigated. See A.B. 197 (Griffith), January 14, 1881, *Assembly Bills, 1881*, vol. 3, and A.B. 16 (Wharton), Jan. 10, 1883, *Assembly Bills, 1883*, vol. 1, California State Law Library.

8. The best history of the convention is Carl B. Swisher's *Motivation and Political Technique in the California Constitutional Convention, 1878–1879* (Claremont, Calif., 1930). Also see Winfield J. Davis, *History of Political Conventions in California, 1849–1892* (Sacramento, 1893); Dudley T. Moorhead, "Sectionalism and the California Constitution of 1879," *Pacific Historical Review* 12 (September 1943); Ralph Kauer, "The Workingmen's Party of California," *Pacific Historical Review* 13 (September 1944): 278–291. The Constitution of 1849 had been tailored to fit a frontier economy heavily dependent on mining and stock raising. Critics claimed that it produced a tax system that overtaxed land used for agriculture because growing crops were assessed along with the land itself. Such a system, according to reformers, encouraged land speculators and cattle barons but discouraged irrigation and intensive agriculture. Farmers also hoped a new constitution would resolve many other problems ranging from high transportation and interest rates to uncertain water rights.

competing groups of water users. Proposals to create a state irrigation system, prohibit future appropriations, and confiscate existing rights (with compensation), were checkmated by champions of vested rights and guardians of the state treasury.[9] Given the nature of the California legislature, those who favored the complete abolition of private rights must have wondered whether such a course might not play into the hands of the very interests they hoped to defeat. If all water belonged to the state, then the legislature would have to decide who could use it and under what terms. The state's *entire* water supply, not just the surplus, would become available.

The constitutional convention spent relatively little time discussing the complicated "water question." In February 1879, when the new constitution was ready to present to the public, several disgruntled delegates charged that reform had been stymied by an alliance of the Spring Valley Water Company and hydraulic mining and ditch companies. The *Sacramento Daily Union* noted that the new constitution was "silent as the grave" on such important questions as mining debris, water monopolies, and the wanton destruction of forests that protected the state's watersheds. There was some substance to the charge. Hydraulic miners did worry that the farm block would try to destroy mining by confiscating its water in the name of a higher good, agriculture. Nevertheless, even without the symbiotic relationship enjoyed by mining and municipal water companies, substantial reform was unlikely if not impossible. The *Union's* editorials represented a ritualistic attack on San Francisco capitalists. Newspaper editors from interior towns and cities charged repeatedly that San Francisco's stranglehold over the state's economy would weaken with the expansion of agriculture, allowing such communities as Sacramento, Stockton, and Los Angeles to challenge the "queen city's" commercial supremacy.[10]

The new constitution did little to check the power of water companies. Article XIV provided that "the use of all water now

9. *Debates and Proceedings of the Constitutional Convention*, 1: 81, 86, 95, 101, 143–144, 165, 220, 225, and 272; ibid., 2: 1019–1022, 1024–1031, 1371–1376, and 1472–1473; ibid., 3: 1373.

10. *Sacramento Daily Union*, February 12, 1879; also see the *Union* of November 20, 21, 23, 1878. The *Union* was controlled by the Central Pacific Railroad. Not surprisingly, it charged that the Workingmen's Party preoccupation with railroad issues prevented full consideration of agricultural problems.

appropriated, or that may hereafter be appropriated, for sale, rental, or distribution, is hereby declared to be a public use, and subject to the regulation and control of the State in the manner to be prescribed by law." This section gave the appearance of strengthening state control, and in 1913 became the prime justification for enacting a statute that extended administrative control over the acquisition of *all* new water rights. However, such was not the purpose in 1879. The article went on to require *local* officials, not a state commission such as the railroad commission, to set rates each February for *both* municipal and rural water companies. Thus "public control" over rates masked the forfeiture of state responsibility.[11]

The legislature of 1880 implemented Article XIV by adopting a law requiring boards of supervisors to set maximum rates for irrigation water each year. Section IV of the act permitted any interested person to file suit against any board that refused to exercise this responsibility.[12] In 1885, the legislature absolved the boards from setting rates unless twenty-five or more local taxpayers requested the service, but the new law required rates of not less than 6 nor more than 18 percent of the company's total value. It also broke new ground by requiring water companies to open their books to the boards. Nevertheless, the companies were not required to submit financial statements, and they continued to pad their net worth with such questionable assets as water rights and options on reservoir sites. Moreover, the law did not pertain to water companies that charged annual operation and maintenance fees in lieu of water rates. Virtually all the ditch companies that served the Fresno irrigation colonies fell in this category; apparently, the law was never used in Fresno, Kings, or Tulare counties.[13]

11. The new constitution's section on water is reprinted in *Debates and Proceedings of the Constitutional Convention*, 3: 1472–1473. After it completed its work, the constitutional convention appointed a committee to explain the new document to the voters. As to water, the committee noted: "We provide that when water is offered for sale or hire to the public it should become a public use, and be regulated by law." For the committee report see the *Pacific Rural Press*, 17 (March 8, 1879): 156.

12. *Tulare County Times*, April 10, 1880.

13. *Cal. Stats.*, 1885, 95; C. S. Kinney, *A Treatise on the Law of Irrigation* (San Francisco, 1912), 3216–3220; Arthur Maass and Raymond L. Anderson, . . . *and the Desert Shall Rejoice* (Cambridge, Mass., 1978), 171. An 1881 law (*Cal. Stats.*, 1881, 54) pertaining to municipal water companies required them to submit complete lists of

Despite the legislature's failure to provide effective rate regulation, the late 1870s represented a turning point in the state's
involvement in water planning. The drought of 1876–1878, during
which the average annual precipitation in the southern San Joaquin
Valley fell to less than four inches, crippled the sheep industry and
touched off countless battles over water. In March 1877, the *Tulare
County Times* reported that "armed forces" were guarding local
dams, canals, and headgates. Conflict between rival irrigators would
not end, the editor warned, until a coordinated state irrigation
system had been established.[14] For years, California's shipping,
flood, arid and swampland reclamation problems had begged for
a state water policy. Moreover, during the 1870s, the autonomous
swampland districts had proven incapable of coordinating their
efforts at flood control and reclamation, and hydraulic mining cast
a shadow over the future of the Sacramento Valley's farms and
towns.[15]

Hydraulic mining began in California when miners decided
to tap gold deposits in the gravel beds exposed when streams and
rivers changed their course and carved new channels.[16] Initially,
during the 1850s and early 1860s, tailings were allowed to wash

---

customers and charges to the local agency responsible for setting rates along with
detailed sworn statements describing the amounts spent on construction and maintenance since incorporation. Most companies followed the Spring Valley Water
Company's lead and either ignored the law or submitted inadequate or distorted
reports. The defiance of municipal water companies probably made it easier for
ditch companies to avoid effective regulation.

14. *Tulare County Times*, March 31, 1877.

15. Flood control, swampland reclamation, and irrigation were obviously
related. The failure of the autonomous swampland districts authorized in 1868
underscored the need for state coordination. It also promised the possibility of an
alliance between northern California, where flood control and swampland reclamation were the pressing water problems, and southern California, where irrigation
took precedence.

16. The standard work on California's hydraulic mining controversy is Robert
Kelley, *Gold vs. Grain: The Hydraulic Mining Controversy in California's Sacramento
Valley, A Chapter in the Decline of the Concept of Laissez-Faire* (Glendale, Calif., 1959).
The opening chapter sketches the historical development of hydraulic mining. Also
see Joseph A. McGowan, *History of the Sacramento Valley* (New York, 1961), 293–301;
John Walton Caughey, *The California Gold Rush* (Berkeley, 1948), 249–268; and Rod-
man W. Paul, *California Gold: The Beginning of Mining in the Far West* (Cambridge,
Mass., 1947), 147–170.

into adjoining streams; but advances in technology permitted miners to use a greater volume of water under greater pressure, and by the middle 1850s they began to dump the increasing mass of debris into the deep river canyons. From 1867 to 1870, a series of innovations—ranging from Hoskins's "Little Giant" hose to steam-driven, diamond-tipped drills designed to excavate tunnels—exacerbated the problem. Fortunately, the drought that plagued farmers in most of the state from 1868 to 1874 was a blessing in disguise to communities scattered along the Feather, Bear, Yuba, and American rivers. Though they began to build ever higher levees during the early 1870s, severe flooding did not occur until the middle of the decade. Then, in January and November, 1875, Marysville, located at the confluence of the Feather and Yuba rivers, experienced a devastating flood. Twice its residents dug out from under tons of mud that filled streets, basements, homes, and shops. The damage occurred even though Marysville was virtually a walled city surrounded by levees as high as chimney tops.

Miners claimed that floods were an act of nature, but by the middle 1870s, mining debris had virtually eliminated navigation on the Feather River, and silting had sharply curtailed transportation on the Sacramento River. Moreover, as much as 30,000 acres of prime farmland adjoining the Yuba, Bear, and Feather rivers had been choked with layer upon layer of mud. Some critics blamed shoaling in northern sections of San Francisco Bay on the miners, and predicted that if allowed to continue, navigation would be destroyed in the bay, crippling the economy of northern California. The *Sacramento Daily Union* led the editorial attack on hydraulic mining. In January 1876, it commented:

> Hydraulic mining may be said to be in its infancy to-day, and it is growing with wonderful strides. . . . The mass of sediment carried into the rivers is increasing almost in a geometrical ratio, and the action of the silting process upon the upper waters . . . becomes more marked every month. We believe that if no measures are taken to abate this evil, five years from the present time will see a large proportion of the valley lands desolate, the navigation of the Sacramento and other rivers completely destroyed, and an annual destruction of property exceeding the net value of the products of the

California mines. That the agricultural interest should stand by supinely and witness the ruin of its prospects is not to be expected.[17]

Beginning in 1876, farmers sought relief in the courts and legislature.[18]

The hydraulic mining controversy reflected California's difficult transition from a mining to an agricultural economy, as gold gave way to grain. A correspondent to the *Pacific Rural Press* wrote in 1877: "you have often been reminded of the forlorn and dilapidated condition of the Sierra foothills. . . . When the flush times of mining prosperity began to wane, the interest and industry was transferred to the great San Joaquin valley."[19] However, despite plunging stock values, abandoned mines, and a shrinking population, the mining counties retained a disproportionately large delegation in the state legislature, while the blossoming agricultural counties lagged in representation. The mining industry counted many allies among the San Franciso delegation, which represented the state's "business community." For example, while the agricultural counties voted overwhelmingly to accept the Constitution of 1879, the two strongholds of opposition were the foothill counties and those surrounding San Francisco Bay. The agricultural counties looked to Sacramento and Stockton for leadership, as the hydraulic

---

17. *Sacramento Daily Union*, January 3, 1876.

18. For broad overviews of the environmental impact of hydraulic mining, see the *Reports of the Joint Committees of the Assembly on Mines and Mining Interests, and Agriculture, Relative to the Injury now being done to Lands and Streams in this State by the Deposit of Detritus from the Gravel Mines*, JCSA, 21st sess. (Sacramento, 1876), Appendix, vol. 4; *Majority Report of the [Assembly] Committee on Mining Debris*, JCSA, 22d sess. (Sacramento, 1878), Appendix, vol. 4. This first report included one engineer's prediction that even if the flow of debris did not increase, as valley residents expected, Suisun Bay would fill in 15.5 years and San Pablo Bay in 31.25 years (p. 11). The second report concluded that the beds of the Sacramento River and its tributaries were 4 to 25 feet higher as a result of hydraulic mining, and some of the river canyons contained debris 100 feet thick (pp. 4–5). For newspaper coverage of the controversy during the 1875–1876 legislative session see the *Sacramento Daily Union*, December 24, 28, 1875; January 17, 29, 1876; March 4, April 1, 1876; *Pacific Rural Press* 11 (January 8, 1876): 28; (January 15, 1876): 40; (March 18, 1876): 177; (June 3, 1876): 356.

19. *Pacific Rural Press* 13, (May 19, 1877): 306.

mining controversy added another chapter to the persistent sectional struggle between the "neglected" interior and the "greedy" metropolis. As the *Sutter Weekly Banner* remarked:

> The farmers of the interior have fought this giant evil alone, and their efforts to prevent the ruin of their land and the destruction of our great national highways, have met with little recognition or encouragement. San Francisco has furnished the capital to wash down mountains upon their most fertile lands with charming indifference to the injuries inflicted, so long as a steady stream of gold flowed into her coffers. It is to be hoped that as the destruction has now reached her own domains [e.g. silting in Suisun and San Pablo bays], she will hesitate in the further prosecution of an industry that brings nothing but ruin to the State.[20]

Thus, by 1878 the long-standing political and economic relationship between the mines and urban capital became even more conspicuous.

The mining controversy also contributed to the emerging agricultural rivalry between northern and southern California. Completion of the Southern Pacific's line into Los Angeles in 1876 touched off an agricultural colony boom in southern California that lasted into the late 1880s. But by the middle 1870s, would-be farmers could find little cheap, accessible land in the Central Valley. Northern California newspapers repeatedly asked large landowners to sell small parcels to immigrants at reasonable prices, but with little result. Consequently, the papers began to tout foothill land as a new agricultural eden. The *Sacramento Daily Union* noted that the advantages of foothill farms outweighed the "volumes of word painting of scenery and climate" generated by southern California journals to lure farmers into that part of the state. Plenty of fertile government land remained in the mountain valleys, and the nearby mining camps provided ready markets for farm commodities. The foothills were cooler, secure from floods, and enjoyed heavier rainfall than land in either the Central Valley or southern California.

20. As reprinted in the *Sacramento Daily Union*, June 7, 1877.

Moreover, they contained an abundant supply of timber, and farmers there could avoid the "miasmatic diseases" commonly associated with irrigation in the state's flatlands. Best of all, these lands were already served by an extensive network of flumes and canals. Unfortunately, the mining industry claimed most of the water supply for its own purposes and often refused to share it with farmers for fear that a flood of new settlers would erode its political power. Moreover, since the ownership of farmland seldom included mineral rights, foothill agriculturalists faced the grim prospect of having their land ravished by hydraulic miners. The *Union* noted that the foothills would not be settled until the primacy of mining had ended: "The people of California are continually crying out for more and quicker settlement. The press devotes its space to elaborate demonstrations of the fertility of the foothill lands. . . . We know now that the dependence of California henceforth must be upon agriculture, and that mining can only play a subordinate part in State development."[21]

Drought, flood, and debris convinced the 1878 legislature to create the office of state engineer and to appropriate the then enormous sum of $100,000 "to provide a system of irrigation, promote rapid drainage and improve the navigation of the Sacramento and San Joaquin Rivers." The salary of the state engineer reflected the new office's importance: the $6,000 annual compensation matched that paid the governor. Though flood and debris damage were largely limited to the Sacramento Valley, irrigation could be practiced almost anywhere in the state. The new law ordered the state engineer to locate and map all land capable of irrigation; divide this land into natural drainage districts; designate the best water sources in each district; determine the average annual water supply; prepare plans for irrigation works; and give his "opinion and advice to such parties as may be engaged in irrigating a district, or who may be about to undertake the irrigation of a district."[22]

21. *Sacramento Daily Union*, June 2, 1880. On the potential of foothill lands, also see the *Union* for April 12, 14, 25, 1875; May 1, June 23, August 4, October 6, November 15, 1875; August 16, 1877; January 12, 26, May 29, September 7, 1878; June 12, 1880.

22. *Cal. Stats.*, 1878, 634; *Third Progress Report of the State Engineer to 1883 Session of the Legislature, JCSA*, 25th sess. (Sacramento, 1883), Appendix, 1: 3–4. Also see William Hammond Hall's later testimony before the Stewart Irrigation Com-

Governor Will Irwin, a strong friend of the West Side Irrigation District, appointed William Hammond Hall, the district's chief engineer, as the first state engineer. Hall was born in Hagerstown, Maryland, in 1846, but his family migrated to California and settled in Stockton when he was seven. After attending an Episcopal academy, Hall began his career in 1865 as a draftsman surveying mountains in Oregon for the U.S. Corps of Engineers. Then, from 1866 to 1870, he worked as a surveyor and field engineer for the U.S. Board of Engineers for the Pacific Coast, surveying locations for lighthouses, military posts, and harbors from San Diego to Neah Bay. His topographic survey of San Francisco's Golden Gate Park in 1870 led to an appointment as superintendent and chief engineer of the park in 1871. He held that job until 1876 when he began doing consulting work on irrigation projects in the San Joaquin Valley for the Bank of California and the Nevada Bank. At this time he also became chief consulting engineer for the West Side Irrigation District.[23]

---

mittee in *Report of the Special Committee of the United States Senate on the Irrigation and Reclamation of Arid Lands*, 51st Cong., 1st sess., 1890, S. Rep. 928. 2: 208–218. In *Gold vs. Grain*, pp. 104–106, Robert Kelley suggests that the hydraulic mining controversy was the main reason for creation of the office of State Engineer. However, in his testimony before the Stewart Committee, William Hammond Hall suggested that creation of the office represented a compromise between cattlemen and the Grangers. The stockmen did not want irrigation, while an increasing number of Grangers, stung deeply by the cattle barons' attempts to kill the West Side District, clamored for a centralized state irrigation system. While Hall had no experience designing flood control works, his service as chief engineer for the West Side District gave him some familiarity with irrigation and made him an acceptable appointee to the Grange.

23. William Hammond Hall deserves a biography. Unfortunately, the records from his eleven years as state engineer have been lost or destroyed, with the exception of his field surveys and notebooks, which are held by the California Water Resources Department. The small collection at the California Historical Society in San Francisco is of little value. For a brief but superficial account of Hall's career as state engineer, see Charles P. Korr, "William Hammond Hall: The Failure of Attempts at State Water Planning in California, 1878–1888," *Southern California Quarterly* 45 (December 1963): 305–318. Hall's most important reports have been collected in two volumes at the California State Library under the title *Miscellaneous Reports of the State Engineer, 1880–1888*. For thumbnail sketches of his life see *Who's Who on the Pacific Coast, 1913* (Los Angeles, 1913), 243–244; *Press Reference Library* (New York, 1915), 285; Otto Von Geldern, "Reminiscences of the Pioneer Engineers of California," *Western Construction News* 4 (October 25, 1929): 555–556; *California Water Atlas* (Sacramento, 1978), 23.

Hall sent out five survey parties in 1878. They measured the mining debris deposited by the Yuba, Bear, Cosumnes, American and Feather rivers, and their tributaries. The Sacramento River was mapped and measured as far north as its juncture with the Feather River, and the Feather was surveyed from its mouth to a point eight miles above its confluence with the Yuba River. Hall's lieutenants also surveyed 170 miles of the San Joaquin River as well as parts of the Kings, Fresno, Chowchilla, Mariposa, Merced, Kaweah, and Tule rivers. Irrigation surveys focused on Tulare Lake, Fresno, Kern, Los Angeles, and San Bernardino counties.

The debris problem, drought, and litigation over water rights all contributed to the creation of the state engineer's office, but the hydraulic mining industry was the first to take advantage of the new state agency. In 1878, William Parks, Nevada County's representative in the state senate, proposed that impoundment dams be used to trap mining debris and store water for irrigation. To prevent flooding, he recommended the construction of uniform levees to confine flood-prone streams. Some engineers believed that overflow channels or flood basins offered the best answers to flooding, but Parks assumed that the greater the volume of water a stream carried, the greater its power to "scour out" its channel and carry silt in solution. By forcing a river to recarve its old channel, navigation would be improved and the towns and farms along northern California streams protected. The silt that remained could be flumed into tule basins and aid in reclaiming swampland. Though Parks won widespread support among northern Californians, the legislature wanted to give the new state engineer time to study the debris problem before it approved a particular plan.[24]

In April 1880, the mining companies, assisted by San Francisco's powerful legislative delegation, rushed through a bill creating a special state drainage and debris commission whose members included the state engineer and the surveyor-general. The group was charged with dividing the state into drainage districts and implementing flood control schemes designed by the state engi-

24. For William Parks's influence, see the *Report of the State Engineer to the Legislature of the State of California, Session of 1880, JCSA*, 23d sess. (Sacramento, 1880) Appendix, vol. 5. Hall's recommendations for solving the debris problem are summarized in pt. 2, 13–14. San Francisco's *Bulletin* reprinted excerpts from the state engineer's report on January 22, 1980. Also see Kelley, *Gold vs. Grain*, 99–100.

neer. These schemes could not be built without concurrent approval by district boards of drainage commissioners appointed by the governor from among each district's residents. Four taxes were authorized to pay for the work. Each drainage board could impose a tax of 1/20 of 1 percent of the assessed value of district land; swampland districts whose boundaries overlapped a drainage district could assess a tax equal to the increase in land values incidental to flood control (but no more than $3 an acre); and the state's hydraulic mining companies were required to pay a tax of 1/2¢ per miner's inch of water used in their operations. In addition, the state could collect a uniform tax of 1/20 of 1 percent of assessed value of *all* land in the state. Proponents of the new law claimed that the mixture of taxes fairly distributed the cost of state works. For example, one justification of the statewide property tax was that flood control would improve navigation and benefit the entire state's economy.[25]

Only one district, encompassing the entire Sacramento Valley, was created under the statute, but the new law faced intense criticism. Chico's *Butte Record* charged that the legislation had been passed in exchange for the mining delegation's support for a bill championed by insurance companies. The *San Francisco Chronicle* claimed that the mining lobby had paid up to $1,000 a vote to pass the measure. Many northern California farmers hoped that the legislature would shut down the hydraulic mining industry permanently; others argued that the miners should pay the full cost of protecting the Sacramento Valley from debris damage. The *Weekly Colusa Sun* bitterly complained that the miners "are compelled by a long settled principle of law, to so use their property as that it shall not injure anyone else, and now that we have permitted our property to be injured, they come in and ask us to pay to protect it from further injury." The *Stanislaus County News* called the $5,000 annual tax that county property owners would have to pay a "heavy luxury" because Hall's debris plan was "a scientific experiment for the benefit of another part of the State. . . . The law is wrong in principle, doubtful in expediency, and liable to many abuses." Opposition was particularly strident in Los Angeles County. Los Angeles residents felt slighted because virtually all the state prisons,

25. Kelley, *Gold vs. Grain*, 150–152.

hospitals, and normal schools were located in northern California. Early in 1880, the normal school at San Jose burned down. Subsequently, a bill to establish a teacher's college in Los Angeles passed the senate but stalled in the assembly. Similar bills had been rejected many times during the 1870s. While most of northern California stood to gain directly or indirectly from the drainage bill, flood problems were not serious south of the Tehachapis. The *Los Angeles Herald* declared that "no more iniquitous measure ever passed a California Legislature," charging that the new law represented precisely that class of special interest legislation banned by the new constitution. "It would be just as logical to tax the whole State to pay for a failure of the crops and fleece which have been ruined by drought in the southern counties as to levy a tax to repair the ravages of the debris of mines." Only one southern California senator voted for the bill (ironically, his vote was the price of winning senate approval for the normal school), and Los Angeles newspapers bristled with angry threats that southern California should secede from the rest of the state. The *Los Angeles Evening Express* bitterly complained: "This Debris bill is offensive to us in every respect. It not only takes our money from us without any fair return, but it emphasizes the utter disregard the populous portion of the State has for a section which is weaker politically than the other."[26]

The *Sacramento Daily Union* became one of the new law's strongest defenders. It chided critics of the bill, noting that hydraulic mining benefited the whole state and that the economic health of the entire Pacific Coast depended on keeping San Francisco Bay free of debris. Moreover, the drainage law, by increasing the amount of land under cultivation, might well *lower* state taxes by expanding the tax base. In any case, the principle of using general tax revenue to benefit particular sections was well established. For example, state taxes to support public education were

26. *Los Angeles Herald*, March 21, April 8, 11, 15, 1880; *Los Angeles Evening Express*, April 17, 24, 1880; *Weekly Butte Record*, March 13, 20, April 17, 1880; *San Francisco Chronicle*, April 8, 1880; *Humboldt Times* April 10, 1880; *Weekly Colusa Sun*, April 3, 1880; *Stanislaus County News*, March 19, 1880; *Fresno Expositor*, April 14, 1880; *Bakersfield Californian*, December 18, 1880; January 15, February 5, 19, 1881. On the secession movement touched off by the debris bill also see the *Expositor* of February 9, 1881, and the *Californian* of February 26, 1881.

distributed not according to each county's tax burden, but according to its educational needs. The *Visalia Weekly Delta* was one of the few San Joaquin Valley papers that supported the debris law. It echoed the *Union*'s arguments against sectionalism and suggested that the new law set a valuable precedent. Irrigators could now demand state aid even though irrigation was restricted to a relatively small part of the state.[27]

In April 1880, the legislature asked Hall to design the debris dams. Many years later he insisted that he had privately opposed the drainage act in the legislature because it failed to provide for the construction of an extensive Sacramento River levee system along with the dams, because the act placed no limit on state expenditures, because groups of irrigators could use it to defend their demands for a state irrigation system, and because he thought the federal government should have paid half the cost. He pleaded that the plan carried out in 1880–1881 was largely the brainchild of G. H. Mendell and James B. Eads, whom the governor had appointed to assist Hall. Apparently, time and the painful effects of the debris fiasco on Hall's promising career dulled his memory. In the early months of 1880, the state engineer never voiced his objections publicly, despite many opportunities, and at the time he gladly accepted credit for the debris plan.[28]

In any case, in August 1880, Hall submitted a formal plan to Mendell and Eads for review, and the two men quickly gave their approval. Mendell, who represented the U.S. Army Corps of Engineers, had worked closely with state officials on flood problems since the middle 1870s. Since the legislature hoped to persuade the federal government to pay at least part of the cost of drainage work, his support was essential. Eads had won national recognition by designing a fleet of armored gunboats during the Civil War. Later, he bridged the Mississippi River at Saint Louis and devised a jetty system to prevent shoaling at the mouth of that stream.[29]

27. *Sacramento Daily Union*, November 17, 1877; April 9, 28, Nov. 13, 1880; *Visalia Weekly Delta*, December 10, 1880.

28. See Hall's "An Account of the State Drainage and Debris Work of 1878–1881," December 1904, State Engineer's File, California State Archives. Also see Hall to Governor George Stoneman, November 19, 1882, ibid.

29. William Hammond Hall, *Memorandum Concerning the Improvement of the Sacramento River Addressed to James B. Eads and George H. Mendell* (Sacramento, 1880).

In the fall of 1880, 800 men went to work building two re-
straining dams on the Yuba and Bear rivers. Other workers built
levees along the Yuba, American, and Feather rivers. Apparently,
Eads persuaded Hall to build brush dams rather than earthfill or
masonry structures. Such dams cost less to construct, and their
height could be raised easily to provide additional storage as the
debris accumulated. Since the woven brush mattresses were little
more than dense screens, they did not seriously impede stream
flow. In fact, they were designed to capture the heaviest debris,
such as rocks and timber, even as most of the stream flowed over
the dam's crest. Maintaining swift currents in Sacramento River
tributaries was vital to "scour out" debris that had already washed
into the valley. For this reason, upstream from the Yuba River dam,
nearly two miles of brush and sapling embankments were constructed to maintain the channel behind the dam and to prevent
erosion of the stream banks. After inspecting the completed Yuba
River dam in November, the Marysville *Appeal*'s editor remarked:
"In all its parts the dam is a splendid piece of engineering. The
work has been done thoroughly, and is highly creditable in every
way to both contractors and engineers." Visitors flocked to see the
flood control works. In January 1881, virtually the entire state assembly inspected the Bear River debris works.[30]

The legislature of 1881 convened during one of the rainiest
winters California had experienced since the 1850s. Many legislators grumbled that the dams had cost too much and would not last
through the winter. Despite Governor George Perkins's warning
that debris damage might exceed $160,000,000 unless the state forged
ahead with its drainage work, several bills were introduced in the
legislature to rescind the 1880 law. The mining block, aided by
members of the San Francisco and Sacramento delegations, pre-

---

30. Kelley, *Gold Vs. Grain*, 154, 239; *Annual Message of George C. Perkins to the
Legislature January, 1881, JCSA*, 24th sess. (Sacramento, 1881), Appendix, 1: 18; *Report
of the State Engineer to the Legislature of the State of California, JCSA*, 24th sess. (Sac-
ramento, 1881), Appendix, 3: 31–65; *Report of the Board of Directors of Drainage District
No. 1, Showing Progress of Work to January 1, 1881*, in ibid.; *Sacramento Daily Union*,
September 4, November 11, 15, 17, December 14, 1880; Jan. 22., 1881; *Tulare County
Times*, Aug. 20, 1881; *Weekly Colusa Sun*, January 22, 1881; *Pacific Rural Press* 20 (July
31, 1880): 72, 116; (November 20, 1880): 330; (Nov. 27, 1880): 346; *Stockton Daily
Independent*, January 28, 1881. The Marysville *Appeal* quote is as reprinted in the
Rural Press of November 27, 1880.

vented consideration of the leading repeal bill at the beginning of March by a vote of thirty-nine to thirty-six. Nevertheless, the drainage work continued only into the fall. On September 26, 1881, the California Supreme Court ruled the law unconstitutional, mainly on a technicality. Article IV, Section 24, of the state constitution required that each bill concern only one subject, and that subject had to be clearly revealed in the title. The court ruled that, strictly speaking, channel improvement and debris impoundment were not "drainage works." It also decided that the assessments authorized by the law were unconstitutional because the legislature could not delegate the responsibility of collecting state taxes to boards that represented special districts, and also because the law had provided a system of raising revenue that violated the principle of equal taxation. As if this were not bad enough, a month later fire ravaged the Yuba River dam which, along with the Bear River structure, had already been seriously damaged by floods.[31]

In 1884, a United States circuit court banned hydraulic mining in California, and thereafter the industry's prominent part in the state's economy quickly faded—though the debris it had produced continued to wash down into the Sacramento Valley for over thirty years. Robert Kelley has credited the legislature of 1880 with launching "the state's first large-scale effort at controlling the Sacramento River."[32] Yet the Drainage Act of 1880 had much broader

31. *Stockton Daily Independent,* January 31, 1881; *Pacific Rural Press* 21 (January 15, 1881): 41; *Bakersfield Californian,* March 19, April 19, 1881; *Tulare County Times,* October 8, 1881; *Sacramento Daily Union,* January 8, 14, February 9, 16, 21, 23, 25, 26, March 2, July 12, 13, 29, September 28, October 24, 1881. In all, the state spent over $500,000 on debris work, including $105,000 for the Yuba River dam and $74,000 for the Bear River Structure. For a breakdown of how the money was spent, see *Report of the Assembly Committee on Claims, Twenty-Fifth Session, on Assembly Bill No. 207, JCSA,* 25th sess. (Sacramento, 1885), Appendix, 6: 4, 11. Hall persistently maintained that the brush dams were practical and well designed. Apparently, he warned the Sacramento River Drainage District's directors to keep a large supply of brush, sandbags, and men on hand through the winter to repair breaks in the dam as well as sections that settled. The district itself constructed the dams, and in an October 15, 1881, letter to Governor George Perkins, Hall admitted that they had been built "very hurriedly" and that a widespread fear that the California Supreme Court would overturn the Debris Act prevented the district from buying supplies on credit or even hiring laborers. The state engineer denied that the dams had washed away, arguing that they had "settled" due to lack of maintenance. The letter is in the State Engineer's File, California State Archives.

32. Kelley, *Gold vs. Grain,* 132.

significance as well. It authorized the first substantial state expenditure for water resource development in the arid West and reflected a dawning awareness of the need for centralized resource planning. Unfortunately, passage of the act, and its failure to accomplish the results promised by its sponsors, reinforced the common, complementary assumptions that the legislature was hopelessly corrupt and the state government could not be trusted to build efficient public works. Critics of the law charged that Hall, Mendell, and Eads had designed faulty dams and that the cost of those dams had greatly exceeded what a private company would have paid for the same work. The dam builders had few precedents to follow. But if three such prominent engineers had been unable to solve the debris problem, could any engineers draft, let alone build, a comprehensive irrigation system for the state? Hall and the office of state engineer suffered most of all, because the debris scheme undermined confidence in his judgment and the value of his other work, especially the irrigation surveys. The drainage act revived and intensified powerful rivalries between groups of water users in different parts of the state. Farmers in southern California and the San Joaquin Valley were now even more likely to oppose measures designed to aid northern California water users, just as the mining block became more hostile to bills proposed by representatives from the irrigation counties.

After a heated debate, the legislature of 1881 slashed an appropriation to pay for continuing the irrigation surveys begun in 1878, from $50,000 to $20,000. One disgruntled lawmaker characterized the legislation as a "twin sister of the debris bill." In 1878, Hall expected his investigations to produce a comprehensive state water plan that would reconcile the needs of different groups of users. But the failure of the state debris program polarized these groups, and the irrigation surveys suffered from the same charges of localism as the debris work. Hall maintained that his office had been blamed for events beyond his control. The budget cuts stymied his grandiose plan to publish seven volumes discussing the social, political, legal, physical, and technical aspects of irrigation. Instead, the $20,000 all but eliminated the hydrographic survey work and restricted his staff to finishing a map of the state's irrigated land and preparing a closing report. Hall expected the 1883 legislature to abolish or reorganize his office and complained to a friend: "I must get out of public life; it is no place for a man to make money except by dishonesty and I can't do that. . . . I am now preparing to go out of this office when my report is made two years from now. Nothing would induce me to hold it longer unless I positively could get at nothing else to make a living."[33]

Sadly, William Hammond Hall's greatest accomplishment was all but eclipsed by the debris controversy and the battle between riparian owners and appropriators described in the next chapter. In his 1880 report, Hall noted that "the establishment of a proper water right system will do more to bring about a solution of the . . . problems of irrigation . . . than all else which can be accomplished at this time." He gave several telling examples of inadequate state control over the acquisition of water rights. Eighty-three claims had been filed on the Kings River, but only forty-two listed specific quantities of water, and those exceeded the maximum flow of the stream by 250 percent. On the Kern River, conditions were even worse. Seventy-six quantifiable claims had been filed, amounting to more than twenty times the river's peak flow. Not surprisingly, the result had been litigation and the reluctance of most business-men to invest in irrigation projects. Hall assumed that California's future depended on intensive agriculture and the family farm. Be-fore potential immigrants would consider California as a permanent home, huge estates would have to be divided up and conflicts over water rights reduced or eliminated. The state engineer concluded that the state should not build irrigation works, but "should foster irrigation interests by establishing a business basis for enterprise

---

33. *The Sacramento Daily Union* of April 16, 1881, reprinted the legislative debate over the appropriation. The quote is from Hall to "My dear Yorke," July 17, 1881, Box 1, MS 915, Correspondence 1880–1885, William Hammond Hall Collection, California Historical Society Library, San Francisco. Hall even faced criticism in the irrigation counties, where many farmers considered his work superfluous or re-dundant, and a threat to private water development. For example, the *Bakersfield Californian's* editor proclaimed in the April 23, 1881, issue that the county did not need any state assistance: "If our canal-owners want places for storage they will send their own suveyors and make their own locations, probably ignoring the work of the State Engineer altogether. Let the Legislature devise a law regulating the use of rivers and streams and the irrigation problem will regulate itself. This comprises all the State assistance needed. This indiscriminate appropriation business means nothing more than fat places for parasites who believe the public owe them a living." The effect of the debris controversy may be seen in the fact that the same newspaper had supported state construction of storage reservoirs in its issue of July 15, 1880.

in irrigation projects." This could be done by giving state officials control over the acquisition of water rights, by establishing irrigation districts, by attaching water rights to the land, by restricting the state's irrigable land to the amount of water needed to raise different crops, by subjecting all water rights (except municipal) to condemnation by the state, and by declaring agriculture the highest use of water. Hall urged that the land itself should bear the cost of irrigation, and that the farmers should distribute the water according to state "schedules and regulations." In particular, all diversions from navigable streams should be monitored closely by the state to protect navigation. The state should inspect irrigation systems to reduce waste, but Hall warned that existing water rights "should not be interfered with" otherwise. He requested, and was subsequently granted, permission to prepare water laws for submission to the 1881 session of the legislature.[34] Hall presented his suggestions to the governor in October 1880. He had prepared two bills and the outline of a third. The first was designed to provide a full record of all existing claims to California's water supply; the second provided for the filing of future claims and the allocation and distribution of that supply; and the third offered proposals concerning the formation and operation of irrigation districts.

Hall recognized the need for a complete record of water rights to reduce litigation, to determine the state's surplus supply, and to pave the way for state administrative supervision over allocation and distribution. Since the surveyor-general maintained a full record of land titles, he suggested that this official also be designated "state register of water rights." The process of compiling an accurate record of claims would begin with the county recorder. Each water user, except riparian owners and those served by municipal water companies, would be given a year to record his claim with the county. The claim would indicate when the notice of intent to divert had been posted (if applicable), when actual diversion began, how much water was currently being used, and the location and

---

34. *Report of the State Engineer to the Legislature of the State of California, Session of 1880.* The first quote is from pt. 4, 125. The material on water rights controversies is from p. 4 of the same section. Hall's suggestions regarding water law reform are from his conclusions, pp. 4–5. Also see Hall's letters to Governor George Perkins, January 29, 1880, and to Governor George Stoneman, November 19, 1882, State Engineer's File, California State Archives.

a description of ditches and diversion works. The recorder would relay this information to the surveyor-general who would compare it with information gathered by the state engineer. The surveyor-general would then forward copies of all documents he had acquired to the state attorney-general who, in turn, would order each district attorney to file suit on behalf of the state, in a superior court, against all claimants to a stream. The court would ask each claimant to demonstrate the validity of his water right, on penalty of forfeiture for noncompliance. The final court decree would be recorded in both Sacramento and the appropriate county, so the would-be appropriator would be able to tell easily how much water remained available for future use. Hall emphasized that this bill posed no threat to established rights. But since the courts limited water rights to "beneficial use" (though they often defined the phrase very broadly), most claims that existed only on paper would be quickly eliminated, and other claims might well be reduced. Any such legal action posed a direct threat to water users; the burden of proof was on them, not the state.

The second bill built directly on the first, and Hall urged that the legislature consider them as a package. It would have created a state board of commissioners, consisting of the governor, surveyor-general, and state engineer, to supervise the issuance of new water rights and regulate distribution. Among its responsibilities, the state board would establish uniform standards for measuring water quality and the volume of diversions. Though the board would have broad review and veto powers, most day-to-day administrative tasks would be performed by local administrative units comprising three to five counties. The boundaries of these districts would be drawn by the state board to conform to natural drainage basins. The districts would be directed by boards of local appointees selected by the governor. The state engineer would serve as engineering consultant to each district, and the state surveyor-general would provide local water officials with copies of all water rights, court decrees, and other data collected by state agencies. The local boards could grant surplus water to individuals or companies, though any application approved at the local level could be rejected in Sacramento. In particularly wet years, the local board might also grant temporary rights to use water, but these "rights" would not enjoy any priority or other legal recognition. Local officials would continually monitor stream flow and diversions and could prevent waste by banning faulty water works or uncapped artesian wells. In dry years, they could also prohibit diversions that interfered with riparian rights or threatened shipping on navigable streams. However, all riparian rights would be limited to stock and household uses, and could not be used on land more than one-quarter of a mile beyond the riverbank. Rights acquired by appropriation not used for two consecutive years, or not put to "good use," could be revoked. In October of each year, the local board would publish a schedule of water rights for each stream under its jurisdiction in an appropriate newspaper. The board's expenses would be paid from a tax levied against all land within the district. In Hall's first annual report, filed in January 1880, the state engineer recommended that the state assert the power to condemn *all* water rights, though he realized that the expense involved would make wholesale condemnation actions impractical. Consequently, neither of the bills proposed in October 1880 contained such a provision. Hall doubtless expected that his system would weed out weak, extravagant, or speculative claims anyway.[35]

Had the legislature accepted Hall's proposals, California would have enjoyed the most advanced code of water laws in the arid West. As noted earlier, Utah and California both asserted public control over water used for irrigation during the 1850s but abandoned the principle by 1880. However, in 1879 and 1881 persistent water conflicts on the Cache la Poudre and South Platte rivers prompted Colorado to create a state engineer's office and to divide the state into water districts that conformed to the boundaries of natural drainage basins. These were the most elaborate water laws adopted by an arid state to that date. Any resident of a water district could appeal to a district court to adjudicate all water rights on a particular stream, and each water user had to participate in the proceedings. Unfortunately, the law did not permit the state to participate in the process. The state was not considered a party to

---

35. The bills, along with Hall's explanation of their provisions, were reprinted in pt. 4 of the *Report of the State Engineer to the Legislature of the State of California: Session of 1881, JCSA,* 24th sess. (Sacramento, 1881), Appendix, 3: 11–44. For a concise summary of the proposed legislation, see Hall's *The Irrigation Question in California: Appendix to the Report of the State Engineer to His Excellency George C. Perkins, Governor of California,* in ibid.

the suits, nor was it allowed to collect information to guide the courts. In fact, the courts were not even required to notify the state engineer that an adjudication suit had been instituted. Since court officials did not have the training, time, or resources to measure stream flow or inspect diversion works, they usually accepted the statements of claimants at face value. Hence, most of the early adjudication cases produced decrees which granted far more water than the particular stream carried. Decrees usually granted either the amount of water claimed, or the capacity of a claimant's ditches, not the amount of water actually used or needed. So while a state water commissioner with substantial theoretical powers administered the distribution of water in each district, his efforts were severely limited by the nature of the decrees themselves. Similarly, the laws did not provide complete state supervision over the acquisition of new water rights. The 1881 law required each water user to file a claim and map with the appropriate county clerk within ninety days of the beginning of project construction. But Colorado did not require irrigators or water companies to file with the state engineer's office until 1887, when the legislature provided for a centralized record of claims. In almost every respect, Hall's proposals went far beyond the pathbreaking laws of Colorado.[36]

The debris controversy and heavy rainfall of the early 1880s completely overshadowed Hall's bills; they received little public attention. But even had he proposed them during a drought, the chances of adoption were slim. The state engineer failed to realize that ideals such as planning and efficiency had a very small constituency. His measures were as "impractical" as they were farsighted because they challenged virtually every water user in the state. They struck at monopoly whether exercised in the name of riparian rights or appropriation. And while they rejected a centralized state irrigation system, they also threatened the principle of local control. (Ironically, proponents of local control charged that Hall wanted to build up a "water bureaucracy" in Sacramento, while supporters of a state water system claimed that he pandered

---

36. Elwood Mead, *Irrigation Institutions* (New York, 1903), 222–223, 143–179; A. E. Chandler, *Elements of Western Water Law* (San Francisco, 1913), 57–61. Also see Robert Dunbar, "The Origins of the Colorado System of Water Right Control," *Colorado Magazine* 27 (October 1950): 241–262; idem, "The Significance of the Colorado Agricultural Frontier," *Agricultural History* 34 (July 1960): 119–126.

to powerful "local interests," specifically land and water companies.) Not surprisingly, Hall's two bills were never formally introduced in the legislature. They were thirty years ahead of their
time.[37]

Hall consistently opposed using a state-owned irrigation system to quiet litigation and prevent future water conflicts.[38] He shared
most of the prevailing fears of overlarge government: that it offered
a feeding ground for corrupt and incompetent bureaucrats; that it
promoted waste and inefficiency; that state governments invariably
tried to assume powers and responsibilities better exercised at the
local level; and that they reflected the interests of powerful capitalists and corporations rather than the "public interest." Nevertheless, he also had more immediate reasons for opposing a state
system. The only large-scale attempt at state construction of public

37. In his address to the legislature in January 1881, Governor George C.
Perkins urged the legislature to consider Hall's bills. He noted: "Our present laws
[not] only do not provide for the issuance of definite water rights, but they make
no provision for the prevention of waste, and no provision for the adequate organization to construct irrigation works." *Annual Message of George C. Perkins to the
Legislature [1881]*, JCSA, 24th sess. (Sacramento, 1881), Appendix, 1: 17. The *Pacific
Rural Press* was one of the few journals to notice Hall's bills. See ibid. (January 1,
1881): 1; ibid. (January 15, 1881): 34. Not surprisingly, the appropriation bill to pay
for continuing the state engineer's work drew more attention. For example, see the
Sacramento Daily Union, April 16, 1881; *Stockton Daily Independent*, April 21, 1881;
Pacific Rural Press 21 (April 9, 1881): 262.

38. For examples of support for a state irrigation system, see the *Biennial
Report of the State Board of Agriculture for the Years 1870 and 1871*, JCSA, 19th sess.
(Sacramento, 1872), Appendix, 3: 25–26; *Tulare County Times*, November 25, 1871;
California Mail Bag 1 (December 1871): 33; *Transactions of the California State Agricultural
Society During the Year 1873* (hereafter TCSAS), JCSA, 20th sess. (Sacramento, 1874),
Appendix, 6: 609; *Fresno Expositor*, May 14, June 4, 1873; *Pacific Rural Press* 6 (October
11, 1873): 232; *Sacramento Bee*, October 1, 1876; *Expositor*, January 9, 1878; *Sacramento
Daily Union*, December 13, 28, 1880; January 3, 6, 1881; *Visalia Weekly Delta*, April
9, 1885; *Bakersfield Californian*, July 31, 1886. In the 1870s, support for a state irrigation
system grew out of antimonopoly sentiment, while in the 1880s most supporters
believed that water conflicts would end only when the state exercised absolute
control over water rights. Newspaper support for a state system waned considerably
during the 1880s. This may reflect the increasing power of land and water companies
over local papers, but, in addition, by the 1880s private water companies had
acquired title to so much of the state's water that a state system no longer seemed
as practical as during the 1870s when the first irrigation boom began. By 1887, the
sheer cost and complexity of state condemnation of water rights acquired during
the 1870s and 1880s made such a scheme infeasible. Moreover, many who supported
a state system in the 1870s did not understand the extent, or nature, of riparian
rights. As noted in the next chapter, the first suits over riparian rights grew out of
the drought of 1878–1879.

works, the debris dams, had been thoroughly discredited, and the state engineer could not resolve conflicts over water rights without cooperation from the large land and water companies. Perhaps most important, in the absence of a permanent state bureaucracy, Hall could not expect state employment to last forever. He needed to remain on good terms with private companies because when he quit or was fired, he would need a job. He lacked the professional independence to live up to high ideals, and his opposition to a state system may have been more a measure of his political vulnerability and the impermanence of his office than of philosophical convictions.

Hall could not escape the tug and pull of powerful, rival groups of water users. Not the least of the many dilemmas he faced was how to maintain his independence and initiative when the survival of his office depended on an uneasy pact with the "appropriation party" (whose activities are described in greater detail in the next chapter). At least by 1880 or 1881, and probably long before, Hall wanted to establish a weekly or semimonthly magazine called *Agricultural Engineering* to keep the public posted on the many activities and responsibilities of his office. In 1883, he asked E. W. Hilgard, head of the University of California's College of Agriculture and the best-known soil scientist in the state, to join him. "My object," he wrote, "is to shape public sentiment generally to an appreciation of scientific and professional labor in connection with the work of developing the resources of the country—to build up a sentiment or understanding of such subjects that will support similar works as that I have had in charge." Hilgard argued, however, that existing periodicals, such as *Pacific Rural Press*, provided adequate forums to publicize his work and balked at the financial risks of launching a new publication. Hall responded that he did not want his writings associated "with quack advertisements, wishy-washy local articles, vapid editorials and trash generally." He countered with a proposal for an "official monthly bulletin" issued by the state printing office, a publication whose contributors would include all the state's resource specialists, but Hilgard showed no interest in the suggestion.[39] Hall doubtless hoped such a publication

39. Hall to E. J. Wickson (ed. *Pacific Rural Press*), March 5, 1881, Box 1, MS 915, Correspondence 1880–1885, William Hammond Hall Collection, California Historical Society, San Francisco; Hall to E. W. Hilgard, May 10, 13, 1883, Hilgard Family Papers, Bancroft Library.

would further his career by publicizing his work. Broad popular support for his program, particularly water law reform, would reduce his dependence on the special interests and put his engineering reputation above the battle among hydraulic miners, shippers, towns, riparian owners, appropriators, and other water interest groups.

From 1882 to 1886, Hall limped along on a shoestring budget. Though he managed to sidestep the legislative battles over water rights fought out in 1883 and 1885, his career seemed to have reached a deadend. Nevertheless, the debate over water rights in the 1885 legislature rekindled the hope that Hall's office might be restored to its original glory by the next legislature. He formed a partnership with James Schuyler, the former assistant state engineer, and dusted off plans for a magazine; Schuyler became editor and front-man for *Water and Land*. However, attempts to raise seed money and sell advertisements to San Francisco Bay Area businessmen, the railroad, and the state's large land and water companies failed. Most potential advertisers considered the water rights issue either too controversial or too "academic" for a popular journal. Schuyler promised that the periodical would treat irrigation "as a political and social question, a legislative problem, an agricultural method, and as an art."[40] But Boss W. B. Carr, who masterminded the legislative strategies of the state's large land and water companies in Sacramento, found Hall abrasive and uncompromising. Carr backed off from providing financial support for the journal, perhaps fearing Hall would be too hard to control once he had his own forum. Thus, the state engineer poured more than $1,000 of his own money into the publishing venture, and the first and, apparently, only issue of *Water and Land* appeared in late May or early June, 1886. This issue surveyed litigation over riparian rights. Hall subsequently poured out his feelings of betrayal in a series of letters to J. De Barth Shorb, Carr's manager in the legislature. He fumed that *Water and Land* would have served the appropriation party far better than the bribes freely distributed during the legislative sessions of 1885 and 1886 (discussed in chapter 8).[41]

40. *Bakersfield Californian*, June 5, 1886.

41. On the attempt to launch *Water and Land*, see Hall to James Schuyler January 25, 1886; Hall to J. De Barth Shorb, May 25, 1886; Schuyler to Hall, April 3, May 27, May 29, June 29, 1886, Hall Collection, Box 1, MS 915 (1886–1889), California Historical Society; Hall to Shorb, May 17, June 19, July 2, July 25, August

For all his petulance, for all his wounded pride, Hall survived the 1886 special session. His downfall began in 1887. With public opinion split between proponents of autonomous irrigation districts and a state irrigation system (though far more favored the districts), Hall supported neither and cautioned delay. He continued to believe that no substantial increase in irrigation should occur until water rights matters had been settled. Yet he did not think that the lawmakers could draft effective water laws on their own. So he recommended the formation of a special commission to prepare the laws and present them to the 1889 legislature. A bill creating such a commission passed the assembly early in the 1887 session but stalled in the senate. Many legislators balked at the bill's $50,000 appropriation and thought that the state engineer himself had had more than enough time to gather the needed information. Before adjourning, the legislature called for "completion of all work now in the hands of said engineer." Once again, Hall had urged study while the public demanded action.[42]

21, 1886, J. De Barth Shorb Collection, Incoming Correspondence, 1886, Huntington Library, San Marino, Calif. A copy of the only issue of *Water and Land* is included in the William Hammond Hall Collection, California Department of Water Resources Archives, Sacramento.

In his May 17 letter to Shorb, Hall warned (with the tone of impatience and condescension that often entered his letters): "Mark my word, my dear Shorb, the solution of this [water rights] question is in state control of watercourses. Just what form that control shall take, is a matter of great concern to your people. Just how soon it will come is a matter of greater concern. . . . The thinking people of this state look to me to point the way in this matter. I want to do so without making a jar, or a conflict, or without precipitating a system too soon." In his June 19 letter, he wrote: "I have sounded the real public opinion closer than you have, my dear Shorb, and it is not all your way [pro-appropriation], by a long, long, long shot. . . . Going on as you have, you are creating a universal feeling and opinion that Mr. Haggin's purse is longer than Mr. Lux's, or that he is using it more freely."

42. *Stockton Daily Independent*, January 26, 29, and February 15, 1887; *San Francisco Chronicle*, January 3, 26, February 8, 27, 1887; *Sacramento Bee*, February 11, 1887; *Pacific Rural Press* 33 (January 29, 1887): 81; (February 19, 1887): 81; (February 19, 1887): 148; *Weekly Colusa Sun*, January 29, 1887. For the bills, see A.B. 247 (Brierly), January 24, 1887, *Assembly Bills, 1887*, vol. 3; A.B. 226 (Mathews), January 21, 1887, ibid. The latter bill had been introduced in 1885 as A.B. 544 and S.B. 301. Also see Hall to Governor Washington Bartlett, January 16, 1887, and Hall to George Stoneman, December 22, 1887, State Engineer's File, California State Archives. In his letter to Bartlett, Hall noted: "Watching, as I have for five regular sessions the manner in which it [water law reform] is looked upon by individual members generally, I must say that I have no confidence in the outcome of legislative action at any one regular session. The subject is too big to be handled in sixty days, and interests are too diversified to admit of agreement in that time."

Hall remained the center of controversy during his last two years in office. He had published but two of seven or more promised volumes: a study of irrigation and water law in France, Italy, and Spain, which his critics considered a scholarly extravagance far removed from the state's immediate problems, and a survey of irrigation in southern California. In an October 1888 letter to Governor R. W. Waterman, Hall claimed to have finished studies of irrigation in central California and the southern San Joaquin Valley. He pleaded that the state controller had unfairly withheld his salary "under an understanding that I had voluntarily agreed with him last January to await further payment until the work was published." Publication had been delayed, Hall maintained, only because the legislature had not appropriated sufficient money to print the reports. Nevertheless, a year later—when he left office amid charges that he had carted off documents, notes, and manuscripts belonging to the state—Hall denied that he had any publishable irrigation studies. He suggested to William Irelan, the state mineralogist and ex-officio state engineer, that the state hire him as a consulting engineer to prepare surveys of irrigation in northern, eastern, and central California. If the state agreed to pay him $10,000, the manuscripts would be delivered within two years.[43]

The state controller watched Hall closely. In 1888, he charged that the engineer had paid for lithographic and map work that had not been done. Hall claimed that he had approved large expenditures in June 1887, with the prior approval of Governor Washington Bartlett, because the 1887 legislature had refused to credit funds allocated in 1885 to the new budget. Map work had lagged behind schedule, and the state engineer's office stood to lose nearly half the $20,000 voted by the 1885 legislature had he not acted before the end of the fiscal year. The 1889 legislature appointed a special senate committee to investigate the controller's charges. On March 16, 1889, the committee reported that it had not had time to investigate the charges fully but that "so far as our limited inquiries have gone, we have no evidence of dishonesty on the part of the State Engineer, or on the part of any one connected with that office." Meanwhile, in January 1889, Hall resigned to take a

43. Hall to R. W. Waterman, October 31, 1888, September 3, 1889; Hall to William Irelan, September 3, 4, 1889, Waterman Family Papers, Bancroft Library.

job with John Wesley Powell's United States Geological Survey, which Congress had charged with surveying the arid West's irrigable land and potential reservoir sites.[44] In both 1889 and 1891, the legislature named the state mineralogist as ex-officio state engineer, and it abandoned water planning entirely in 1893.[45]

In his last formal report to the governor and legislature, William Hammond Hall penned a swan song filled with bitterness and disillusionment:

> This department was set up as sort of a compromise medium between the powerful couplets of unreasonable and selfish contending interests: The Hydraulic Mining and Anti-debris couple, and the Appropriation and Riparian couple. It is but natural that under such circumstances it should suffer in efficiency and popularity, and that those should come to the surface willing and ready to accomplish their private ends, or vent their personal animosities, by making use of whatever popular prejudice or misunderstanding there might be on the subject. . . . Now there is a reason for this, outside of any personality or the outcome of the respective works. It is this: To be acceptable and popular before the public, every procession must be headed by a band. A mere individual worker, no matter how efficient or how much multiplied in the public parade, cuts no figure unless there be acceptable popular music to which the appearance is made. . . . No technical or scientific man can study his subject, attend to the duties of such a department, and at the same time make the appearance and music necessary to popularize his efforts. . . . The State Engineering Department needs a Board—a Board of Directors, or Trustees, or Consulting Engineers, call them what you will—to do the popular things, and secure appropriations, while the State Engineer . . . does the work.

44. John P. Dunn (state controller) to George A. Johnson (attorney-general), March 16, 1888; Hall to Governor R. W. Waterman, March 19, 1888, State Engineer's File, California State Archives. Also see the *Biennial Report of the State Controller for the Thirty-Eighth Fiscal Year Ending June 30, 1887, and the Thirty-Ninth Fiscal Year Ending June 30, 1888* (Sacramento, 1888), 29; *The Journal of the Senate During the Twenty-Eighth Session of the Legislature of the State of California,* (Sacramento, 1889) March 16, 1889.

45. *Cal. Stats,* 1889, 328.

Hall also issued a prophetic warning: "When, as is sure to come, the State is forced to take control of her streams for irrigation, arterial drainage, and reclamation regulation, it will be found that the time has passed in which alone the data might have been acquired necessary for intelligent action, both in an engineering and political way." California would wait for a decade before new champions of water law reform picked up the flag and for more than two decades before that reform movement bore fruit.[46]

Hall worked for the U.S. Geological Survey until July 1890 when he returned to private practice, devoting his energies to water supply and irrigation problems in California and Washington state. Then, in 1896, he went to South Africa and designed water supply systems for the mines around Johannesburg, in the Transvaal. He also drafted a new water code for the Cape Colony and built irrigation systems in Russia's Transcaucasus region and in Central Asia. In 1900, he returned to California "where until the present time," a 1915 press guide noted, "he has been engaged chiefly in the management of properties for investment and development."[47] In short, Hall gave up engineering for land speculation, a pursuit he had followed ardently even as state engineer.[48]

In the early years of the twentieth century, the city of San Francisco turned to the Sierra Nevada to supplement its inadequate water supply. In addition to filing claim to a large share of the

46. *Report of the State Engineer to his Excellency R. W. Waterman, Governor of California, for the Year and a Half ending December 31, 1888, JCSA*, 28th sess. (Sacramento, 1889), Assembly, 1: 9–10, 8.

47. *Press Reference Library* (New York, 1915), 2: 285.

48. For example, in 1887, Hall urged Francis G. Newlands, manager of the Sharon estate, to invest in an irrigation project in San Bernardino County. "Now, my dear Mr. Newlands," he wrote, "I do not present this matter to you as a land agent or speculator; but I speak of things that I know professionally."
In the following year, he urged Governor R. W. Waterman to invest in a project to water land between Stockton and the Stanislaus River, wheat land which he claimed could be acquired for 25¢ to 50¢ an acre: "I think, Governor, that this is a chance not only to make a great deal of money, but to make a host of friends, and a very valuable public capital." Hall asked for a chance to get "some small share of the enterprise." Since he did not expect his state job to last beyond 1889, he also hoped to support himself by designing and managing the irrigation system. See Hall to Francis G. Newlands, March 26, 1887, Sharon Family Papers, Bancroft Library; Hall to R. W. Waterman, July 28, 1888, Waterman Family Papers, Bancroft Library.

Tuolumne River, in July 1901 the city claimed the water of Lake Eleanor on one of the stream's tributaries. One year later, Hall also filed on Lake Eleanor and began to locate reservoir sites there as well as on Cherry Creek, another tributary. He formed the Sierra Ditch and Water Company with the intent of damming the lake and selling water to the Turlock and Modesto Irrigation Districts in the Central Valley. Hall claimed that the city did not need Lake Eleanor because the surplus flow of the Tuolumne was sufficient for its needs. He also maintained that the city had not shown diligence in putting the water to use by pushing ahead with the construction of its new water system. After years of bitter controversy, the city bought out Hall, whose speculation bordered on extortion. The former state engineer also invested in hydroelectric power companies and secured water rights on other Sierra streams. J. Rupert Mason, one of California's leading irrigation district bond dealers, commented years later: ". . . whether he should be looked upon as a *pro bono publico* or a *pro bono Hall*, somebody else should try and offer that judgment."[49] When Hall died at age eighty-eight in 1934—the same year the first Tuolumne River water reached San Francisco—the event went all but unnoticed. San Francisco's leading dailies printed his obituary notice among dozens of others and made no mention of his career or service to the state.[50]

A career filled with promise foundered in the 1880s and after both because Hall was ahead of his time and because of flaws in his character. He was a person of extremes: a keen mind, an uncompromising idealism, and bold vision blended with a colossal ego, boundless vanity, tactlessness, political naiveté, and, perhaps, simple greed. One example speaks volumes. In 1892, San Francisco's Board of Supervisors appointed George Davidson, G. H. Mendell, and Irving Scott to design a new sewage and drainage system for the city. Davidson, who served on the University of California faculty as well as headed the Coast Geodetic Survey, was one of

49. On the Lake Eleanor, Cherry Creek, Tuolumne River matter, see William Hammond Hall, *In the Matter of Water Storage and Utilization on the Tuolumne River, California* (San Francisco, 1907), Bancroft Library; Hall to James D. Phelan, November 7, 1907, Bancroft Library; Hall to Marsden Manson, January 1, 1912, Phelan Family Papers, Bancroft Library. J. Rupert Mason's comment is from his oral history transcript at the same library, p. 138.

50. *San Francisco Examiner* and *San Francisco Chronicle*, both October 17, 1934.

the state's best-known scientists with a vast knowledge of water systems. Mendell, who helped Hall design the debris dams in 1880, was the ranking Army engineer on the Pacific Coast. Nevertheless, Hall wrote to Davidson suggesting that the three-man commission was, in effect, incompetent. After complaining that most of California's civil engineers made little more than "the bare living of a clerk," he urged Davidson to resign his commission:

> Has your study and practice in connection with astronomical determinations, great triangulations, coast line topography and sea and bay hydrography, made of you a sanitary engineer? Has anything in your practice given you experience specially applicable and useful to be applied in the work of a sewerage commission? I know that I am injuring myself by writing this letter. I shall be perfectly satisfied to see this work go to others, provided it falls to those who are fitted for it and who have a right to it.

Whether Hall's motives were self-serving or designed to aid his profession, the letter demonstrated a serious lack of judgment, a shortcoming exhibited many times during his career.[51]

In considering Hall's work, strength must be balanced against weakness. On the positive side, the state engineer's office served as a training ground for several of the state's most artful hydraulic engineers, including James Schuyler, Marsden Manson, and C. E. Grunsky. The skills they learned in the field were put to use building some of California's largest water projects in the twentieth century. Hall's office also perfected techniques for measuring stream flow and gathered an immense amount of information concerning irrigation and water laws which proved of great value to later water planners, particularly champions of water law reform during the Progressive Period. Most important, Hall recognized a fundamental truth: no comprehensive water planning was possible without new water laws. He did more than anticipate the later legal reforms proposed by Elwood Mead in the Wyoming constitution of 1889. Like John Wesley Powell, he took a broad view of western devel-

51. William Hammond Hall to George Davidson, March 21, 1892, Incoming Correspondence, George Davidson Collection, Bancroft Library.

opment. Settlement should occur as part of a planned use of natural resources, not willy-nilly. Irrigable land should be surveyed and mapped, streams gauged, reservoir sites located, existing conflicts over water quieted, the needs of different groups of users harmonized or reconciled, and procedures for the acquisition of future rights clearly established—all of which needed to be done before new dams and canals were constructed. Unfortunately, few Californians had the foresight to recognize the boldness of Hall's vision. Many more remembered that as an engineer his judgment often failed, as when he overlooked the salinity of Tulare Lake while serving as the West Side District's chief engineer, or in his ineffective brush debris dams.

In the years from 1878 to 1888, the State of California spent $259,023.70 on the State Department of Engineering, including $58,500 on Hall's salary. It spent about the same amount on Josiah D. Whitney's California Geological Survey in the years from 1860 to 1874, and the histories of the two agencies, as well as the characters of their leaders, bear striking similarities. Given the persistent expectation that all state expenditures should produce immediate practical benefits, the public perceived Hall and Whitney as far too academic. Both appeared visionary and impractical; for example, Hall never provided any plans for dams and canals. Both were also resource experts who scorned the give and take of politics, were highly sensitive to public criticism, and found their years as state officials filled with controversy and disappointment.[52]

Hall clearly anticipated the experts whom Samuel Hays claims led the Progressive conservation movement at the federal level during the years from 1890 to 1920.[53] Yet the needs of hydrography were little understood. For example, Hall's critics complained that he had stalled in publishing the survey data his office had collected.

52. Nash, *State Government and Economic Development*, 103. Nash comments: "Whitney's lack of tact further contributed antagonism to his enterprise. A dedicated scientist and a perfectionist, he was contemptuous of all who criticized his work. At the same time he was extremely blunt and had little patience with those who failed to understand the nature of scientific inquiry. His sharp tongue and biting sarcasm widened the gap between the advocates of theoretical research and those favoring applied science" (p. 101). Whitney shared Hall's lack of faith in the legislature.

53. Samuel P. Hays, *Conservation and the Gospel of Efficiency: The Progressive Conservation Movement, 1890–1920* (Cambridge, Mass., 1959).

In response, he pointed out that by the mid-80s much of the irrigation data gathered from 1878 to 1881 was hopelessly anachronistic. The legislature considered Hall's job temporary, while the state engineer recognized that resource surveys had to be ongoing.

By 1887, Hall had little enthusiasm for what was left of the broad responsibilities entrusted to him in 1878. He might have taken the Wright Act as a new opportunity, even as a justification for continuing his office and making it permanent. He could have gathered impartial information for district promoters, including stream flow data, soil quality, and reservoir sites. He might also have suggested district boundaries, reviewed district plans for irrigation works, and certified district bonds. But neither the legislature nor Hall wanted him to assume that job. In his nine years in office, Hall had been called upon to solve problems to which there were no easy solutions, problems far beyond the financial resources of the state. He had also been saddled with many other distracting responsibilities, ranging from a plan on how to use and develop Yosemite State Park to the design of sewage and fire control systems for public buildings and state asylums.[54] He bristled at the public's and legislature's lack of respect for his knowledge and ability, forgetting that prophets rarely win recognition in their own time. That he failed to win recognition later, when Californians better understood the need for long-range water planning, was due to the ambiguity in his character and temperament. In the end, Hall's speculative nature and his haughty, stubborn, and abrasive bearing were qualities of the individualistic, entrepreneurial engineers of the nineteenth century, not of the well-heeled organization men of the twentieth.

54. *First Biennial Message of Governor George Stoneman*, January 5, 1885, *JCSA*, 26th sess. (Sacramento, 1885), Appendix, 1: 16.

8

Lux v. Haggin:
The Battle of the
Water Lords

In the late 1870s, litigation over water rights erupted throughout the San Joaquin Valley, particularly in the drier counties of Tulare, Kings, and Kern. By 1874, the Southern Pacific had completed its line through the valley, and the settlers who followed in its wake touched off a canal-building boom. This strained the region's limited water supply, especially when drought returned in 1877–1878. For the first time, water users in large numbers resorted to the courts. The battle between riparian owners and appropriators culminated in the famous *Lux* v. *Haggin* suit, probably the most important water case decided in the nineteenth-century West. It also forced the California legislature to focus full attention on irrigation and water rights for the first time. *Lux* v. *Haggin* had repercussions throughout the West, not just in its theory of the origin of water rights—which came to be called the "California doctrine"—but in paving the way for the Wright Act adopted in 1887.

The principals in this drama included Henry Miller, James Ben-Ali Haggin, and William B. Carr—some of the most powerful and colorful figures in California history. Miller, an austere, hardworking, strong-willed German immigrant, began his long business career in 1850 as a butcher in San Francisco. Recognizing the value of vertical consolidation, he soon moved into the livestock business and formed an alliance with Charles Lux, his chief competitor in San Francisco. In 1857, Miller secured options on ranches in the northern San Joaquin, Santa Clara, and Salinas valleys, and leased the land to graze his herds. He did not buy the land until 1863–1864, after losing two-thirds of his cattle to drought. In 1890, he recalled: "From that time on, the people settled in around the country and utterly wiped out the free range and then the question arose what should we do—should we keep less cattle or buy more land? So we commenced to buy a little land and a little more." In addition to buying up such San Joaquin ranchos as the Santa Rita, Buri-Buri, Salispuedes, Juristac, La Laguna, Bolsa de Felipe, Las Lomarias, Muertas, San Antonio, San Lorenzo, Las Animas, and Tesquesquito, Miller used the Swamp Land Act of 1850 to acquire a 100-mile-long block of land adjoining the San Joaquin River, beginning northwest of Fresno and extending to the confluence of the San Joaquin and Merced rivers south of Modesto.[1]

Miller moved into the Kern Valley in the early 1870s, again in search of cheap grazing land. In 1868, James C. Crocker began to piece together the land and herds that became the nucleus of the Kern County holdings when he purchased the Temblor Ranch, adjoining Buena Vista Slough. Whether Crocker bought the land on his own or for Miller and Lux is not clear, but by 1871 he was clearly the cattle barons' chief agent in the county, and by 1873 the three men owned a vast tract of land south of Tulare Lake at the end of the Kern River, roughly twenty miles west of Bakersfield. Most of the land was swamp, to which titles were not secured until 1876 and 1878. Miller and Lux's holdings in Kern County climbed from 61,969 acres in the early 1870s to 80,073 acres in 1881, to 120,587 acres in 1890. This was a vast estate by anyone's standards; still it paled compared with the 201,307 acres owned by the two stockmen in Merced County, or with their 180,150 acres in Fresno County.[2]

---

1. William D. Lawrence, "Henry Miller and the San Joaquin Valley" (M.A. thesis, University of California, Berkeley, 1933); Henry Miller autobiographical dictation, H. H. Bancroft Collection, Bancroft Library, University of California, Berkeley; Edward F. Treadwell, *The Cattle King* (Boston, 1950).

2. Lawrence, "Henry Miller and the San Joaquin Valley," 60; Herbert G. Comfort, *Where Rolls the Kern* (Moorpark, Calif., 1934), 115; *Kern County Weekly Courier,* November 22, 1873; *Bakersfield Californian,* July 18, 1878; Charles Lux et al., v. James B. Haggin et al., Brief for Respondent, filed November 6, 1883 (San Francisco, 1883), Lux v. Haggin File, California State Archives, Sacramento. See the map of Miller and Lux landholdings on p. 18 of the brief.

James Ben-Ali Haggin's character was in stark contrast to Miller's. While Miller rose from humble beginnings and shunned the public eye, Haggin descended from a prominent Kentucky family and moved freely in Calfornia's "high society." He often traveled in a posh personal railroad coach, with an attached dining car, or on his private steam yacht. He won recognition as a horse breeder and assembled the best racing stable on the Pacific Coast. His exotic middle name came from his maternal grandfather, a Turkish physician. After practicing law for four years in New Orleans, Haggin emigrated to California in 1850, and in the following year formed a long-lived partnership with his brother-in-law, Lloyd Tevis, initially as moneylenders or fledgling bankers. Subsequently, the two set up a successful law practice and engaged in a variety of business ventures, most notably as leading investors in Wells, Fargo & Company. They had much in common, particularly their genteel southern background and marital ties.[3]

Miller, Haggin, and Tevis are well known to students of California history, but William B. Carr has been all but forgotten. "Billy" Carr was a fascinating character, the most powerful man in California politics during the late 1870s and 1880s. His crude, brash, flamboyant nature was forged on the frontier, and perfectly complemented the more refined personalities of Haggin and Tevis. Born in Indiana in 1830, Carr came to California as an argonaut but soon turned to the lucrative business of digging mining ditches in El Dorado and Sacramento counties. Subsequently, California's rapid growth fueled his career in construction. In the mid-1850s, he built most of the levee system surrounding Sacramento, and during the 1860s, after he moved to San Francisco, his factory supplied most of the brick used by the city in its public buildings. Gradually, Carr established the San Francisco political connections that made him as dominant in his time as Christopher Buckley and Abraham Ruef would be in theirs.[4]

In nineteenth-century California politics, no city had more corrupt political institutions than San Francisco, and no political

3. Alonzo Phelps, *Contemporary Biography of California's Representative Men* (San Francisco, 1881), 325–329; James Burnley, *Millionaires and Kings of Enterprise* (Philadelphia, 1901), 265–270.

4. Phelps, *Contemporary Biography of California's Representative Men*, 309–310.

figure was more vilified than the shadowy Boss Carr. As in the rest of the nation, political parties in California served many purposes, but none was more important than distributing the booty of office—and San Francisco's rapid growth offered plenty of opportunities. Modern historians have argued that political bosses filled an "institutional gap" as the nation's towns gave way to sprawling metropolises. In particular, the absence of a class of "professional" politicians and experts trained in solving the new range of urban problems worked against the development of responsible party leadership. Men such as Carr, who skillfully turned San Francisco's Republican party into his own private preserve, provided some measure of continuity. However, their "talents" were rarely appreciated by the public. As early as 1873, San Francisco's *Bulletin* declared that "no man who has a particle of self-respect cares to be found in [Carr's] company." Four years later, in a scathing indictment of the Republican Party, San Francisco's *Argonaut* charged that Carr and his associates were "a ring of mercenary bandits who steal to get office, and who get office to steal. . . . Through this man *Carr* alone the honors of the party, its offices, its patronage and its emoluments must be dispensed." The *San Francisco Chronicle,* one of Billy Carr's most persistent critics, revealed the extent of the boss's political power:

> The most "influential" politician among us is a man who has no idea of politics apart from the money he can obtain by the business. Coarse and ignorant, he cannot appreciate the higher aspirations of gentlemen, but measures man and principles by dollars and cents. He is a power in the primaries; he designates our public officials; he makes and unmakes laws at the State Capital; he essays to elect Congressmen and United States Senators; he orders them to vote as he chooses upon public measures, and they must allow him to name the men who are to fill the Federal offices in this State. No matter how high a character or how good the qualifications candidates may have there is no chance for them except to "see Billy Carr" and pass the ordeal of his approval. . . . From the highest to the lowest and all along the line the commanding influence of this mighty potentate is felt and feared. If a man shows any independence, he is put down at once.

Later in the same month, the *Chronicle* charged that United States Senator Aaron A. Sargent "belonged body, mind and soul, to Billy Carr."[5]

In 1874, Haggin selected Carr to serve as his land agent in Kern County. Not only did the boss have substantial political skills and influence, but as a lobbyist for the railroad, he could help acquire odd-numbered railroad sections. Haggin had already purchased the 52,000-acre Gates Ranch in 1873. Carr acquired at least 33,000 acres of railroad land in 1874, along with the Bellevue (or Belle View) and McClung ranches west of Bakersfield. Subsequently, he also bought thousands of acres of swampland. By early 1876, the Haggin-Carr team held close to 100,000 acres. The land stretched in a nearly solid block between the Kern River and Buena Vista and Kern lakes fifteen miles to the south.[6] The biggest block of land was secured through the artful use of the Desert Land Act, passed by Congress on March 3, 1877. That act permitted settlers to purchase a section (640 acres) of land for $1.25 an acre if the claimant irrigated all the land within three years of filing. Some California periodicals, notably San Francisco's *Argonaut* and *Chronicle*, charged that the bill had been drafted and introduced by Senator Aaron Sargent at the insistence of Haggin and Carr. True or not, in the early weeks of April 1877, the pair, through surrogates, filed on more than 100,000 acres of even-numbered sections of desert land adjoining the new Southern Pacific line north of Bakers-

5. The *Bulletin's* description of Carr was reprinted in the *Stockton Daily Independent* of December 4, 1873. The other quotes are from the San Francisco *Argonaut* 1 (June 30, 1877): 4; *San Francisco Chronicle*, May 10, 22, 1877. Apparently, Carr also had close ties to the railroad, which stood to profit from expanded irrigation in the southern San Joaquin Valley, both from freight traffic and land sales. The *Kern County Weekly Courier* of January 17, 1874, described Carr as the *former* "political Napoleon of the Railroad Company." But on May 9, it reported that Carr and Leland Stanford had visited Bakersfield together and that Carr was "the right-hand man of the railroad company, entrusted with nearly every matter of delicacy and difficulty, requiring skill, finesse, shrewdness, and untiring perseverance that is to be managed on this coast. He is the Napoleon of lobbyists and intriguers, and is never known to fail in any point he attempts to carry." Carr's connection to the railroad may have been through Lloyd Tevis, who had been a vice-president of the Central Pacific.

6. Norman Berg, *A History of Kern County Land Company* (Bakersfield, Calif., 1971); *Kern County Weekly Courier*, January 9, 1875; *Bakersfield Californian*, May 25, 1876; February 15, 1877.

field. This gave Haggin and Carr a solid block of land along the Calloway Canal.[7]

The entries were made hastily, before federal officials publicly announced the availability of the land. The *Chronicle* bitterly noted:

> The President's signature was not dry on the cunningly devised enactment before Boss Carr and his confederates were advised from Washington that the breach was open. It was on Saturday, the 31st of March. The applications were in readiness, sworn and subscribed to by proxies, for taking up the intervening sections of the railroad grants through the Kern valley. All that Saturday night and the following Sunday the clerks in the Visalia Land Office were busy recording and filing the bundles of applications dumped in upon them by Boss Carr, although it was not until several days after that the office was formally notified of the approval of the Desert Land Act.

Most of the dummy entrymen were residents of San Francisco, employees of the United States Mint, the U.S. Customs House, or Wells, Fargo & Company. They received $1 to $5 apiece for their signatures. Haggin, Tevis, and Carr paid neither filing fees nor the 25¢ an acre required as down payment under the Desert Land Act. Moreover, their vast land grab superseded the claims of many bona fide settlers who had preempted 160 acres and planned to pay the government $2.50 an acre.[8]

---

7. The Desert Land Act actually dates to 1875, when Congress passed a law providing for the reclamation of desert land in Lassen County, California. The major difference between that law (*U.S. Statutes at Large*, 18: 497) and the Desert Land Act is that the first statute gave settlers two years to irrigate their land rather than three. Congress regarded the Lassen County Act as a legislative experiment, capable of application to other arid lands. However, Haggin and Carr may well have been responsible for extending the original legislation. On the Desert Land Act, see Paul W. Gates, *History of Public Land Law Development* (Washington, D.C., 1968), 638–643; John T. Ganoe, "The Desert Land Act in Operation, 1877–1891," *Agricultural History* 2 (April 1937): 142–157; Stanley Roland Davison, *The Leadership of the Reclamation Movement, 1875–1902* (New York, 1979), 75–78; Roy Robbins, *Our Landed Heritage: The Public Domain, 1776–1970* (Lincoln, Neb., 1976), 219–220. Gates notes (p. 639) that by June 30, 1877, 467 entries had been filed in California covering 166,665 acres.

8. *San Francisco Chronicle*, September 28, 1877, January 21, 1878; Margaret Aseman Cooper Zonlight, "Land, Water and Settlement in Kern County, California"

The Haggin-Tevis-Carr syndicate had far from finished its expansion. In 1878, it acquired the 17,600-acre Mexican land grant of San Emidio, about thirty-five miles south of Bakersfield, and in the following year thousands of additional acres from the Livermore-Redington tracts. Meanwhile, many small farmers, hit hard by drought and monopoly, sold out to the group and fled the valley. In 1882, John Hittell reported that Haggin and his associates owned 300,000 acres in Kern County, only 40,000 of which were under irrigation. This represented about 75 percent of the county's irrigated land.[9]

The Haggin team also moved to secure a water monopoly in the valley. In 1873, groups of independent farmers controlled Kern County's six major canals, and only 5,000 acres were under irrigation. Carr urged the farmers to incorporate and issue stock as a way to pay for expensive weirs and headgates. Once they did, he bought up most of the stock to insure control over the new companies. (He also held out the promise that the price of land would soar when the capitalists consolidated the ditches into a unified,

(M.A. thesis, University of California, Berkeley, 1954), 75, 143, 157. Zonlight's very valuable, though polemical, study was republished by Arno Press in 1979, retaining the same pagination.

9. Zonlight, "Land, Water and Settlement in Kern County," 75; John S. Hittell, *Commerce and Industries of the Pacific Coast* (San Francisco, 1882), 406. Zonlight challenges the assumption that monopoly stimulated the expansion of irrigation—or promoted economic growth. She notes that while the population of Kern County doubled from 1870 to 1877, following passage of the Desert Land Act migration into the county dried up. The population of Kern County did not increase from 1880 to 1886, during which period Fresno County's population grew by 89 percent and Tulare County's by 66 percent. Moreover, she claims that monopoly destroyed community life. Average school attendance in the county dropped from 649 children in 1879, to 246 in 1886 (pp. 259, 316) because so many families fled the valley. The population remained stable because many single, male renters or tenants replaced the family farmers. According to Zonlight, the Desert Land Act destroyed any chance that the state or federal governments, or private irrigation districts, would undertake irrigation projects in Kern County. And the Haggin-Carr/Miller-Lux monopolies set a pattern for the future: "The trends observed in land use, the development of a seasonal labor force, the slowing and retardation of population growth, the poverty of the social, educational, recreational and other community facilities were determined to a large extent when Haggin's enormous desert land acquisition was given the Governmental seal of approval" (pp. 174–175). Unfortunately, Zonlight never fully explained what Haggin and Carr wanted to achieve in Kern County. Were all their motives self-serving? Were they any more greedy than the speculative "small" farmers they replaced? She also ignores the important effects of the *Lux* v. *Haggin* suit on migration into and out of the county.

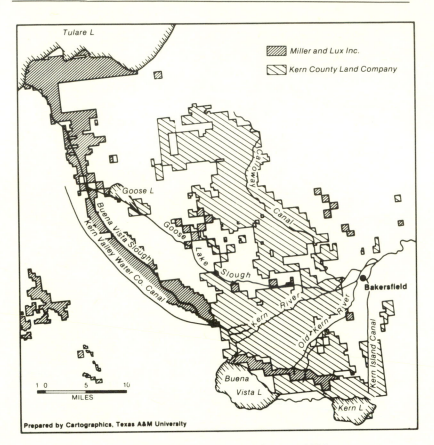

Major Landholdings in Kern County, 1890.

efficient system and constructed new canals to serve the Haggin estate.) Almost immediately, the San Franciscans hired dozens of local men to dig ditches and tend their ranches. This was another way they won support from local farmers. Then, on May 4, 1875, they claimed 3,000 cubic feet per second (74,000 miner's inches) from the Kern River, about three times more water than the stream had ever carried, an amount roughly equivalent to the water then used for irrigation in Los Angeles County. This was a prelude to

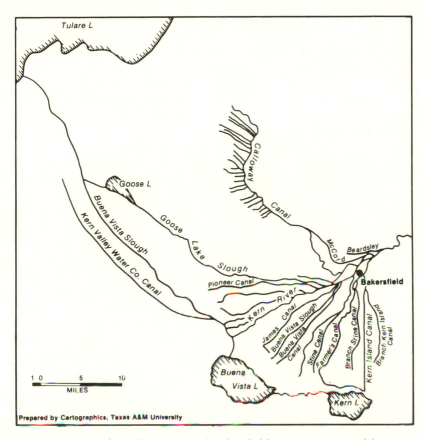

Irrigation Canals in the Vicinity of Bakersfield, Kern County, California, 1890.

the formation of the Kern County Land and Water Company two weeks later. The company dammed the Kern north of Bakersfield, then built the thirteen-mile Kern Island Canal to serve more than 13,000 acres south of the town. The largest canal built during the 1870s, the Calloway Canal was started by rival land promoters in May 1875, but the company acquired this ditch in June 1877 and expanded it to more than thirty miles in 1879. The structure had the capacity to serve 70,000 acres. The company's diversion dam, constructed in the fall of 1877, provoked the *Lux* v. *Haggin* suit.[10]

The Kern County Land and Water Company used a variety of tactics to intimidate uncooperative farmers. It reduced or cut off water supplies, an act that often forced settlers to sell out to the company. It filed over 100 separate suits against Kern County farmers who did not buy water from the company, in an attempt to eliminate diversions by individuals or small groups of farmers. And, because Boss Carr dominated Kern County politics, it occasionally used local officials, ranging from judges to tax assessors, to harass recalcitrant settlers. Other forms of control were more subtle. For example, farmers whose families depended on their off-season job constructing irrigation ditches for Haggin, Tevis, and Carr were not likely to rock the boat. Even so, the Kern County Grange and Workingmen's party bitterly assailed the land and water monopoly and repeatedly called for repeal of the Desert Land Act. And in January 1878, the Bakersfield Grange asked the State Senate Committee on Irrigation to introduce a bill to permit farmers to elect their own boards of water commissioners to parcel out the water. They charged that the land syndicate had wasted water and shortchanged many irrigators.[11]

10. Zonlight, "Land, Water and Settlement in Kern County," 300; Wallace Smith, *Garden of the Sun* (Los Angeles, 1939), 447–449; Charles Lux et. al., v. James B. Haggin et al., Brief for Respondent, filed November 6, 1883, 12–15; Victor M. Cone, *Irrigation in the San Joaquin Valley, California*, U.S.D.A., Office of Experiment Stations, Bulletin (hereafter OESB), no. 239 (Washington, D.C., 1911), 21–22; *Transactions of California State Agricultural Society, 1900* (Sacramento, 1901), 89–95; *Fresno Expositor*, December 29, 1875; *Bakersfield Californian* April 29, May 13, 1875; June 29, July 6, 1876; February 14, 1885; *Kern County Weekly Courier*, June 13, 1874; July 3, December 11, 1875; *Tulare County Times*, July 20, August 17, 1878. F. H. Newell, in his *Report on Agriculture by Irrigation in the Western Part of the United States at the Eleventh Census*, 1890, reprinted in 1896, H. Misc. Doc. 340. (serial 3021), 53–54, suggests the effectiveness of the Haggin-Carr water monopoly. There were thirty-one canals constructed during the 1870s and none in the 1880s. Three canals were located in 1870, one in 1871, three in 1872, nine in 1873, five in 1874, four in 1875, two in 1876, one in 1877, and the final canal in 1878. By the latter year virtually all the ditches had fallen under the direct or indirect control of the Kern County Land and Water Company.

11. *San Francisco Chronicle*, October 19, 1877; January 29, 1878. Haggin and Carr were the largest employers in Kern County. The *Bakersfield Californian* reported in its issue of May 3, 1877, that 325 men worked on their ranches and construction

During the 1870s and 1880s, the state's periodicals painted the motives of Haggin, Carr, and Tevis, and of Miller and Lux, in black and white. More often than not, at least until 1885, the Kern County Land and Water Company appeared in print as the champion of irrigation, friend of the small farmer, and agent of economic progress, while Miller and Lux became archmonopolists bent on preserving Kern County as one huge pasture. The *Bakersfield Californian* (or *Kern County Californian*) was virtually a company paper, and may well have been subsidized by Haggin and Carr. It applauded every activity of the company: the trees planted, the artesian wells sunk, the generous lease terms, the barns and houses offered to settlers, and the brick dairy constructed in Bakersfield. It claimed that monopoly promoted an efficient use of land and water and would, ultimately, lure more residents into the valley than the speculative efforts of independent small landowners. The fact that Haggin and Carr bought out so many small farmers could be easily explained. As the *Californian* commented in September 1883, the company had wasted a great deal of water trying to convey it to distant and widely separated tracts of land: "One reason that Mr. Haggin has bought out so many of the owners of small tracts was his unexpected inability to supply them with water."[12]

Since Haggin did not reside in Kern County, rarely visited his vast estate, and seldom talked to reporters, the *Californian* had a hard time explaining his actions or intentions in his own words. It frequently referred to a statement written by Haggin in 1877 at the height of the controversy over his Desert Land Act claims. The statement, accompanied by affidavits from dozens of residents of the county which pronounced Haggin a public benefactor for promoting irrigation, was published and widely circulated in the county. "My object has not been, nor do I wish to monopolize large bodies

---

projects and that their cash expenditures averaged more than $1,000 a day. The *Californian* of September 10, 1881, claimed that the Kern County Land and Water Company's payroll stood at 600 workers.

12. *Bakersfield Californian*, September 1, 1883. For other examples of strong editorial support for Haggin and Carr, see the issues of March 1, 1877, and May 13, 1880.

of land," Haggin wrote, "but I desire to make valuable and available that which I have by extending irrigating ditches over my lands, and when these lands are subject to irrigation to divide them up and sell them . . . in small tracts with the water rights necessary for irrigation." Nevertheless, a vexing question emerged by the 1880s: If the Kern County Land and Water Company wanted to encourage the immigration of small farmers, why did it not launch colonization schemes similar to those undertaken in the Fresno area? Like so many California newspapers, the *Californian* did not want to appear pessimistic, or to gain the reputation of being a "knocker," or to encourage the growth of rival agricultural communities, so it scarcely noticed Fresno's success. Instead, the paper blamed the railroad for discouraging settlers with high fares and freight rates: "We can see no drawback on the prosperity of this country but the railroad monopoly," it commented in May 1880. Of course, an even more persuasive argument emerged in the early 1880s when Haggin and Carr seemed to have abandoned all efforts to promote irrigation: "Were it not for the blackmailing water suits that have been brought against them," the editor commented in September 1881, "they would be making extensive improvements and their expenditures would probably be three times as large."[13]

These arguments have the ring of truth when the Haggin-Carr efforts in Kern County are viewed in isolation. But the Kern County estate was simply one part of a vast "grasslands empire" which included 1,400,000 acres in Arizona, New Mexico, and Kern County. Irrigation was valued not as a way to raise high-value crops like those cultivated around Fresno, but to produce alfalfa for stock. W. H. Hutchinson has noted: "They combined their agricultural production in California with their livestock production in the Southwest, via railroad linkage on both the Southern Pacific and Santa Fe lines, to make Bakersfield important in meeting the ever-growing market for meat in ever-growing Los Angeles." In

13. *Bakersfield Californian*, May 13, 1880; September 10, 1881. James B. Haggin's statement is from *Desert Lands of Kern County, California: Affidavits of Various Residents of Said County* (San Francisco, 1877), as reprinted in the *Californian* of May 20, 1880, available at the Bancroft Library. The *Tulare County Times* of November 2, 1878, reprinted an interview with W. B. Carr which echoed Haggin's statement.

the early 1880s, Haggin and Carr grazed 70,000 sheep, and as many cattle and horses, and imported breeding stock from all parts of Europe. While subdivision of the Kern County land would have produced substantial profits—had Haggin and Carr tried seriously to resolve the water conflict with Miller out of court—it would also have destroyed their rapidly expanding stock business by driving up taxes and creating a potentially "intractable" community of small farmers. As it was, Haggin and Carr had a hard time controlling Kern County politics.[14]

In short, the motives and objectives of Miller-Lux and Haggin-Carr-Tevis had much in common. Certainly, Miller and Lux were as committed to irrigation as their rivals—when it furthered their interests. They were among the first Californians to raise alfalfa, rice, and cotton, all three of which required irrigation. By 1881, they had nearly doubled the length of the aqueduct of the San Joaquin and Kings River Canal and Irrigation Company. In 1880, the canal served 29,000 acres, including 12,000 planted to grain and another 5,000 to alfalfa (though it was capable of serving about 120,000 acres). In the 1870s, Miller joined with W. S. Chapman to build a thirty-mile ditch from the San Joaquin River to Chowchilla Slough, south of Merced. By the second decade of the twentieth century, this San Joaquin Valley irrigation system could water 340,000 acres. Though Miller sold and rented thousands of acres to farmers, he was not a land promoter and did not welcome the prospect of irrigation transforming the San Joaquin Valley into a haven for small farmers. Still, the cattleman was a wily business-man capable of adapting to new circumstances. His most critical biographer claims Miller had a plan to "divide his vast domain in an orderly way." "His plan," William D. Lawrence has written, "was for a gradual transformation of his vast holdings into business enterprises, such as banks, stores, etc. As he gradually sold off his best lands to small settlers he would buy up business enterprises to get the trade of the settlers, in the meantime buying up cheap

14. W. H. Hutchinson, *California: Two Centuries of Man, Land, and Growth in the Golden State* (Palo Alto, Calif., 1969), 198. On the Haggin stock business, see the *Bakersfield Californian*, July 16 and November 26, 1881. Haggin's ranches rivaled in size the 750,000 acres Henry Miller owned in California, Nevada, and Oregon.

land to continue his stock business. In this way his firm would be a constantly growing and expanding institution."[15]

Unfortunately, Haggin and Carr have not left a substantial body of private papers, so their objectives remain unclear. Certainly, even the illusion of dividing up their estate won substantial political support in the valley and allowed them to exercise greater control over local officials. Moreover, they may have hoped that leasing irrigated land to small farmers would strengthen their water rights in court. Judges and juries were likely to consider the irrigation of wheat, alfalfa, and other cash crops as a higher use of water than the flooding of fields to produce natural, or even seeded, pasture grasses.

In any case, a court test came soon enough, though not on the Kern River. The drought of 1877–1879 dried up many streams in the San Joaquin Valley. In October 1877, the *Fresno Expositor* warned that "those owning farms along the [Kings] river shores are possessed of certain riparian rights. What these rights are, we understand, the stockmen propose to determine before the Courts, by instituting suits against the ditch-owners for damages sustained by reason of the cattle dying." The battle between riparian owners and appropriators was an extension of the no-fence law controversy of the 1870s in that it usually pitted farmers against stockmen. But the battle lines had changed. By the late 1870s, many livestock owners knew that irrigation had come to stay and realized that the new institution presented them with economic opportunities. They began to claim the right to irrigate using riparian rights, and sometimes the additional right to sell water to other irrigators. Usually, riparian owners did not oppose irrigation per se, as they had a decade earlier. Instead, they complained that upriver diversion

15. See the unpublished typescript biography of Charles Lux, prepared by W. H. White from an interview with Lux, H. H. Bancroft Collection, Bancroft Library. Also see Treadwell, *The Cattle King*, 187; Lawrence, "Henry Miller and the San Joaquin Valley," 64–65, 109–110, 141; *Fresno Expositor*, October 15, 1873. Treadwell's biography must be read with caution. He served as Miller's attorney, and became one of California's leading experts in water law. As a persistent defender of the riparian doctrine and the rights of private property, he pictures Miller as a beleaguered businessman threatened by the "mob rule" of the 1870s. Consequently, his book gives a fragmentary, distorted view of the water conflicts of the 1870s and 1880s.

dams often completely dried up a stream when farmers needed only a small fraction of the flow for irrigation. Also, by altering the course of a stream, appropriators threatened to eliminate downstream riparian rights entirely. This fear was particularly acute in the southern San Joaquin Valley where creeks and rivers frequently disappeared in the desert and changed course on their own.[16]

The first important case erupted in April 1877 in Tulare County. J. M. Creighton sued Dudley Evans and others, asking for a perpetual injunction to prevent the defendants from diverting water from Elk Bayou, a tributary of the Kaweah or Tule River. He also asked for $2,500 in damages to compensate for the loss of cattle, hogs, fruit trees, grapevines, and a field of alfalfa. After one jury failed to agree, in September a second panel ruled against Creighton's claim, arguing that he had not proven any extensive loss of property. Creighton then appealed to the Supreme Court, which reviewed the case in the following year. That court commented:

> . . . the [Tulare superior] court further instructed the jury that if defendant [Evans] diverted a portion for a useful purpose, and . . . enough water was left in the stream for the use of the plaintiff for watering his stock and for domestic purposes, and if the plaintiff was not damaged by the diversion, the verdict should be for the defendant. This . . . was erroneous as a matter of law. So far as appears on the record, defendant was not entitled to divert the water for any purpose.

Even though the lower court had found that Creighton had not suffered from the diversion, the higher court argued that "the right of a riparian owner to have the water of a stream run through his land is a vested right, and any interference with it imports at least nominal damages, even though there be no actual damages." The

---

16. *Fresno Expositor*, October 21, 1877. Conflicts among rival ditch companies were as common as those between riparian owners and appropriators. For example, the mutual water companies at Mussel Slough challenged upstream diversions from the Kings River at Fresno, and the Washington and Central colonies also clashed over their rights. See Maass and Anderson, . . . *and the Desert Shall Rejoice* (Cambridge, Mass., 1978), 236; Virginia E. Thickens, "Pioneer Colonies of Fresno County" (M.A. thesis, University of California, Berkeley, 1939), 89–96.

Visalia Weekly Delta solemnly declared that the court had undertaken "to sweep away our whole irrigating system. . . . The supreme court of California seems willing to defy public opinion and the statute law of California."[17]

A year later, the supreme court dropped a second bombshell in the case of *Pope* v. *Kinman*. In this San Diego County contest, the owners of the Muscupiabe Ranch diverted the entire flow of Lytle Creek and claimed an exclusive right based on priority. The riparian owners of the San Bernardino Rancho downstream filed suit. Once again the county court upheld appropriation, only to have its arguments rejected by the state supreme court. The latter court ruled:

> It is not necessary, in this case, to define in detail the precise extent of the riparian right as existing in this country; it is enough to say that under settled principles, both of the civil and common law, the riparian proprietor has an usufruct in the stream as it passes over his land. The judgment of the Court below deprived the plaintiffs of this usufruct.

The supreme court went on to say that the defendants had no right to Lytle Creek except as riparian owners with land adjoining the stream. They could use water, but not to the detriment of other riparian owners. The *Los Angeles Daily Commercial* described the judgment as "unjust and outrageous," recognizing that it might influence suits pending in Kern and Fresno counties.[18]

A third important case involved the lower Kings River in Tulare County. John Heinlen owned a 4,400-acre ranch near Lemoore in Mussel Slough at the sink of the stream. In 1879, he filed suit against Moses J. Church's Fresno Canal and Irrigation Company, one of many suits filed against Church during the 1870s and 1880s. Heinlen charged that diversions at Fresno killed 500 of his

17. Creighton v. Evans, 53 Cal. 55 (1878); *Tulare Weekly Times*, February 9, 1878; *Visalia Weekly Delta*, May 27, 1881. In 1882, Creighton filed suit to prevent diversions by the Kaweah Canal and Irrigation Ditch Company. For this suit, see the *Delta*, June 2, 1882.

18. Pope v. Kinman, 54 Cal. 5 (1879); *Los Angeles Daily Commercial*, April 22, 1881. Also see the *Los Angeles Evening Express* editorial on *Pope* v. *Kinman* as reprinted in the *Fresno Expositor*, June 29, 1881.

cattle in 1877, 300 in 1878, and another 200 in 1879. Church did not contest Heinlen's rights or challenge the riparian doctrine directly. Instead, he maintained that since the diversion began in 1871, and continued for seven or eight years without interruption, his company's water right had ripened through prescription. However, Heinlen denied that he knew about, or had condoned, the diversions, and Judge W. W. Cross awarded him $11,000 in damages and a permanent injunction against future diversions by the company.

The injunction did Heinlen little good because individual farmers took over the operation of the canals in 1883, defying the cattleman to file suit against each water user. In a March 1883 letter to the *Visalia Weekly Delta*, Bernard Marks, chief promoter of the Central Colony, charged that Heinlen wanted to use his riparian right to help subdivide his land and sell it at a handsome profit to new settlers. Since Heinlen was the last riparian owner on the stream, he could use the water without fear of reprisal from other riparian claimants, promising land purchasers a guaranteed supply. Though the California Supreme Court overturned Judge Cross's decision in 1885, it did so on a technicality, ruling that much of Heinlen's land was too far from water to be considered riparian.[19]

These cases clearly suggested that California's highest court considered riparian rights supreme and inviolable.[20] For this reason, and to avoid the enormous expense and bitterness produced by litigation, many appropriators decided to try to settle with riparian owners out of court. For example, in February 1883, appropriators and riparian owners in the Mussel Slough area met at Hanford's Palace Hotel to discuss their differences. One faction proposed that riparian owners tolerate diversions in wet years, when streams ran full, in exchange for a promise that appropriators

19. *Fresno Expositor*, January 19, 1880; February 16, June 8, 1881; *Bakersfield Californian*, September 3, 1881; January 3, April 14, 1883; November 29, 1885; *Visalia Weekly Delta*, March 17, 1882; March 2, 1883; November 26, 1885. The right of riparian owners such as John Heinlen to irrigate their land using riparian rights was first established in Anaheim Water Company v. Semi-Tropic Water Company, 64 Cal. 185 (1883).

20. For lower court rulings in support of riparian rights see the *Bakersfield Californian*, January 8, 1880; *Fresno Expositor*, April 7, 1880; *Visalia Weekly Delta*, April 9, 1880; *Tulare County Times*, April 30, 1881.

refrain from irrigating in dry years. Others argued that this depended too much on good faith, and that conflicts over water rights would end only when the state bought up all rights, riparian and appropriative, and parceled out the water itself. The expectation that the legislature would find a way to resolve the conflict worked against settlements at the local level, at least until after 1887.[21]

As in other parts of the state, on the Kern River, legal conflicts between competing ditch companies which relied on appropriative rights were nearly as common as disputes between riparian owners and appropriators. For example, *Lux* v. *Haggin* was preceded in 1877 by a bitter feud pitting Haggin and Carr against their chief rivals, J. H. Redington and H. C. Livermore. The struggle became so intense that during the summer of that year, twenty-five men armed with Sharps rifles patrolled the Kern Island Canal and its diversion dam.[22]

In 1878 and 1879, new diversions from the rapidly expanding Calloway Canal coincided with, or contributed to, the death of 16,000 cattle, most owned by Miller and Lux. The animals had grazed on two parcels of riparian land in Buena Vista Slough, one near Tulare Lake, fifty miles below the head of the Calloway ditch, and the other twenty miles from the diversion dam. The riparian owners offered upstream appropriators 75 percent of the river's flow if the diverters promised to permit the remaining water to reach the slough. However, the appropriators, confident they could defend their rights in court, refused.[23] Consequently, Miller and Lux, along with J. C. Crocker and six other riparian owners, filed suit to block the diversions on May 12, 1879. As defendants, the suit named James B. Haggin, the Kern County Land and Water Company, and 117 others. In previous years, "flood water" from the rapidly melting Sierra snowpack reached the slough early in

21. *Visalia Weekly Delta*, February 9, 1883. On the attempts of riparian owners and ditch companies in the Mussel Slough region to resolve their differences, also see the *Delta* of February 9, August 2, 9, 1883.

22. Wallace Morgan, *History of Kern County California* (Los Angeles, 1914), 83; Comfort, *Where Rolls the Kern*, 150–151; *Bakersfield Californian*, February 14, 1885; *Kern County Weekly Courier*, April 3, 1879.

23. *Sacramento Bee*, July 28, Aug. 20, 1886; *Bakersfield Californian*, May 7, 1881.

the spring. But in 1879, the extra water, which flowed out onto the plains and produced lush pastures, did not arrive until June 10 and the flow lasted only twenty days. In the spring and summer of 1880, over seventy additional suits were filed in the Kern County Superior Court, listing nearly every appropriator and ditch company in the county. Ditch owners feared that Miller and Lux would prosecute the other suits in a series if they lost their main contest against the Calloway Canal.[24]

The trial did not begin until April 15, 1881, in part because 1880 was an unusually wet year, in part because the riparian owners were preoccupied in 1880 with an unsuccessful effort to win a hearing in the San Francisco federal court.[25] Shortly before the trial opened, the *Fresno Expositor* acknowledged that "if the present line of Court . . . decisions hold, the farmers are completely at the mercy of a few cattle rancheros, who settled along the principal streams in early times." The *Bakersfield Californian* predicted that if the court ruled in favor of Miller and Lux, "a green and smiling territory" would revert "to the condition of a hard, forbidding desert." The *Californian* tried to convince its readers that the suit was much more than a personal battle between two breeds of water monopolists. If the riparian interests prevailed, most of the county's ditches and canals would be worthless. Property values—and not just the value of land under ditch—would plummet, and the many jobs provided by Haggin and Carr would disappear.[26]

In light of the Supreme Court's demonstrated sympathy for riparian rights, Haggin and Carr's team of lawyers—John Garber, George Flournoy, and Louis T. Haggin—decided not to challenge the doctrine directly. Instead, they spent most of the trial trying to demonstrate that Miller and Lux were not riparian owners. As the *Kern County Weekly Courier* commented: "All the parties in the suit seem to consider the riparian question settled, so the question

---

24. *Bakersfield Californian*, April 16, 1881, published a full list of the suits. However, in late 1880 or early 1881, all legal complaints except those against the Kern County Land and Water Company were dropped.

25. Ibid., March 25, 1882.

26. *Fresno Expositor*, February 23, 1881; *Bakersfield Californian*, April 16, 1881.

. . . is whether the Buena Vista Slough is or is not a natural channel of the Kern River."[27] The defense called many witnesses who swore that the slough constituted the sink of the Kern River, not a part of the river itself. These included G. H. Mendell, head of the U.S. Army Corps of Engineers in California, and Assistant State Engineer James Schuyler. In most places, the slough had no discernible banks and changed its course from year to year. In other places the water often flowed in two directions at once. And in some years—notably in 1871, 1876, 1877, and 1878—the slough was no more than a swamp, with no current at all.

The trial wore on for seven weeks. The prosecution called twenty-three witnesses solely to demonstrate a continuous flow in the slough. On June 4, 1881, the *Californian* characterized the trial as "dull, dry, laborious" and "monotonous," as well as "exhaustive." The defense found plenty of ammunition. It argued that in dry years the streambed was so porous and evaporation so severe that the water disappeared even without diversions. It also pointed out that in wet years Miller and Lux had themselves blocked the flow of water into the slough to aid in reclaiming their swampland. This complied with state reclamation laws but violated the essential principle of riparian rights—that a stream be permitted to follow its own natural course. Garber, Flournoy and Haggin also claimed that Miller and Lux did not receive formal title to the riparian land in question until after construction began on the Calloway Canal. And since the cattle barons did not file suit until 1879, the lawyers charged that the riparian owners' silence in 1876, 1877, and 1878 constituted a "lache," or tacit legal acceptance of the diversion. Many other arguments were introduced, though few were fully developed. For example, the defense argued that Haggin and Carr, who owned land adjoining the outlet to the canal, were themselves riparian owners with a right to divert water for irrigation. It also argued that the Kern River had at one time followed the course of the canal, implying a sort of "residual" riparian right, and that the canal aided in reclaiming swampland that the defendants owned. The lawyers for Miller and Lux—Hall McAllister and R. E. Houghton—tried to prove only two points: that the slough was a water-

27. *Kern County Weekly Courier* May 26, 1881.

course, if not a river in the conventional sense, and that riparian rights included actual ownership of water out to the middle of a channel.[28]

Judge Benjamin Brundage issued his decision on November 3, 1881, after spending several months in San Francisco hearing final arguments, pondering the evidence, and studying related cases. He might simply have ruled in favor of Haggin and Carr by denying that Miller and Lux had provided sufficient evidence to prove riparian status. Instead, he went far beyond any earlier decision of either a superior court or the state supreme court. He denied that Buena Vista Slough was part of the Kern River or constituted a separate running stream. By nature, he argued, swampland could not contain rivers. But if a river existed, then *all* the swampland in the slough, as well as the water itself, constituted the bed of the stream. The only riparian owner, if one existed at all, was the federal government, which owned the dry land adjoining the swampland granted to the state. Though Brundage suggested that Miller and Lux had forfeited their right to sue by permitting more than $100,000 to be spent on construction of the Calloway Canal after 1875, his most persuasive arguments developed the implications of swampland laws and the Desert Land Act. He noted:

All the Acts and all the legislation, both State and Federal, concerning public lands, must be construed together, and as establishing one great system, and expressing one comprehensive policy, and so construed it cannot be believed that either the State or the United States intended the laws made for the sale of that land to contain anything destructive of that system.

28. See Lux v. Haggin: Transcript on Appeal, Lux v. Haggin File, California State Archives, Sacramento, as well as the many other briefs in the same file. For the trial itself see Lawrence, "Henry Miller and the San Joaquin Valley," 116–132; Berg, *A History of Kern County Land Company*, 13–15; Maass, *. . . and the Desert Shall Rejoice*, 226–231; *The Kern County Californian*, April 9, 23, 30, May 7, 14, 21, 28, June 4, 11, 1881; *San Francisco Chronicle*, April 16, 1881; *Daily Alta California*, April 21, 27, 29, November 5, 1881; *Stockton Daily Independent*, April 18, 1881; *Pacific Rural Press* 22 (November 19, 1881).

Just as the state would not have passed laws providing for the reclamation of swamplands if it intended settlers to flood them as pasture, the national government would not have enacted the Desert Land Act without tacitly reserving a sufficient supply of water to irrigate and improve the land made available by that law. Brundage even argued that the Kern County Land and Water Company had inherited the federal government's original riparian rights and enjoyed the right to irrigate for that reason alone. He did not reject the riparian doctrine entirely but claimed it was subject to the special needs and conditions of life in the arid West.[29]

Though the *Fresno Expositor* applauded Brundage's decision as the "entering wedge" by which the riparian doctrine would be destroyed in California and the *Bakersfield Californian* promised that the "sense of insecurity felt . . . by every owner of real estate will now disappear," the judge narrowly lost in his fall 1884 reelection campaign. Subsequently, Brundage joined the legal staff of James B. Haggin. In 1881, the *Californian* assured its readers that the ruling would survive an appeal to the supreme court, but this was far from certain. Late in 1880, in Tulare County, Judge W. W. Cross—in *Heirs of Jefferson Janes v. South Side Tule River Ditch Company*—ruled that the sink of the Tule River constituted a natural watercourse even though upstream appropriators maintained that the river disappeared ten miles above the riparian lands of the plaintiffs.[30]

This decision, as well as Judge Cross's consistent defense of riparian rights in Tulare County, doubtless helped persuade Miller and Lux to appeal, and the California Supreme Court heard the case in 1883 and 1884. Ironically, the heavy rains of early 1884 produced a new conflict; now each side tried to divert damaging

29. Judge Benjamin Brundage's decision was published in full in a supplement to the November 5, 1881, issue of the *Bakersfield Californian*. The main points in the ruling were summarized in many newspapers including the *San Francisco Chronicle*, November 4, 1881; *Kern County Weekly Record*, November 10, 1881.

30. *Fresno Expositor*, November 9, 1881; *Bakersfield Californian*, November 5, 1881; *Tulare County Times*, November 27, 1884. In the same fall election Judge Brundage lost, Judge Cross won relection easily in Tulare County. This demonstrates that public opinion in the two counties was more evenly divided than the large land and water companies would admit. On the Jefferson Janes case see the *Visalia Weekly Delta*, December 20, 1880, January 14, 1881; *Tulare County Times*, December 18, 1880.

flood waters onto the lands of the other, besieging the Kern County Superior Court for injunctions and counterinjunctions. Nevertheless, the arguments presented to the supreme court did not substantially change, however much they were honed and polished. The artfulness of the arguments was the way in which they were piled on top of one another: Henry Miller was not a riparian owner, and could not be, but if he was, he had sanctioned the Calloway Canal by remaining silent for so long. The riparian doctrine did not pertain to lands owned, or at one time owned, by the state or federal governments, but if it did the Kern County Land and Water Company had inherited the federal government's original right as a "supra-riparian proprietor." The various water commission acts passed in 1854 and after had superseded or destroyed the riparian doctrine, but if they had not, then the new water code adopted by the legislature in 1872 had been intended as a warning that no new riparian rights could be acquired after the code took effect. (One section of that law promised protection to *existing* riparian rights, suggesting that only appropriative rights could be secured in the future.) The Miller-Lux lawyers were less artful, but more consistent. As they reasoned in a brief filed in February 1884:

> The whole contest here is just this: Shall this defendant, and the individual parties interested with this defendant, succeed in their speculation of carrying the water of Kern river from its appropriate natural channel, where God and nature have placed it, and carry it thirty miles off to make a garden of what nature has made a desert, and to make a desert of plaintiffs' swamp land, which nature has enriched with this plentiful affluent, pouring down annually from the mountains? That is the question. By what right, by what reason, shall men, who own this land and who own this water, be deprived, by a simple notice of appropriation placed at some point along the Kern river, for which not one dollar was paid, of property for which they, the plaintiffs, have paid the State forty thousand dollars? . . . There is no possible advantage; there is no possible benefit; there is no possible justice; there is no possible ground of public policy; why these waters should be diverted to this desert land, thirty miles from the natural channel in which God has appointed them to flow.

Haggin and Carr had violated not just property rights, but a law of God and Nature.[31]

The court issued its brief decision on October 27, 1884, supported by a narrow four-to-three vote. It did not address the major arguments presented by either side. The two most important findings in the decision were that swamplands carried riparian rights like any other land grant from the state or nation and that Brundage had ignored evidence that Miller and Lux owned most of their land long before work began on the Calloway Canal. Certificates of purchase, argued the court, were as binding as formal land patents. In effect, the ruling suggested that only appropriations made *prior* to the purchase of riparian land from the state or federal government were valid. Perhaps because of the fragmentary nature of the decision—which many attributed to the age or senility of Judge J. R. Sharpstein, who wrote the opinion—late in November the court agreed to rehear the case. One of the wisest comments on the ruling came from John Norton Pomeroy, the foremost scholar of riparian rights in the nineteenth-century arid West. Pomeroy, a strong defender of the riparian doctrine, noted:

> The conclusions thus reached by the court may have been unexpected by many, and may be disappointing to some parties whose supposed interests are affected. In our opinion, however, no other conclusion was possible, unless the court should directly overrule an unbroken line of decisions, reaching from the earliest period of the State's history down to the present time, which had uniformly recognized, declared and upheld the common law riparian rights of private owners.

Interestingly enough, Pomeroy argued that riparian rights gave only "natural and primary rights" to water. Once the needs of

---

31. For a summary of the Kern County Land and Water Company's arguments, see Charles Lux et al., v. James B. Haggin et al., Brief for Respondent filed November 6, 1883 (San Francisco, 1883), 4–5. The quote is from *Charles Lux et al., v. James B. Haggin et al., Brief for Appellants in Reply* filed *February 28, 1884,* 9, 19, *Lux* v. *Haggin* File, California State Archives.

riparian owners had been "reasonably supplied," he concluded, appropriators could use the remaining supply.[32]

While the *Los Angeles Times* commented that the decision destroyed the value of the 200,000-acre Haggin-Carr estate at the same time it increased the value of the 100,000 acres owned by Miller and Lux by "eight or ten times," the opinion elicited little newspaper comment because the decision to rehear the case followed the verdict so swiftly. However, in light of the court's decision, public attention shifted to the forthcoming meeting of the California legislature.[33]

The legislature had been grappling with riparian rights since 1880. Both Haggin and Carr and Miller and Lux had turned to the legislature at the same time they filed suit. In 1880, Senator John W. Satterwhite introduced a bill to amend Section 1422 of the Civil Code. That section read: "The rights of riparian owners are not affected by the provisions of this title" (meaning the preceding sections concerning the acquisition of appropriative rights). Satterwhite wanted to add: "A riparian proprietor is one through whose land a natural stream flows, or whose land borders on a natural stream, and he has only the right to the flow of sufficient water . . . for culinary and household purposes, and to water the livestock which he owns and keeps on his premises. Such rights are subject to condemnation for irrigation purposes." The "natural stream" obviously excluded Buena Vista Slough and the other swampland sinks at the end of streams flowing into the San Joaquin Valley. At the same session, however, Senator Benjamin F. Langford introduced a bill to protect riparian owners. It provided that when a riverbank owner believed that he had been deprived of his fair share of water, the county sheriff would measure any upstream

---

32. *Bakersfield Californian* published the supreme court's 1884 decision in a supplement to its issue of November 1, 1884. John Pomeroy's opinion is reprinted in the *Tulare County Times*, November 20, 1884. The lawyers for Miller and Lux argued that the two men had not actually inspected the Calloway Canal until shortly before they filed suit and that J. C. Crocker, who joined them as a plaintiff in the case, did not examine the canal until the fall of 1880. Some of the riparian land had been purchased in June, 1874, before the formal notice of appropriation for the Calloway Canal was posted in the following year.

33. *Los Angeles Times*, October 29, 1884.

diversions and remove dams and other obstructions that prevented the landowner "from receiving the same proportion of water that his or their proportion of frontage on the stream bears to the whole frontage of the stream within the county." The bill also provided that the superior court would appoint a watermaster to supervise distribution on the petition of a "respectable minority" of riparian owners. Langford's bill would have virtually destroyed appropriative rights because any riparian owner would have been able to challenge *any* appropriative right no matter how long the water had been used, even if the diversion had never been questioned in the past. Neither bill, nor a third which protected vested riparian rights at the same time it prohibited the acquisition of new ones, made any headway. Significantly, the Langford bill demonstrated that the major riparian owners did not want an absolute right to the entire flow of a stream. They were willing to compromise, as Miller had been, but the large water companies refused, confident that one day the riparian doctrine would be overturned.[34]

As noted in chapters 6 and 7, the legislation considered in 1881[35] and 1883 was no more successful. Northern Californians feared that any limitation or abridgement of riparian rights, however justified, would make it easier for hydraulic miners to pollute Sacramento Valley streams, so they sided with the large riparian owners in the San Joaquin Valley. Then, too, rival bills tended to cancel each other out. In 1883, Assemblyman Wharton introduced a bill to repeal Section 1422 on grounds that riparian rights had not been recognized in Mexican California, but Assemblyman G. E. Whitney proposed to prohibit the acquisition of appropriative rights in the future and turn the regulation of existing rights over to each county's board of supervisors.[36]

34. S.B. 211 (Satterwhite), January 28, 1880, *Senate Bills, 1880,* vol. 3; S.B. 257 (Langford), February 4, 1880, ibid. The third bill was A.B. 525 (Sayle), March 15, 1880, *Assembly Bills, 1880,* vol. 4, California State Law Library, Sacramento. Also see *Fresno Expositor,* April 28, November 24, 1880; *Visalia Weekly Delta,* February 13, 20, 1880; *Bulletin,* February 6, 17, 1880; *Sacramento Daily Union,* February 10, 17, April 14, 1880; *Alta California,* February 11, 1880; *Los Angeles Evening Express,* February 14, 1880; *Stockton Daily Independent,* July 8, 1880.

35. S.B. 279 (Rowell), January 26, 1881, *Senate Bills, 1881,* vol. 2, California State Law Library, Sacramento. Also see the *Fresno Expositor,* February 23, 1881.

36. A.B. 365 (Wharton), February 1, 1883, *Assembly Bills, 1883,* vol. 3; S.B.

The Wharton bills introduced in 1883 had been endorsed, if not sponsored, by such friends of appropriation as Haggin and Carr, Moses J. Church, and the 76 Land and Water Company. But in 1884, the land and water companies decided to "go public" in an effort to win broad popular support. From 1879 to 1883 they had concentrated on the courts and legislature; now they tried to mold public opinion by orchestrating and financing large irrigation conventions perceived by many Californians as spontaneous protests of small farmers against the iniquities of "riparianism." The first major convention met in the middle of May 1884 at Riverside. The keynote address declared riparian rights "repulsive, dangerous and ruinous to California. . . . Let our courts in an evil hour give preference and sanction to this principle and the spectacle of decadence in Los Angeles and San Bernardino counties inside of five years would be mournful, aye hideous, to contemplate." Some who attended the convention doubted that either the courts or the legislature could overturn riparian rights, but the delegates agreed to meet again in Fresno in December to draft legislation for the 1885 legislature.[37]

A new sense of urgency confronted the friends of corporate irrigation who met at Fresno in December 1885. The supreme court's recent ruling in *Lux* v. *Haggin* was fresh in their minds, and the convention's executive committee issued an "Address to the Legislature" which argued that since eastern states had frequently modified the common law to suit local conditions, California should do the same:

> The conclusion must be that, by the Act of 1850, we adopted only such portion of the common law of England as was applicable to our condition, and whatever we did take of the common law included a power and duty existing in Judges and Courts exercising common law jurisdiction to modify the common law when demanded by common necessity.

---

87 (Whitney), January 16, 1883, *Senate Bills, 1883*, vol. 1, California State Law Library, Sacramento.

37. *Pacific Rural Press* 27 (February 16, 1884): 145, 178; (March 8, 1884): 221; (Mar. 29, 1884): 293; (May 24, 1884): 512, 514, 516; (June 14, 1884): 588, 590; 28 (July 19, 1884): 50. The quote is from the May 24 issue. Also see the *Visalia Weekly Delta*, April 24, May 29, November 14, 1884.

The men who wrote these words were not disinterested idealists. Their economic interests and home counties spoke volumes. The group consisted of Will Green of Colusa County, D. K. Quinwalt and E. D. Ruggles of Tulare, J. De Barth Shorb of Los Angeles, Richard Hudnut of Kern, J. T. Wharton (who led the appropriation interests at the 1883 legislature), and H. S. Dixon of Fresno. There was no representation from such important irrigation counties as Napa, Yolo, or Monterey, nor did any of the members represent business, transportation, or other economic groups. Shorb, chairman of the group and its most active lobbyist before the legislature, was an attorney and civil engineer from Maryland who founded the San Gabriel Winery on the present site of San Marino, California; his farm contained 1,500 acres of vines. He also owned thousands of additional acres of potential farmland in southern California and Arizona. Without an adequate water supply, of course, such land had little value. As mentioned earlier, Green owned a princely estate near the Sacramento River. When the legislature was considering the bills drafted by the legislative committee, Green candidly admitted: "if the bills pass, the present year will see the inauguration of an enterprise to water 450 square miles of land in the Sacramento Valley." He was referring to the Stony Creek Canal Company which he had formed early in 1883 in the hope of tapping that stream as a water supply for his dry land in Colusa County.[38]

38. "Address to the Legislature of the State of California by the Legislative Irrigation Committee of the State Irrigation Convention held at Fresno in December, 1884," *Pamphlets on California*, vol. 17, Bancroft Library; *Pacific Rural Press* 28 (December 27, 1884): 556; 29 (February 7, 1885): 125; *Sacramento Daily Union*, February 20, 1884. The *Sacramento Bee* of July 22, 1886, reported that Will Green's company claimed virtually the entire flow of the Sacramento River during the irrigation season, or 500,000 miner's inches of water. For details of the scheme see the *Weekly Colusa Sun*, March 3, 24, 31, 1883; March 1, 15, June 7, 28, 1884; January 10, 24, 1885. J. De Barth Shorb's papers at the Huntington Library, San Marino, California, are filled with discussions of speculative land and railroad ventures. Shorb was very prominent in the Democratic party and, doubtless, hoped his efforts on behalf of irrigation would help win the gubernatorial nomination to succeed his friend, and next-door neighbor, George Stoneman. Instead, the ill-fated special session of 1886 killed his political career, though he later served on the State Viticulture Board and as Los Angeles County treasurer. To his credit, Shorb helped build local rail lines and citrus marketing cooperatives and brought a unique method of condensing wine back from Europe. Like most nineteenth-century land speculators, he believed that his efforts ultimately benefited the public as much as himself.

On the eve of the new legislature, Wharton wrote to Shorb, emphasizing the need to "satisfy" G. E. Whitney, a powerful Republican state senator from Alameda County who was expected to lead the opposition to appropriation as he had in 1883. Wharton also urged that the executive committee prevent the introduction of any water bills other than those sponsored by the committee: "It will only tend to confuse and destroy that harmony of action that is so essential to the passage of any irrigation laws."[39] The "Fresno Bills," as they came to be called, constituted a comprehensive package of water legislation. S.B. 210 and A.B. 410 limited riparian owners to the amount of water they actually used and granted appropriators the right to condemn any riparian claims that limited their diversions. They also confirmed *all* existing appropriative claims.[40] S.B. 37 and A.B. 170 borrowed from William Hammond Hall's proposals of 1880, but without challenging any established appropriative rights. They required each appropriator to file a formal claim on forms prepared by the state engineer. Once a complete list had been compiled, the state attorney-general would file suit to quiet titles on each stream, though the individual superior courts would simply confirm existing diversions and establish chronological priorities. The resulting decree would be conclusive, though disgruntled claimants could appeal. Each year, water users would be required to update claims by providing the county recorder with information concerning the location and extent of diversions. Anyone who claimed water in the future would have to file the same annual statement, but the state was not given the power to evaluate new claims or to distribute the water.[41] S.B. 38 and A.B. 171 provided for the formation of irrigation districts administered by local boards of water commissioners, upon petition from the owners of half the land within the proposed district, whose boundaries would be designated by the state engineer. The

39. J. F. Wharton to J. De Barth Shorb, December 28, 1884, Box 1, Incoming Business Correspondence, Shorb Collection, Huntington Library. Also see Wharton to Shorb, December 10, 1884.

40. S.B. 210 (Reddy), February 2, 1885, *Senate Bills, 1885*, vol. 3; A.B. 410 (Weaver), February 2, 1885, *Assembly Bills, 1885*, vol. 5. The four "Fresno Bills" were discussed in the *Pacific Rural Press* 29 (January 24, 1885): 69; (February 7, 1885): 125.

41. S.B. 37 (Reddy), January 19, 1885, *Senate Bills, 1885*, vol. 1; A.B. 170 (Weaver), January 19, 1885, *Assembly Bills, 1885*, vol. 2.

commissioners would issue bonds and purchase or condemn established water rights, and a district board of trustees would determine the assessed value of all district lands and apportion taxes to retire the debt according to the value of the land and anticipated benefits from irrigation.[42] Finally, a proposed constitutional amendment would have allowed district officials, as well as county boards of supervisors, to set the rates charged by private water companies. However, the amendment promised the companies a minimum return of 7 percent per year on capital invested. The proposed legislation was designed, in part, to meet the needs of irrigation districts that chose to allow private water companies to build their irrigation works.[43]

Many California newspapers, including the San Francisco *Alta California* and *Chronicle*, the *Fresno Expositor* and *Republican*, the *Los Angeles Herald* and *Evening Express*, and the *Colusa Sun*, supported the legislation.[44] However, considerable opposition surfaced in Sacramento and San Francisco, as well as in the mining counties. Sacramento's *Bee* and *Union* both opposed the Fresno Bills. The *Bee* argued that any increase in diversions from northern California streams would reduce the volume of those streams, limiting both navigation and their capacity to scour out mining debris. It also argued that riparian proprietors had rights that could not be ignored; the legislature should not "rob one set of men for the benefit of another." The *Union* wondered if the limitation on riparian rights embodied in S.B. 210 had been engineered by the mining interests. Any restriction on riparian rights might limit the liability of hydraulic mining companies: "Better that the deserts remain unreclaimed and the parched lands go unwatered, than that the law

42. S.B. 38 (Reddy), January 19, 1885, *Senate Bills, 1885*, vol. 1; A.B. 171 (Weaver), January 19, 1885, *Assembly Bills, 1885*, vol. 2.

43. *Pacific Rural Press* 29 (February 7, 1885): 125.

44. Appended to the "Address to the Legislature" cited above were eighty pages of newspaper editorials supporting the Fresno Bills. The editors blamed a wide variety of interest groups, including cattlemen, miners, and even the railroad, for blocking the legislation. Many of the editorials assumed that the Fresno Bills addressed the needs of small farmers. For example, the *San Francisco Chronicle*, February 21, 1885, claimed that a group of senators were prepared to kill the bills through a filibuster. The filibuster was "doubtless in the interest of the capitalists, who foresee that legislation will defeat their hopes of securing a monopoly of water." Also see the *Visalia Weekly Delta*, January 29, 1985.

be blotted out, which is to-day the safeguard of the people against the unnatural descent of mining debris and slickens, and which, if unchecked, would render the fertile regions along the Sacramento river uninhabitable and utterly destroy the navigability of the chief free highway of the State." San Francisco's *Bulletin* noted that the proposed constitutional amendment guaranteeing a 7 percent return to water companies was unrealistic because it did not require those companies to open their books. Nor could any city or county dependent on a water monopoly afford to penalize a company that failed to comply with the law. The *Bulletin* also charged that this proposal had been inspired by the Spring Valley Company. A prominent correspondent of the *Pacific Rural Press* noted that public control over water rates was of limited value in any case:

> Under those [Fresno] bills a man that secures possession of water can do just what he pleases with it. He is not bound by law to divide with any one, pay or no pay. What does Haggin & Carr care that the Board of Supervisors shall fix a water rate? They are under no obligation to furnish water to others, even if they get all [the] Kern river. Under these bills then a water owner can dry out his neighbors, and buy for a song all the land his water will cover. He even is not bound in his yearly statement to show whose land was irrigated (Sec. 15). These bills should be entitled: Acts to monopolize the land and water of the arid portions of the State in the quickest and most effectual manner.

The state's former surveyor-general, James W. Shanklin, commented: "The vice of nearly all the bills on irrigation, is the taking of the patrimony of the people from them, and giving it over to private and corporate ownership." Shanklin maintained that public ownership offered the only safe alternative to monopoly.[45]

---

45. *The Sacramento Bee*, February 25, 28, March 3, 12, 1885; *Sacramento Daily Union*, February 24, 26, 28, 1885; *Bulletin*, January 30, March 11, 1885; *Pacific Rural Press* 29 (February 14, 1885): 142; (May 23, 1885): 492; *Tulare County Times*, November 27, 1884; February 19, 1885. The *Union* quote is from the issue of February 27, the excerpt from the letter to the *Pacific Rural Press* is from the issue of February 14, 1885; James W. Shanklin's statement is from the *Union* of February 28, 1885. In its December 4, 1884, issue, the *Times* called confiscation of riparian rights "communism

By mid-February, P. D. Wigginton, who helped Shorb lobby for the Fresno Bills, privately expressed confidence that "we now have full control of the legislature." The Shorb forces succeeded in having two hours a day set aside in the senate exclusively for discussing irrigation legislation. Wigginton expected that as the sixty-day session wound to a close, opponents of the bills would be forced to accept them to get the time to enact their own pet legislation. Still, he warned against overconfidence, declared that the irrigation bloc was too "independent," and urged that its members trade support with powerful lawmakers from the mining country.[46]

The Fresno Bills passed the assembly by a four-to-one margin on February 24, 1885. However, in the senate they encountered a strong coalition of senators who represented the hydraulic mining and riparian interests. Hydraulic mining had been outlawed by federal court decree in 1884, and the mining section's representatives in Sacramento desperately looked for a way to revive the industry. The mining bloc, led by Nevada County's Senator Cross, pushed two pet bills. One would have authorized the payment to contractors of $260,000 for the debris dams and levees constructed in 1880 and 1881, the amount left unpaid after the supreme court overturned the Debris Act of 1880. The second bill would have permitted mining companies to condemn land and build debris impoundment dams on their own; in exchange, they would have been absolved of all responsibility for future damages to towns or farms in the valleys. Three men led the senate debate over the irrigation bills: Reddy, who had introduced the Fresno legislation; Cox, a lawyer and riparian owner with strong ties to Miller and Lux; and Cross. Debate over the legislation often focused on technical legal questions—for example, whether riparian rights originated in federal or state laws. But Cross's forces opposed the water bills because the agricultural interests had strongly opposed the

---

agrarianism." "If you take the waters from the streams to-day and divide them without any reference to rights, either natural or acquired," the editor fretted, "and if in ten, twenty or fifty years from now a million more people are here than live here now, will it not be equally proper to re-divide them?"

46. P. D. Wigginton to Shorb, February 13, 15, 1885, Incoming Business Correspondence, Box 2, J. De Barth Shorb Collection, Huntington Library.

Debris Bill, and still refused to back any legislation favorable to mining. When the irrigation bills reached the senate, many miners hoped that a tradeoff could be arranged. For example, the Downieville *Mountain Messenger* commented on February 28, 1885: "we expect it [the bill limiting riparian rights] to give our fellows a chance to trade votes, and trade them often, to the end that the dam bill of Senator Cross, or one of kindred import, may become a law, and remove from our backs the old man of the valley, who is riding us to death." However, when Cross's bill came up for a senate vote in February or early in March, it won only one vote from outside San Francisco and the mining counties. When the irrigation bills reached the floor, the Nevada County senator had his chance to retaliate: "Let the people of the southern counties and their representatives feel the wrong we have felt; may they suffer as we have suffered, and may the God who distributes his sunshine equally on our mountains and their plains hear their every cry in vain for relief, until they award us that mercy they themselves claim."[47]

Poor floor management also accounted for the failure of the irrigation bills. Senator Reddy had served two previous terms in the legislature, but he could not match the parliamentary skills of Cross. He waited nearly a month to introduce the bills and consequently, they were buried under a mound of proposed legislation. Moreover, the bill limiting riparian rights ended up in the general file, rather than the much shorter special file. Reddy also took on too many jobs. He spent a week championing a bill to create a home for the blind, and squandered both energy and political support pushing several unpopular measures, including one to abolish voting requirements for jurors. Since senate rules prohibited spending more than an hour and a half per day on any particular bill or set of bills, the time lost was critical. A majority of the senate supported the legislation; but the floor leaders—first

47. *Pacific Rural Press* 29 (February 28, 1885): 187; (March 7, 1885); *The Sacramento Bee*, February 17, March 5, 1885; *Bulletin*, February 20, March 2, 11, 12, 1885; *The Mountain Messenger*, February 28, March 14, 28, 1885; *San Francisco Chronicle*, February 25, 28, March 5, 1885. The *Sacramento Daily Union* published a supplement on March 11, 1885, which printed critical parts of the debate in the legislature. The quote is from Senator Cross's speech in opposition to the Fresno Bills as reprinted in the *Alta California*, March 9, 1885.

Reddy, then R. F. Del Valle—could not muster the two-thirds vote required to consider the Fresno Bills out of order. The *Sacramento Daily Union* noted that many opportunities to bring the bills forward in the senate had been ignored. It suggested that Reddy and Del Valle had dragged their feet in the hope of exploiting the irrigation controversy in their next election campaigns.[48]

Ironically, the railroad also helped kill the Fresno Bills. Its representatives in Sacramento tendered only lukewarm support for the legislation even though it stood to gain from an expansion of irrigation and though Lloyd Tevis and Billy Carr had close links to railroad leaders. Apparently, the Southern Pacific agreed to support the irrigation bills in exchange for the irrigation bloc's endorsement of a constitutional amendment—the so-called Heath Amendment—limiting state taxes on the company to a flat 2.5 percent of gross income. Unfortunately, some of the Democratic leaders of the irrigation forces had also taken strong stands against the railroad in the past. To maintain its iron grip in Sacramento, the railroad apparently decided to punish its enemies and prevent the emergence of a new legislative clique that might challenge the supremacy of the transportation monopoly. San Francisco's *Chronicle* commented: "The proof of the fact lies on the surface. The monopoly controlled the Legislature. It was omnipotent. It could pass what measures it wanted and defeat the measures it did not want. If it had wanted the irrigation measures to pass, all it had to do was raise its hand, and Vrooman, McClure, Perry, Kellogg, Cross and the rest would have passed them at once. . . . A single word from Vrooman would have insured their becoming laws, the word was not spoken and they are dead."[49]

In the end, the legislature of 1885 accomplished little more than its predecessors. Several other water bills were introduced, but they offered little new and foundered in the wake of the Fresno Bills.[50] Near the end of the session, an angry J. De Barth Shorb issued a pathetic, melodramatic appeal to the lawmakers:

48. *Sacramento Daily Union*, March 14, 1885; *Pacific Rural Press* 31 (April 10, 1886): 349, 360; *Los Angeles Times*, February 22, March 4, 1885; *Bulletin*, February 17, March 7, 1885.

49. *San Francisco Chronicle*, March 17, 1885. Also see the *Chronicle* of March 5, 1885; *Bakersfield Californian*, March 21, 28, 1885.

50. The most important bill aside from the Fresno legislation was S.B. 50,

. . . turn not a deaf ear to the supplications of the thousands and tens of thousands whose now happy homes may be made desolate by non-action on your part. . . . You know that thousands of people went upon the arid deserts of California because they had not the means to purchase land elsewhere, and under what they thought the laws of the State, diverted the waters of the streams upon them; that they lived in flimsy huts, affording insufficient shelter from the burning sun of summer, and the cold blasts of winter; that they went poorly clad, and lived on the coarsest food—not enough, in many cases, to properly support life—while they were digging ditches and waiting for the vine and fruit trees to grow, and that thus beautiful homes have been made, and large and prosperous communities built up. . . . But now comes an interpretation of law that these diversions of water are wrong and illegal; that three or four cattle kings, who happen to have swamps at the ends of some of the streams, have the right to have the swamps remain swamps; that the desert shall no longer bloom, but that it shall be a desert. . . . We have spoken thus far in behalf of our present population; but we add to this a prayer for the future of California. We beg that you will not, by non-action, destroy for years the bright future of our State.

Many members of the legislature were moved by the appeal. A majority favored a special legislative session so that the lawmakers could focus their entire attention on irrigation. However, a vocal minority argued that the supreme court, which was in the process of rehearing *Lux* v. *Haggin*, might soon reverse its earlier decision. Critics of a special session also wanted to wait two years for an entirely new legislature. That way the senators who had opposed

introduced by Senator Whitney of Alameda County on January 19, 1885. Whitney had aided Senator Cross in blocking the irrigation bills in the senate. His bill declared California's water supply "the common property of the people of the State" and "forever inalienable." See *Senate Bills, 1885*, vol. 1. Though the constitutional amendment to guarantee water companies at least a 7 percent return failed, a law was enacted (*Cal. Stats.*, 1885, 95) which promised the companies at least 6 percent, but not more than 18 percent annually, on their investment.

irrigation legislation could be thrown out of office, breaking the logjam.[51]

In the months that followed, the water controversy continued to burn at white heat. Even such a staid literary journal as the *Overland Monthly* joined the debate.[52] The State Grange, whose political influence had faded considerably since the middle 1870s, remained the most vocal pressure group to call for greater public control over water. For example, its Committee on Irrigation reported to the thirteenth annual meeting of the Grange in October 1885:

> Your committee knows of no safer or better plan to accomplish this desirable object [of promoting irrigation] than through a general system of irrigation which shall be under complete state control. Of course, in order to carry out this aforesaid system the State will have to exercise its power of eminent domain as well on riparian owners, so called, as upon the owners of existing water ditches, who may have acquired vested rights; in either case we deem that ample compensation should be made for any losses sustained.

However, many Grange members apparently perceived a threat to their own water rights in such a scheme, and the memory of the debris debacle of 1880–1881, as well as the sordid power play of the mining delegation at the 1885 session of the legislature, convinced them that comprehensive state control posed untold dangers. Consequently, the conferees rejected the recommendation. Instead, they adopted a resolution supporting a constitutional amendment

> declaring that all water courses and lakes, except those barely sufficient to water half a section of land, and except such

51. *Los Angeles Times*, March 3, 1885. The Shorb statement is from the *Times* of March 8, 1885. The *Bulletin*, March 12, 1885, reported that Shorb and Will Green saw little chance for a special session.

52. George W. Haight, "Riparian Rights," *Overland Monthly*, 2d ser., 5 (June 1885): 561–569; John H. Durst, "Riparian Rights from Another Standpoint," 6 (July 1885): 10–14; A. A. Sargent, "Irrigation and Drainage," 8 (July 1886): 19–32.

water as is claimed by the United States for navigation purposes, are the inalienable property of the State, and that no
diversion of water from the basin of a lake or a channel of a
stream is lawful without the permission of the public authorities, and that no lease of water obtained from the State
authorities should extend beyond a term of ten years.[53]

The Immigration Association of California also joined the call for
water law reform, noting that northern California was lagging far
behind the growth rate of southern California. The group, dominated by northern California boosters and businessmen, pointed
out that irrigation had contributed to a doubling of taxable property
in Los Angeles from 1881 to 1885.[54]

Whatever the opinions of literary journals, farm organizations, or booster clubs, the state supreme court still had the last
word on California water law. While the 1884 decision had been
brief and fragmentary, the 1886 ruling ran 200 printed pages and
still stands as the longest opinion ever issued by the court. Justice
J. R. Sharpstein wrote the 1884 opinion; Justice E. W. McKinstry
did the job in 1886. E. F. Treadwell, himself a leading student of
California water law, has left a highly laudatory judgment of
McKinstry:

> A man of higher standing, morally and intellectually, could
> not well have been found, but if the whole world had been
> sought for a man of unimpeachable character, who was still
> the best suited to uphold the riparian doctrine . . ., no one
> could have done better than to have selected Justice Mc-
> Kinstry for the part. . . . He was an old man already and lived
> to be much older. He was born and bred in the common law,
> and knew his Blackstone forwards and backwards. He be-
> lieved honestly and sincerely in the omnipotence of the law
> and the infallibility of the common law of England to solve
> every human dispute.

53. The first quote is from the *Pacific Rural Press* 29 (October 31, 1885): 356;
the second is from the *Press* of October 24, 1885, 336.

54. *Fifth Annual Report of the Immigration Association of California, 1886* (San
Francisco, 1887), 4–6.

What part the legislative debacle of 1885 played in McKinstry's decision is unknown, but his ruling was thorough and conclusive, an extraordinary explication and defense of the riparian doctrine.[55]

McKinstry methodically disposed of each argument that had been offered by the Haggin-Carr lawyers. He ruled that the law required only that Buena Vista Slough contain a discernible channel at the time Miller and Lux filed suit; the attempts of both sides to prove that a channel had or had not existed in the years before and after the trial began were irrelevant. Nor did the judge think that a river always had to have readily identifiable, unchangeable banks, easily distinguishable from the bed. Like Sharpstein, McKinstry held that for most legal purposes certificates of purchase constituted just as solid proof of landownership as final patents. However, he considered the issue irrelevant because riparian rights had existed from time immemorial and were not dependent on provisions of land laws. Similarly, since such rights did not date from the purchase of land, they could not be limited by laches; the fact that Haggin and Carr had spent a vast sum building the Calloway Canal was beside the point. Though Miller and Lux had not complained when the canal was surveyed, when construction began, or even when the first diversion began in the fall of 1875, so long as they filed suit within five years, the date they filed suit did not matter. Besides, McKinstry pointed out that Brundage had ignored one simple fact. The diversion dam at the head of the slough had a headgate that could permit water to reach the riparian lands. Obviously, Miller and Lux did not know how much water would be diverted when the first diversion was made. It was reasonable that they would wait until actual damage was done before filing suit. A court could not issue injunctions on the basis of anticipated injury.[56]

The judge also addressed many broader legal issues. He denied that either the federal laws of 1866, 1870, or 1877, which sanctioned appropriation, or the California Civil Code of 1872, had limited or excluded riparian rights. Moreover, he maintained that without the check of riparian rights, appropriation would produce

55. Treadwell, *The Cattle King,* 92.

56. *Lux v. Haggin,* 69 Cal. 255 (1886), pp. 267–287, 296–298, 409–421, 421–439.

a monopoly "by comparatively few individuals, or combinations of individuals controlling aggregated capital, who could either apply the water to purposes useful to themselves, or sell it to those *from whom they had taken it* away, as well as to others. Whether the fact that the power of fixing rates would be in the supervisors, etc., would be a sufficient guaranty against over-charges would remain to be tested by experience." The court could not deprive citizens of vested rights without good cause, and certainly not because another group of citizens considered their needs a higher good. Unlike appropriative rights, riparian rights were inseparable from the land itself, and those rights could not be confiscated without destroying the value of that land. However, for the first time the court formally acknowledged that irrigation was a "public use" of water. In doing so, the justices confirmed the right of irrigation companies and other organizations of farmers, such as irrigation districts, to condemn riparian rights on condition of proper compensation. The case also clarified several other characteristics of water rights that had been widely accepted but not confirmed by the highest court: disuse did not destroy riparian rights; riparian farmers could use a reasonable quantity of water to irrigate their land, or sell water to nonriparian landowners, even if these diversions reduced a stream's volume (assuming the forbearance of others who owned land adjoining the stream); and water rights acquired through adverse use and prescription were not subject to challenge by riparian owners. Thus, condemnation, purchase, and adverse use offered hope to irrigators. The court also confirmed the *primacy* of appropriation on the public domain. Though the ruling encouraged the legislature to provide "further legal machinery" to facilitate the condemnation of riparian rights for "public use," deciding on fair compensation would remain a job for the courts.[57]

---

57. *Lux et al. v. Haggin et al., Majority and Minority Opinions of the California Supreme Court, October 27, 1884 and April 26, 1886* (Sacramento, 1886). The quote is from pp. 66–67. On the *Lux* v. *Haggin* suit, see S. C. Wiel, *Water Rights in Western States* (San Francisco, 1905), 34–37; Treadwell, *The Cattle King*, 78–94; Morgan, *History of Kern County*, 98–109; Wells Hutchins, *The California Law of Water Rights* (Sacramento, 1956), 52–53; Harding, *Water in California*, 38–39. Had the challenge to riparian rights involved the orchards of Los Angeles or San Bernardino rather than the wheat fields of Kern County, the court might have ruled differently. Citrus fruit *required*

Apparently, the appropriators knew about the ruling in advance because several weeks before formal issuance of the decision on April 27, 1886, the friends of the Fresno Bills, now organized as the "State Irrigation Convention," began to create antiriparian clubs throughout the state. To join one of these clubs, prospective members had to endorse the Fresno Bills, without reservations, and reject riparianism. "Organized, you have a potent force in the selection of judges and legislators," the renamed executive committee said in a broadside. "It is under your power to crush the threatening evils of riparianism. Fire the hearts of the people with the justice of your cause. Show political parties that you have the strength and will to enforce what you demand." Many of the appeals issued by the convention emphasized that the riparian doctrine had retarded California's population growth and overall economic development. The group reported that only 284,000 people lived in the Central Valley, a population density of about 5 to the square mile. This compared, according to the antiriparian organization's figures, with 92.6 per mile in the Merrimac Valley; 56.5

---

irrigation to develop, while wheat could be raised in many years, even in the southern San Joaquin Valley, with normal rainfall.

B. Marks, in "The Riparian Decision in Interior California," *Overland Monthly* 9 (February 1887): 145–162, warned would-be appropriators that the condemnation privilege confirmed by the court offered little consolation. For example, since there were sixty-four riparian owners along the Kings River, each might require restitution for *all* the stream's water. And if a river changed its course, as often occurred especially in southern California, a whole new batch of riparian owners would be created along with a new crop of suits. For a defense of the decision see Warren Olney, "The Present Status of the Irrigation Problem," *Overland Monthly* 9 (January 1887): 40–50.

On April 27, 1886, the day following the issuance of the ruling in *Lux v. Haggin*, Judge McKinstry wrote Shorb pointing out that while the final vote was 4 to 3, only Judge M. H. Myrick "has dissented from the general propositions of the opinion." Myrick denied that the Common Law adopted by the legislature on April 13, 1850, included the riparian doctrine because climatic conditions in California were so different from the eastern United States or England and the basic assumption behind the common law was that it should conform to specific conditions. Judge E. M. Ross, who wrote the longest dissent, conceded that the riparian doctrine applied to land within Spanish or Mexican grants, but not to land within the public domain. As McKinstry put it: "R. thinks that grants from the *government* of lands—by some species of reservation—didnt [sic] carry water." McKinstry assured Shorb that nothing in the decision would "interfere with proposed legislation [presumably the Fresno Bills] as I understand it." He summed up his decision in two words: "It's right."

per mile in the Connecticut Valley; 173 per mile in the Hudson Valley; and 109.7 per mile in the Miami Valley. It estimated that the Central Valley could support 11,000,000 people, "a population which would make San Francisco the most desirable business city in the world, and the mart of an immense commerce, as varied in the products of which create it as the globe gleaned trade of London." The message was clear: urban businessmen stood to gain as much from the destruction of riparian rights as farmers.[58]

Many proponents of the Fresno Bills sincerely believed riparianism had hurt California's economy; they considered their self-interest and the public good one and the same. But behind the scenes, much of the work of the convention was orchestrated by Billy Carr. For example, Carr entrusted the job of organizing Los Angeles County's antiriparian clubs to Shorb, urging his front-man to "make the organization as thorough and numerous as possible. . . . Of course we will take care of any expense there is." The boss bought space in newspapers for the convention's press releases and articles as well as editorial support for the proposed irrigation legislation. Those newspapers that refused to accept the party line were singled out for punishment. In one letter to Shorb, Carr remarked that the *Pomona Times-Courier* had attacked the Fresno Bills, and the boss ordered: "turn the Pomona Times Courier over or silence it."[59]

The interest of the appropriators in winning support from San Francisco financial interests was evident in the choice of that city for the antiriparian convention held on May 20, 1886, at the Grand Opera House. One member of the executive committee claimed that the antiriparian clubs organized in the spring now had 20,000 members. However, aside from the notable presence of the president and cashier of the Bank of California, the San Francisco delegation gave little evidence that the water companies had won over the city's business community. Virtually all the delegates were from

58. *Pacific Rural Press* 31 (April 10, 1886): 349. Also see the *Press* for April 3, 1886, 321; and April 10, 1886, 352. The executive committee of the Fresno Convention had recommended the formation of antiriparian clubs as early as December, 1884, but apparently none were formed until Haggin and Carr promised financial support.

59. W. B. Carr to J. De Barth Shorb, April [?], 17, 19, 29, 1886, Incoming Business Correspondence, Box 2, Shorb Collection, Huntington Library. The quote is from the letter with the illegible date.

southern California or the San Joaquin Valley. Most were canal company lawyers, and fourteen were employees of Haggin's Kern County Land and Water Company. Carr was "elected" vice-president and given a seat on the executive committee. Although Colusa County, in the Sacramento Valley, contained three antiriparian clubs, Will Green—who was listed as a member of all three—doubtless fathered and nurtured the groups. Green drafted a "Declaration of Principles" clearly inspired by the water companies. And when some delegates refused to sign the creed, they were prohibited from participating in the convention.[60]

Boss Carr had already decided that the time was ripe for a special session of the legislature. He mapped out his strategy in a candid letter to Shorb dated May 5, 1886. One of the Fresno Bills limited riparian owners to the amount of water actually used, but in 1885 no attempt had been made to abolish all riparian rights. Nevertheless, the San Francisco convention supported the complete eradication of such rights without compensation, and Carr put this objective at the top of his agenda. His second goal was enactment of the constitutional amendment requiring county boards of supervisors to fix water rates at no less than 7 percent of the total assets of individual water companies. "If we get these . . . propositions," he noted, "the water is secured." Carr recognized the importance of electing "sound" judges "who will sustain instead of undoing whatever we may accomplish at the extra session," and he hinted that he favored reorganizing the supreme court—though the letter contained no specific reorganization plan. That Carr cared little for water law reform or the needs of farmers, beyond his immediate objectives in Kern County, was evident in his recommendation that the irrigation district and water rights

60. Thomas E. Malone, "The California Irrigation Crisis of 1886: Origins of the Wright Act" (Ph.D. diss., Stanford University, 1965), 137–138; *Pacific Rural Press* 31 (May 29, 1886): 528–529; (June 5, 1886): 555, 556, 558; (June 12, 1886): 578–582; (June 19, 1886): 602–609; (June 26, 1886): 626–636; *Bulletin*, May 20, 1886; *Call*, May 20, 21, 28, 1886; *The Visalia Weekly Delta*, June 3, 1886. In a letter to Shorb, January 17, 1886, Richard Hudnut, a member of the Fresno Convention's executive committee, urged another large irrigation convention and noted: "Several of the canal companies have promised to furnish money." Doubtless, the San Francisco convention in May received financial support and direction from the same sources. Also see Hudnut to Shorb, February 19, April 12, 19, 1886, Incoming Business Correspondence, Box 2, Shorb Collection, Huntington Library.

adjudication bills be shelved until the next regular session. They were, he argued with delicious irony, too "controversial."[61]

After the San Francisco convention adjourned, Carr traveled the state getting the signatures of twenty-one of the 1885 legislature's thirty-eight senators and sixty-four of seventy-eight assemblymen on a petition urging the governor to call a special session on grounds that before the next regular session "thousands of citizens and their families may be ruined, and millions of property may be destroyed." The petition pledged the signators to support the Fresno Bills, which virtually all had voted for in 1885. Governor George Stoneman, who owned a large citrus farm adjoining Shorb's extensive vineyards in San Gabriel, strongly favored that legislation. He had organized California's first formal irrigation convention in 1873, preached the need for water law reform in his 1885 message to the legislature, and was a prominent delegate to the San Francisco irrigation convention in May. But the governor, Carr knew, was reluctant to convene the legislature without some assurance of success; the chief executive had been roundly criticized for calling an unsuccessful special session to curb railroad rates in 1884. Historian Wallace Morgan claimed that Carr, "reinforced by a stalwart bunch of his friends from Kern county and elsewhere," presented the petition to the governor in a San Francisco hotel room, apparently on July 15, 1886. That night, in Morgan's words, the governor "distinguished and endeared himself . . . by consuming without a quiver more mint julips than any other man in the crowd from below the Mason and Dixon line could carry off." Despite the governor's well-earned reputation as a hard drinker, the feat was all the more prodigious given his service in the Union Army. We do not know whether the governor's advanced state of intoxication had been planned or anticipated by Carr, but toward the end of the evening (morning?), the congenial chief executive signed the executive order calling for the extra session to convene on July 20.[62]

61. W. B. Carr to Shorb, May 5, 1886, Incoming Business Correspondence, Box 2, Huntington Library.

62. Morgan, *History of Kern County*, 108. Haggin and Carr probably also feared that since the next session of the legislature would select a U.S. senator, the lawmakers would have less time to devote to irrigation legislation. The *Sacramento Daily Union* reprinted the petition for a special session in its issue of August 20, 1886.

In his appeal for the legislators to return to Sacramento, Stoneman suggested that a state of emergency existed in California which could lead to armed conflict between appropriators and riparian owners:

> The majority of the judges of the Supreme Court have announced that any riparian proprietor may obtain an injunction against any person not a riparian proprietor, to prohibit him from appropriating, diverting or using water from the stream above his land. Under this ruling the ditches and canals, which are the arteries of the agricultural life of the State, may be closed by writs from the courts, and, too, upon *ex parte* application, without notice or warning or opportunity of being heard until after irretrievable damage has been done. Many such suits are now pending. Writs of injunction have been asked for and in some cases obtained, but have not been obeyed. Should an attempt be made to enforce them and others which are likely to issue, as is apprehended, serious trouble may ensue, because the people may resist to prevent the desolation of their homes, farms, vineyards and orchards.[63]

The governor exaggerated the emergency, just as he incorrectly assumed broad popular support for the proposed legislation. The demand for a special session arose for a number of reasons, not all of which were directly related to *Lux* v. *Haggin*. The *Sacramento Bee* mentioned that the federal government had recently canceled

---

The number of signatures suggested about the same level of support for the Fresno Bills as prevailed in the 1885 legislature. The thin majority in the senate boded particularly ill for the appropriation party.

On the lobbying prior to the special session also see Carr to Shorb, May 16, June 30, July 2, 1886, Incoming Business Correspondence, Box 2, Shorb Collection, Huntington Library. Carr's heavy-handed political style can be seen in several of his letters to Shorb. For example, on May 16 he wrote: "I have made arrangements to have some gentlemen dine with me this evening, and I wish you to join us. Don't fail to be on hand. It is important." Carr's long association with the railroad, whose monopoly Governor Stoneman had bitterly assailed both as a member of the railroad commission and as governor, probably increased his dependence on Shorb. Shorb was a farmer, but his lack of ties to large water companies or other special interest groups inspired public confidence in his motives.

63. *Pacific Rural Press* 32 (July 24, 1886): 82–83.

fraudulent timber entries in California and, since Haggin and Carr's dummy entries in Kern County continued to provoke public criticism, the two men may have taken these cancellations as a warning that the Interior Department might soon revoke their titles to Desert Land Act claims. They could strengthen their title by irrigating as much land as possible. "But if they can secure the water," the *Bee*'s editor wrote, "the recovery of the land would be a mere matter of time. This consideration may be one reason of their great outlay for the extra session." Moreover, Stoneman himself doubtless hoped to strengthen his chances for reelection in November. The irrigation crisis offered a ready-made political opportunity both for him and the Democratic Party.[64]

It was a risky opportunity. While the legislature seemed committed to reform, most of the members cultivated by Carr did not represent counties where irrigation was common. And since they faced little direct pressure from constituents, their support was fragile. It depended much more on vote-trading and bribes than on the fear that riparian rights would destroy irrigation and imperil the state's economic future. Thomas E. Malone, in his excellent study of the 1886 session, noted that within the Central Valley only 19 percent of all farmers practiced irrigation, and only 2 percent of the farmland was under ditch. South of the Tehachapis, 55 percent of all farmers used irrigation but, still, only 12 percent of the farmland was watered. Moreover, the counties of Los Angeles, San Bernardino, San Diego, Fresno, Tulare, and Kern elected only 11 of the legislature's 120 senators and assemblymen.[65] Reform was made doubly difficult because of the three-way split between champions of appropriation, riparian rights, and "state control." The state control party included interest groups ranging from mining companies to the Grangers. Finally, any constitutional amendment to abolish riparian rights required a two-thirds vote of the legislature.[66]

Most of the state's newspapers supported the Fresno Bills in

64. *The Sacramento Bee*, July 26, 1886; *Los Angeles Times*, July 18, 1886.

65. Malone, "The California Irrigation Crisis of 1886," 16, 18.

66. On the composition of the legislature see *The Sacramento Bee*, July 26, 1886; *Stockton Daily Independent*, July 23, August 5, 13, 1886; *Sacramento Daily Union*, July 21, 1886.

1885, but that support all but disappeared by the beginning of August 1886. The *Stockton Daily Independent* commented: "Private ownership and control of the waters of the state is far more to be dreaded than the objectionable law of riparian rights, which is altogether inapplicable to California. . . . The idea of recklessly giving away public property and then paying individuals for the use of that property cannot be entertained." After an exhaustive survey of county records, San Francisco's *Chronicle* prepared a stream-by-stream list of the state's leading appropriators. Not surprisingly, the names of the champions of the pending legislation appeared frequently. The newspaper concluded that only remote Inyo County, on the eastern side of the Sierra, contained any unclaimed water, and most streams had been claimed many times over. Hence, any law that simply confirmed existing appropriative claims would subject future water users to a corporate monopoly, a monopoly that would enjoy a perpetual profit of at least 7 percent a year.[67]

Nevertheless, as the lawmakers poured into Sacramento, Carr expressed confidence that they would pass his bills and adjourn within a week. He was joined in Sacramento by several able lobbyists, including Haggin's personal secretary and J. De Barth Shorb. The trio was frequently seen buttonholing lawmakers in the lobby of the Golden Eagle Hotel. The *San Francisco Chronicle* noted that they were no less active in the "arena" itself:

> That it is a Haggin and Carr fight, pure and simple, for the waters of Kern river is to an unprejudiced observer indisputable. An hour in either chamber would be sufficient to convince any one. At any hour in the day J. De Barth Shorb can be found in the Assembly chamber, while the Senate chamber is perpetually alarmed by the presence of W. B. Carr. Henry C. Dibble, or Judge Dibble as he is popularly called, and who has been here since the opening day of the session, divides his time between the two chambers, consulting and advising with Carr and Shorb and communicating their instructions to their followers.

67. *Stockton Daily Independent*, July 28, 1886; *San Francisco Chronicle*, August 15, 1886. Also see the *Chronicle* of July 18, 1886; *Pacific Rural Press*, 32 (July 24, 1886): 82, 83; (July 31, 1886): 97; *Tulare County Times*, July 22, 1886.

Charles Lux and a host of prominent San Francisco politicians—including the city's Democratic boss, Chris Buckley—ably represented the other side. Though the appropriators dominated the assembly, Miller and Lux controlled a substantial minority in the senate. The San Francisco *Evening Post* published a list of "senatorial cattle" which it claimed "will be seen to bear the brand of Miller & Lux when the final rodeo is made." The statement, indiscreet but fully warranted, resulted in the correspondent's expulsion from the senate floor. Charges of vote buying appeared frequently in the press. For example, the *Stockton Daily Independent*, claimed that Carr's forces paid $300 to each assemblyman who voted for the constitutional amendment abolishing riparian rights, and promised an additional $600 payoff if the amendment cleared the senate.[68]

Many other bills and constitutional amendments won some attention from the 1886 legislature besides those in the Carr-Shorb package. Partisans of state control offered several measures declaring California's unappropriated water public property, and two more providing for a state irrigation system. Senator John Days's plan would have retired state bonds through assessments against the land benefited; Senator Cross's bill would have put the issuance of bonds up to a statewide vote, then used general tax revenue to pay them off. Cross, one of Carr's strongest critics in the senate, called for the creation of dams "for irrigation and other beneficial purposes." Clearly, he hoped that such structures would capture debris and permit a revival of the hydraulic mining industry. Moreover, he probably hoped to form a new alliance with northern California farmers. (His bills were reintroduced by Anthony Caminetti of Amador County in 1889, but the mining counties could no longer count on strong political support from the San Francisco delegation.) With the additional exception of Senator Whitney's irrigation district bill, most of the remaining legislation pertained

68. The *San Francisco Chronicle* editorial is from the issue of August 12, 1886. The *Post* comments were reprinted in the *Sacramento Bee*, July 27, 1886. Also see the *Bulletin*, August 19, 1886; *Los Angeles Times*, July 20, 1886; *Stockton Daily Independent*, July 20, 21, 23, 28, August 13, 1886. The *Times* of March 17, 1887, reported that W. B. Carr had been sued by one of his agents who had bribed legislators using his own money, only to have the boss refuse to pay him back when the irrigation bills bogged down in the senate. Norman Berg claims that Henry Miller confided to a friend during the special session that "plenty of money makes a good politician." See *A History of Kern County Land Company*, 14.

to water rights—specifically how to limit the riparian doctrine. This legislation, dwarfed as it was by the Fresno Bills, had little chance of winning broad support.[69]

Despite almost daily editorial blasts against the warmed-over Fresno Bills and corruption in the legislature, Carr's substantial payroll and lobbying skills might have saved the day had the cause of irrigation not been closely linked to a supreme court reorganization plan. Publicly, Governor Stoneman justified reorganization on grounds that two of the justices had been incapacitated by physical illness, or perhaps even "mental incompetence," for months, and that the court was underpaid. However, these reasons were a transparent cover. Carr and Shorb wanted a new court not just to insure the success of their legislation but also as a "fall-back position." Even if the appropriators failed at the special session, they could hope that a remodeled court would overturn the verdict in *Lux v. Haggin.*

The governor, aided by Carr, Shorb, and company, pushed a bill to reduce the court's size from seven to five members, and increase salaries from $6,000 to $12,000 a year. This would have permitted Stoneman to "retire" justices Robert F. Morrison and John R. Sharpstein. Senator Grove Johnson of Sacramento County claimed that the two judges were totally incapacitated and had not

---

69. S.B. 11 (Days), July 26, 1886; S.B. 7, 8, 9 (Cross), July 22, 1886. For surveys of the legislation considered in 1886 see the *Pacific Rural Press,* 32 (July 31, 1886): 101–102; (August 14, 1886): 139–140; (August 21, 1886): 161. Cross also introduced S.B. 13 on July 26, one of several bills designed to pave the way for the condemnation of riparian rights. But the Cross bill was unusual in that it also provided a process for quieting title to all water through comprehensive state suits against all water users, county by county. Among the other noteworthy bills were S.B. 1 (Del Valle), July 21, 1886; S.B. 3 (Reddy), July 21, 1886; S.B. 4 (Lowe), July 21, 1886; S.B. 6 (Whitney), July 21, 1886; S.B. 10 (Whitney), July 22, 1886; S.B. 12 (Saxe), July 26, 1886; S.B. 15 (Reddy), July 28, 1886; S.B. 18 (Kellogg), August 6, 1886; A.B. 2, August 5, 1886; A.B. 3 (DeWitt), July 21, 1886; A.B. 6 (McJunkin), July 26, 1886; A.B. 7 (Walrath), July 29, 1886; A.B. 8 (Goucher), July 30, 1886. All these bills were bound in a special volume, *Bills, Resolutions, Constitutional Amendments of the Special Session of 1886,* housed with the other bills in the California State Law Library.

Many members of the "riparian block" at the special session hid behind demands for complete state control of waterways, including state purchase of riparian rights and construction of a state irrigation system. Such proposals had no chance of passage, so the riparian forces took no risk. However, their strategy did help undermine the charge that riparian owners were archmonopolists. See the *Bakersfield Californian* editorial, September 18, 1886.

written an opinion for months. He reported widespread support for reorganization among the legal profession. The campaign received welcome help from former Chief Justice David S. Terry, famed for his 1859 duel with Senator David C. Broderick. Terry had served as Sarah Althea Hill Sharon's attorney in her divorce suit against William Sharon, the silver baron and U.S. senator from Nevada, and later married her. The supreme court, led by justices Morrison and Sharpstein, had rejected Terry's claims in the property settlement. In response, the attorney urged the legislature to investigate the fitness of the two judges.[70]

Public reaction to the court plan was overwhelmingly negative. On July 26, 1886, Senator Cox of Sacramento County presented the legislature with a petition bearing the signatures of 20,000 to 25,000 residents of San Francisco and Alameda counties. Many of the petitioners were lawyers and businessmen who originally supported the Fresno Bills. Now they urged state control over all water rights. The San Francisco Bar Association warned that the scheme would render the court subordinate to the legislature and make conformity to public opinion the essential test in evaluating the soundness of high court decisions. Many Grangers also joined the battle. The Pomona chapter considered reorganization a plot by which one group of citizens sought legal permission to steal the property of another group of citizens. It protested any "policy that would assassinate the independence of our judges and strike a fatal blow at the stability and integrity of property—a policy essentially lawless, revolutionary and communistic; a policy that seems to be an open declaration of war against the basis and fundamental principles upon which our Government is grounded." Less than a week after the special session convened, the assembly judiciary committee unanimously rejected Stoneman's reorganization proposal,

70. Malone, "California Irrigation Crisis of 1886," 167; *Los Angeles Times,* July 20, 1886; *Sacramento Daily Union,* July 22, 1886; Theodore H. Hittell, *History of California* (San Francisco, 1897), 4: 696–697. Ironically, while Sharpstein had written the 1884 opinion in *Lux* v. *Haggin* and sided with the majority in 1886, Judge Morrison dissented on both occasions. On August 5, 1886, an outraged Francis G. Newlands wrote Shorb after talking with Morrison: "He [Morrison] said that he understood the movement was a bargain between the Irrigationists and Terry by which they got his support and that their real aim was to reach Sharpstein." Newlands urged Shorb to abandon the supreme court scheme, warning that it had "greatly weakened" the irrigation cause. Shorb Collection, Incoming Correspondence, 1886.

and two weeks later a special assembly committee concluded that the charges of incompetence directed against Morrison and Sharpstein were "groundless." Apparently, both judges had recovered their health by the time of the hearing.[71]

Many supporters of the Fresno Bills in 1885 considered the supreme court reorganization plan an impudent power play, and they quickly soured on the remaining legislation. By early August, Carr and Shorb faced almost unanimous opposition from the state's press, and while they still dominated the assembly, they stood no chance of winning in the senate. On August 20, the governor prorogued the legislature until September 7, claiming that the lawmakers needed time to campaign and nominate candidates for the fall elections. The Republicans had scheduled their state convention for August 27 and the Democrats for August 31. Stoneman's refusal to dismiss the legislature contributed to the atmosphere of suspicion. Senator Lynch of San Francisco, as well as many skeptical newspaper editors, charged that Carr had engineered the break so that he could use his substantial power at the convention to cajole recalcitrant Republicans. In Lynch's words, the boss hoped that "some of the doubtful ones might see the vision of a golden light and . . . return converted, consoled and—compensated."[72]

Perhaps the most damning public revelation concerning the special session occurred soon after the legislature reconvened on September 7. The *Sacramento Bee* got hold of a confidential letter

71. The declaration of the Pomona Grange was reprinted in the *Pacific Rural Press* 32 (August 14, 1886); 132. Also see the *Sacramento Bee*, July 22, 23, 27, 1886; *Sacramento Daily Union*, July 28, 1886; the *Stockton Daily Independent*, August 3, 11, 1886; *San Francisco Chronicle*, August 7, 10, 11, 1886.

72. Senator Lynch's statement is from the *San Francisco Chronicle*, August 14, 1886. Also see the *Chronicle* for August 21, 1886; the *Stockton Daily Independent*, August 21, 1886; *Bulletin*, August 20, 1886; *Los Angeles Times*, August 21, 1886.

Both parties made strong statements against water monopoly in their planks on irrigation. The Republican platform opposed acquisition of any appropriative rights that could, in the future, block establishment of state control over watercourses. It also urged that all private rights should be subject to condemnation for public purposes. The Democratic party platform endorsed irrigation districts as the best way to prevent monopoly, but also adopted a plank which read: "The State may at any time assume control of the diversion, use and distribution of water under general laws enacted for that purpose; provided, the State shall in no event be called upon by taxation or otherwise to construct irrigation works." Both documents were reprinted in the *Bakersfield Californian*, September 11, 1886.

dated August 17, 1886, from Paul Oeker, an agent of the California Immigration Association (which spoke for San Francisco's largest merchants) to Bernard Marks, the San Francisco land salesman and promoter who had served as a lobbyist for the Carr team. Both the *Bee* and the *Stockton Daily Independent* published the letter. Oeker noted that the cause was lost and that the "money spent already [in the legislature] would have built reservoirs in the mountains." He urged that the discredited state irrigation convention disband:

> Instead of this unfortunate combination which was weighed down by the firm of Haggin & Carr, local organizations should be formed. . . . Only a few ditch owners came to Sacramento to plead their cause, but no small irrigators were heard even by letter. What good has been all those 60,000 signatures obtained here from merchants and bankers? If you ask them to-day, ninety-nine out of every 100 will tell you they gave their signatures under a misapprehension. If these people had attached any other meaning to their name, they would have come in large delegations to Sacramento before the committee. The bankers and members of big firms, the Board of Trade and Chamber of Commerce of San Francisco, Los Angeles, San Diego, Stockton, Sacramento, etc. would have called special meetings to indorse your . . . cause, and in the face of that the Senate would not have dared to go back on their pledge. . . . Unfortunately, Messrs. Shorb & Carr and the committee stand alone in the fight with the majority of the press of the State against them and general distrust in their noblest motives. Unless you people can, in the eleventh hour of this ill-spent session patch up some possible compromise with their opponents and carry through some half-measure to make a beginning, nothing else will be left for Haggin & Carr but to either buy out Miller & Lux in Kern county, even at a sacrifice, which will come back to them, or make another sort of compromise with them for the water, so as to keep part of the same sure in any case. . . . Any Constitutional amendment ever indorsed by Haggin & Carr will be defeated at the polls. Any candidate for Judge indorsed by them will not be elected. Miller & Lux have again triumphed.

Oeker warned that proappropriation votes would fetch a very high price in the next legislature and suggested that the money would be better spent in buying a settlement with Miller and Lux. The legislature formally adjourned on September 12. Shortly before the legislators left Sacramento, they rescinded Section 1422 of the Civil Code which acknowledged the primacy of riparian rights. This was purely a symbolic act, and the actors in the drama knew it.[73]

The sordid special session did much to discredit the doctrine of appropriation and riparian rights but little to further state control. In pleading for his bill to create a comprehensive state irrigation system, Senator Days argued that

> where the public controls, through its constituted authorities, all the waters and great canals in the interest of the people, the rights of future generations are reserved, the cost of water to the consumer is infinitely less than when it passes through the hands of the feudal lord appropriators; the revenue goes to defray governmental expenses instead of into the pockets of said appropriators; there is less waste of water, more system, health is attended to and malaria prevented.[74]

Many Grange chapters supported such a system in the hope that the state could prevent monopolies and use general tax revenue to build dams and canals, thus reducing the cost of water to farmers.[75] Unfortunately, the vexing questions that had haunted partisans of a comprehensive state system in the past remained: How could such a system be paid for when irrigation was restricted to limited areas of the state, most of which were sparsely populated? How could sectional rivalries be surmounted in the legislature? How could the needs of farmers be reconciled with the interests of other water users? How could a statewide system pre-

---

73. As reprinted in the *Stockton Daily Independent,* September 10, 1886.

74. *Bulletin,* August 19, 1886.

75. The Grange chapters were virtually unanimous in calling for greater state control over the allocation and distribution of water. However, only a few chapters favored state construction of irrigation works. For samples of the many resolutions sent to Sacramento by local Granges during the special session, see the *Stockton Daily Independent,* July 28, 1886; the *Sacramento Bee,* August 2, 1886; *Call,* August 4, 1886; *Pacific Rural Press* 32 (August 7, 1886): 116; (August 21, 1886): 152.

Thomas Baker
One of many San Joaquin Valley
speculators who profited from the
state's prodigal land policies in the
first three decades after statehood.
*(Courtesy California State Library,
Sacramento)*

Dr. Oliver Wozencraft
His bold plan to tame the Colorado
Desert (Imperial Valley) was 40
years ahead of its time. *(Courtesy
California State Library, Sacramento)*

Will S. Green
He hoped that irrigation would lure family farmers into the Sacramento Valley and break up the baronial nineteenth-century wheat ranchos. *(Courtesy California State Library, Sacramento)*

John Bensley
A versatile San Francisco capitalist, he carried the vision of irrigated fields home from a trip to Chile and formed the San Joaquin and Kings River Canal Company in 1866. *(Courtesy Bancroft Library, University of California)*

William C. Ralston
President of the arid West's largest bank in the 1870s, the Bank of California, he was the first capitalist to propose building a comprehensive water system to serve the entire Central Valley. *(Courtesy California State Library, Sacramento)*

Robert M. Brereton
An English-born engineer, he designed the San Joaquin Valley's first major irrigation system in the early 1870s. *(Courtesy California State Library, Sacramento)*

George Davidson
Scientist and member of the federal irrigation commission of 1873-1874, he recognized the dangers as well as benefits of reclamation through private enterprise. *(Courtesy Bancroft Library, University of California)*

William Hammond Hall
A prophet before his time, Hall's farsighted efforts to promote an efficient, equitable allocation of the state's water won little support during his tenure as California's first state engineer (1878-1889). *(Courtesy Bancroft Library, University of California)*

James Ben-Ali Haggin
In the 1870s and 1880s, he built the largest irrigation system in the arid West, but Henry Miller blocked his attempt to secure a monopoly over the Kern River. *(Courtesy California State Library, Sacramento)*

William B. Carr
Master of political intrigue, "Boss" Billy Carr served as the architect and defender of the Haggin-Tevis Kern County empire. *(Courtesy Bancroft Library, University of California)*

Henry Miller
The "Cattle King" who skillfully manipulated the California legislature to protect his interests and rights. *(Courtesy California State Library, Sacramento)*

William Ellsworth Smythe
Philosopher of irrigation, champion of federal reclamation, and sometime proponent of water law reform. *(Courtesy California State Library, Sacramento)*

Elwood Mead
Like William Hammond Hall 20 years before him, at the turn of the twentieth century Mead recognized the need for California to modernize its water laws but lacked the political support to achieve his objectives. *(Courtesy California State Library, Sacramento)*

George Pardee
As head of the California Conservation Commission, he helped draft California's first comprehensive water code, approved by the legislature in 1913. *(Courtesy California State Library, Sacramento)*

Hiram Johnson
The famous Progressive governor put water law reform at the top of his legislative agenda in 1913.*(Courtesy California State Library, Sacramento)*

Robert B. Marshall
He kindled public interest in a comprehensive state water plan during the early 1920s. *(Courtesy Bancroft Library, University of California)*

serve some measure of local control over irrigation? And how could the irrigation works be built without providing a new arena for fraud and corruption?[76]

At the end of his long life, Henry Miller estimated that he had amassed $100,000,000 in property and spent $25,000,000 in legal fees defending his empire.[77] The amount he and his rivals spent on the California legislature from 1883 to 1886 must have been staggering. The two sides had tried to work out a compromise for years, but the final decision in *Lux* v. *Haggin* and the abortive special session helped produce an agreement by the end of 1886.[78] The pact was not formally ratified until July 28, 1888, when thirty-one ditch companies and fifty-eight individuals agreed to guarantee Henry Miller and his riparian neighbors exclusive use of the Kern River from September to February, and also from March through August when the stream carried less than 300 cubic feet per second. The remaining water was divided in the ratio of two-thirds to the Haggin interests, one-third to the riparian owners. To augment the existing supply, the two sides also agreed to share the cost of damming Buena Vista Lake and building new canals and levees. Arthur Maass has noted that in the wake of *Lux* v. *Haggin*, similar agreements were worked out between appropriators and riparian owners on the Kings River.[79]

76. For typical editorial reservations concerning a state irrigation system, see the *San Francisco Chronicle*, July 27, 1886; *Sacramento Daily Union*, July 28, 1886.

77. Lawrence, "Henry Miller and the San Joaquin Valley," 64.

78. The *Bakersfield Californian* of August 14, 1886, reprinted a long statement by Carr dictated at the headquarters of the Irrigation Convention's executive committee in Sacramento on August 2, 1886. Many efforts to compromise had already been made. As mentioned earlier, Miller and Lux first offered to give Haggin and Carr 75 percent of the Kern's water, but the two sides could not decide where to measure the flow. Miller and Lux subsequently suggested that Buena Vista and Kern Lakes be turned into storage reservoirs for use by the Kern County Land and Water Company, but this suggestion was refused because Carr thought the reservoirs would flood too much good farmland. According to Carr, his company offered to pay half the cost of building storage reservoirs in the mountains, at the headwaters of the Kern, exchange good grazing land for that owned by Miller and Lux in Buena Vista Slough, or build a storage reservoir at Tulare Lake for the use of Miller and Lux, but each offer was refused.

79. "Contract and Agreement between Henry Miller and others of the first part and James B. Haggin and others of the second part, July 28, 1888," Bancroft Library; G. H. Baldwin, "Water Rights on Kern River: Report for the California State Water Commission (1918)," appended to J. B. Lippincott's "Report on the

Once the agreement had been ratified, the Kern County Land and Water Company began to sell part of its vast holdings. During the 1880s, Haggin had been approached many times by European and American land agents who wanted to establish irrigation colonies in Kern County. In each case, however, they ultimately "declined to buy into law suits."[80] The first 7,000 acres of agricultural land was auctioned off in brisk selling in February and March, 1889. However, the largest land sales did not occur until after formation of the Kern County Land Company in September 1890. That company's historian claims that Carr opposed "any type of land sales," particularly colonization, and strongly resisted the sale of the rich land near Bakersfield served by the Kern Island Canal. His disagreements with S. W. Fergusson, a real estate promoter hired to take charge of colonization, over the high cost of preparing the land for new settlers, led the boss—now sharply reduced in political power and influence—to sell out to Haggin in 1889 or 1890. Lloyd Tevis became the first president of the new land company, and Haggin took one of the four board of directors' chairs.

The new company established four irrigation colonies divided into 10- and 20-acre farms covering 45,000 of the firm's 300,000 to 400,000 acres. The land sold for $60 to $100 an acre, and settlers had ten years to pay for their farms. Among other amenities, the promoters established four experimental farms to help farmers unfamiliar with the techniques of desert agriculture. However, the land company continued to operate fourteen ranches and cultivated 97,000 acres—most of it alfalfa needed to feed the company's 40,000 cattle, 40,000 sheep, and 2,000 horses. Still, the company suffered from the economic depression of the 1890s and from the competition for settlers provided by the large number of irrigation districts formed in the late 1880s and early 1890s. While land monopoly in Fresno County stimulated agricultural growth and diversification

---

Miller and Lux Ranch, Southern Division on Kern River, California," File 36–1, J. B. Lippincott Collection, California Water Resources Archives; *Visalia Weekly Delta,* December 30, 1886; Treadwell, *The Cattle King,* 94; Harding, *Water in California,* 39; Maass, . . . *and the Desert Shall Rejoice,* 372–374. The Miller-Haggin agreement was embodied in a court decree issued in August, 1900.

80. *Bakersfield Californian,* August 23, September 13, 1884; June 12, 1886. Also see Paul Oeker to Shorb, May 18 and 31, 1886, Incoming Correspondence, 1886, Shorb Collection, Huntington Library.

during the 1870s and early 1880s, colonization failed in Kern County. Twenty years after formation of the Kern County Land Company, the elaborate network of canals built and bought by Haggin and Carr—capable of irrigating 200,000 acres—served only 116,500 acres. Twelve thousand were pasture, 70,000 were planted to alfalfa, 34,000 to grain, and only 2,000 acres produced fruits and vegetables.[81]

The effects of *Lux* v. *Haggin* were profoundly felt throughout the arid West. Despite litigation over riparian rights and the supreme court's recognition of the supremacy of riparian rights, the area of irrigated land increased four or five times during the 1880s. Virtually all California historians, including those who have specialized in the water story, condemn riparian rights. For example, the riparian doctrine has been described as "a serious obstacle to the development of irrigation in the West" and a "potential death sentence" on nonriparian rights.[82] Such judgments are unfair and encourage historians to portray appropriation as a distinctly preferable alternative. Carey McWilliams observed in 1935: "The doctrine of appropriation was obviously the fairest and most economical and the fullest use of an inadequate water supply. It was based on an equitable idea and a practical consideration." McWilliams, a passionate enemy of land monopoly and agribusiness, ignored the fact that the doctrine of appropriation offered little protection to small farmers in most parts of the state because water rights were not attached to the land or restricted to "reasonable" (not just "beneficial") use. Litigation was inherent in the whole system of water rights in California; it was not exclusively a by-product of riparian rights. Haggin and Carr used the courts to destroy rival appropriators who refused to sell out to the Kern County Land

81. *Bakersfield Californian*, January 26, February 9, 16, March 22, 23, 24, 1889; Berg, *A History of Kern County Land Company*, 18, 21–22; William Ellsworth Smythe, "A Study of Two Modern Instances," *Irrigation Age* 5 (Sept. 1893): 112–113; Cone, "Irrigation in California," U.S.D.A., Office of Experiment Stations, Bulletin no. 237 (Washington, D.C., 1911), 53. During the 1890s, despite the nationwide depression, irrigated land in Fresno County increased from 105,665 acres to 283,737 acres. At the same time, Kern County's irrigated land decreased from 154,549 to 112,533 acres. See William L. Preston, *Vanishing Landscapes: Land and Life in the Tulare Lake Basin* (Berkeley, 1981), 97.

82. The first quote is from J. A. Alexander, *The Life of George Chaffey* (Melbourne, 1928), 35, and the second from Irwin Cooper, *Aqueduct Empire* (Glendale, Calif., 1968), 409.

and Water Company just as riparian owners used their rights to beat down diverters. Each doctrine promoted monopoly. The climatic and geographical diversity of the state, the variety of uses to which water was put, the substantial economic and political power of land and water companies, the speculative nature of California agriculture, and the inadequate legal system all insured that California would produce an abundant crop of water suits with or without the riparian doctrine.[83]

Neither riparian rights nor appropriation provided an ideal system of water rights, but the riparian doctrine had some clear advantages. Prior to the construction of large storage reservoirs in the West during the first decades of this century, the surface water supply irrigated no more than a small fraction of potential farmland. The most productive land was usually the alluvial soil adjoining rivers; and the irrigation of this riparian land did not require long canals (which were expensive and exposed the limited water supply to great losses through evaporation and seepage). In an 1872 Montana case, that territory's chief justice argued that water in Montana ought to belong to the land as it did in the humid half of the nation: "Is it not the true policy of this Territory to erect such a system of laws here as shall distribute our short supply of water to the best advantage to all our people? The common law applied to this country is ample and sufficient to secure this much desired end." Thirty-one years later, the Nebraska Supreme Court similarly declared: "we doubt whether a more equitable starting point for a system of irrigation law may be found [than the riparian doctrine]."[84]

Riparian owners could monopolize water as a class, but they did not own water as property. The common law assumed that water was too precious a resource, and subject to too many different uses, to be sold and traded like land or precious metals. So the riparian doctrine recognized no priorities at all. When conflicts arose, riparian owners usually settled their differences out of court,

83. Carey McWilliams, *Factories in the Field: The Story of Migratory Farm Labor in California* (Santa Barbara, Calif., 1971), 33.

84. Thorp v. Freed, 1 Mont. 651 (1872); Crawford v. Hathaway, 67 (Neb. 365 (1903). Also see Benton v. Johncox, 17 Wash. 277 (1897).

through informal agreements. During the first few decades after statehood, prior appropriation was a useful economic tool because it attracted investors and developers. However, it proved less suited to the needs of densely settled rural communities where justice demanded that the water supply provide the greatest good to the greatest number. Theoretically, prior appropriation allowed the first water user to claim the entire supply within the basin, though in California this was limited by riparian rights. Ironically, then, the system of water rights best suited to the 1850s was anachronistic in the 1880s. To serve the largest number of water users, each doctrine required modification, and some modifications were enacted. For example, towns and cities could condemn both appropriative and riparian rights, a privilege extended to irrigation districts by the Wright Act in 1887. But the legislature never seriously considered modifying the riparian doctrine to fit conditions in California. The courts expanded the doctrine to permit riparian owners to irrigate, even when they reduced the flow of a stream, but never tried to stretch the definition of *riparian* to include all the land within a river basin. This was one way to eliminate the evils of both systems, even though such a change would have required a complete overhauling of water laws. Since all riparian rights were "correlative" rather than absolute, the legislature clearly underestimated the riparian doctrine's potential to serve as the foundation for a system of public control.

The effects of *Lux* v. *Haggin* were felt far beyond the borders of California. The state had operated under a dual system of water rights since the 1850s, but the 1886 decision sanctified that arrangement. The supreme court had not overturned the doctrine of appropriation or denied its applicability to California; in fact, *Lux* v. *Haggin* gave appropriative rights acquired before the acquisition of government land by riparian owners clear precedence. But to justify the primacy of riparian rights—which had not existed in California before 1850—the court reasoned that the doctrine had originated in the federal government's original sovereignty over the public domain. As the "original proprietor," the nation had owned not just land, or the minerals under the soil, but also the water. In the West, natural conditions reinforced this interpretation because most of the agricultural land was worthless without water. When the nation transferred land titles to the states, as in the swamp tracts acquired by Miller and Lux, title to water passed along with title to the land as a condition of landownership. Later, when the national government sanctioned and encouraged appropriation, it did not extinguish its original title to the water. By the turn of the twentieth century, this interpretation of the origin of western water rights became known as the "California Doctrine," and it prevailed in Kansas, Montana, Nebraska, North and South Dakota, Oregon, Texas, Washington, and Oklahoma, as well as in California. The California Doctrine built directly on the assumptions in *Lux* v. *Haggin*.

By contrast, seven states—Colorado, Utah, Wyoming, Arizona, New Mexico, and Idaho—followed the "Colorado Doctrine." This legal theory assumed that all water rights derived from the individual territories or states. Under the California Doctrine, appropriation was rooted in acts of Congress (the laws passed in 1866, 1870, and 1877). However, under the Colorado Doctrine, when the federal government acquired new territory—or at least at the time of the formation of new territories and states—full sovereignty over water passed to the states. Proponents of this idea claimed that Congress had implied the transfer of sovereignty by giving the states administrative control over the acquisition of water rights, and explicitly recognized state sovereignty by ratifying the constitutions of Colorado and Wyoming, both of which contained provisions declaring state ownership of water.[85]

Historians need to study the influence of *Lux* v. *Haggin* on the evolution of western water laws. Five of the nine states that follow the California Doctrine entered the Union after 1886: Montana, North and South Dakota, and Washington in 1889, Oklahoma in 1907. With the exception of Montana, each of these states, along with California, Kansas, Nebraska, and Texas, has humid as well as arid sections. This climatic diversity, not shared in such states as Nevada and Arizona, helps explain the popularity of the dual

---

85. The California and Colorado doctrines have generated an enormous body of legal literature. For a good summary of the two legal theories see C. S. Kinney, *A Treatise on the Law of Irrigation and Water Rights and the Arid Region Doctrine of Appropriation of Waters* (San Francisco, 1912), 2: 1093–1124.

system of water rights. Whether the evolution of western water law would have followed the path it has even without *Lux* v. *Haggin* is an open question.

Of course, all of this was far in the future in 1886. After the comic special session, filled with the kind of corruption, bargaining, and deceit that confirmed the worst fears and expectations of the state's voters, newspapers were virtually united in their condemnation of large land and water companies as well as of riparian owners. They turned to the irrigation district with new hope that the state's farmers could find a way to pay for irrigation themselves. The great days of private water companies were over.

9

The Wright Act, 1887–1897: Promise Unfulfilled

Lux v. *Haggin* left irrigators and land and ditch companies with several choices. By refusing to hear the case in 1879, northern California's federal court implied that water rights disputes were matters for the states to settle. And since the state supreme court's opinion was well argued, exhaustive, and conclusive, no attempt was made to challenge the judgment before the United States Supreme Court. As mentioned in the last chapter, many appropriators settled with riparian owners out of court. However, nonriparian irrigators had three other choices. First, they could push for construction of storage reservoirs in the mountains, a particularly popular choice in the late 1880s when some Californians thought that congressional appropriations for the survey of reservoir sites and designation of irrigable land would be followed by federal construction of dams. As the *Visalia Weekly Delta* noted in 1885: "And so long as there continues a limited and insufficient supply, so long will there be discontent, ill-will and strife; and no law, and no administration of law can prevent it. The surplus water of wet years must be stored and distributed as needed; and in this feasible and practical plan, rests not only the easiest and readiest solution of present difficulties, but also the indefinite increase in population and wealth of the whole San Joaquin Valley." This hope depended not just on finding the money to build dams but also on the legal assumption that "flood water" could not be claimed by riparian owners as part of normal stream flow.[1] The second option was the use of underground water. In the 1880s some southern California and San Joaquin Valley farmers discovered that pumping water often cost less than constructing and maintaining irrigation ditches. Some boosters claimed that artesian wells could be found almost anywhere—at the right depth—and that this source of water would one day virtually eliminate the reliance of irrigators on surface streams. Underground water was purer, not subject to seasonal fluctuations in volume, and seemingly inexhaustible. Most important, no laws regulated its use. The farmer could draw as much water as he wanted when he wanted with no fear of litigation.[2] Both of these alternatives promised to sidestep water conflicts by tapping largely unused sources. But the third option, the irrigation district, offered an entirely different choice. It promised to redistribute the existing supply—and provide a bonanza of economic opportunities for speculators, bond investors, and businessmen, as well as farmers.

When the California legislature met in 1887, irrigation and water rights attracted far less attention than in 1885 or 1886. Since the Miller-Lux and Haggin-Carr forces had resolved their differences, the irrigation question was overshadowed by the offer of hydraulic mining companies to build debris-restraining dams if the legislature would limit or waive their responsibility for property damages.[3] Still, Stockton's *Independent* warned that "unless some-

1. *Visalia Weekly Delta,* July 16, 1885. Also see the *Bakersfield Californian,* April 17, 1886. The *Sacramento Daily Union* led the fight for construction of storage reservoirs in 1886.

2. See, for example, the *Visalia Weekly Delta,* January 26, 1884; March 10, 1887; Richard J. Hinton, *Irrigation in the United States,* 49th Cong., 2d sess., 1887, S. Misc. Doc. 15, (serial 2450), 16. Not until 1903, in Katz v. Walkinshaw, 141 Cal. 116, did the state supreme court provide a rule of law to settle disputes between rival underground users. It simply extended the riparian doctrine to include subterranean water. The court ruled that each landowner deserved a "fair and just proportion" of the supply, and that local courts could issue injunctions if the greed of one user led to "irreparable and substantial injury" to his neighbors.

3. On the debris controversy and the legislation considered by the 1887 legislature to correct the evil, see the *San Francisco Chronicle,* January 18, February 4, 6, March 5, 1887. The senate killed the debris bill on March 4. On March 19, 1887, the *Mountain Messenger* of Downieville charged that large landowners in the Sacramento Valley, almost none of whom irrigated their land, paid $1,500 a vote to block the legislation.

thing is done of an immediate and practical character bankruptcy will follow in hundreds of instances in sections [of the San Joaquin Valley] where irrigation is needed and demanded. Capital will not seek investment unless some law is passed to protect it in building ditches so long as the Supreme Court decision stands as it now does." Four hundred and twenty landowners in Stanislaus County urged the legislature to approve an irrigation district bill introduced by Assemblyman C. C. Wright, who had gone to Sacramento pledged to the proposed legislation. Their petition argued that "the local character of this bill, its recognition of existing rights, and its provision for the equal distribution of the waters, where equal burdens are borne and equal benefits conferred, renders the system particularly adapted to our wants, and as we believe equally adapted to other localities subject to irrigation." The *San Francisco Chronicle* came out in favor of the Wright Bill when it discovered that Haggin and Carr, as well as Miller and Lux, opposed the legislation:

> Whatever may be the defects of the Wright bill, it will have the effect of shutting off all schemes for the wholesale seizure of the running water of the State under any claim of law whatever. If it will accomplish this, as it will do, it will preserve the water for the use of those who need it, and the next Legislature, having had two years experience under the system of this bill, can remove any crudities that may be found in it. . . . The Wright bill is all that stands between the rich water monopolist and the poor farmer.

The measure passed the assembly by a vote of sixty-five to zero on February 18, 1887, and though it was slightly modified to meet the objections of the riparian block in the senate, easily won approval in the upper house on February 28. On March 7, after word reached the assembly chamber that the governor had signed the legislation, applause swept over the floor. The lawmakers subsequently approved a companion bill to give superior court judges the power to deny or suspend injunction suits against appropriators if the defendant posted an indemnity bond. This law was designed to prevent riparian owners or appropriators from using injunction suits to block the construction of district irrigation works.[4]

The Wright Act was conservative in tone, carefully drafted, and thorough. Part of its appeal was the close and obvious kinship it bore to the familiar, localized 1872 and 1876 district laws. None of its provisions were revolutionary, though the statute provided a much more detailed description of the process of district elections, the issuance of bonds, the collection and assessment of taxes, and the nature of the condemnation power, than any bill proposed from 1872 to 1885. Wright assumed, as had the draftsmen of the West Side Act, that irrigation districts were "political divisions" in which property qualifications for voting were prohibited by both the state constitution and the civil code. So while the new law required at least fifty "freeholders" to petition their county board of supervisors before an election to form a district could be scheduled, all eligible voters could participate in the elections themselves, whether to select district officials or to issue bonds. A two-thirds vote was required to form a district, but the initial bond issue required the approval of only a simple majority. As in the 1876 statute, the Wright Act required districts to tax *all* property within their boundaries, including town lots and buildings; taxes would be levied according to property values, not according to specific benefits received by individual parcels of land. Tax revenue could be used to pay any district expenses, including the purchase of land, existing irrigation works, or riparian rights. The bonds could not be sold at less than 90 percent of face value.

Each district would be governed by a board of five directors representing different geographical sections within the district. Their responsibilities included purchasing or condemning water rights and rights-of-way, supervising construction of dams and canals, and distributing the water supply. All water would be apportioned

4. The first quote is from the *Stockton Daily Independent*, February 15, 1887; the second from "Petitions of Citizens of Stanislaus County urging the Passage of Assembly Bill No. 12 Relating to Irrigation," of the California Senate and Assembly (hereafter *JCSA*), 27th sess. (Sacramento, 1887), Appendix, vol. 8; and the third from the *San Francisco Chronicle*, January 17, 1887. Also see the *Chronicle* of January 15, March 14, 1887; *Stockton Daily Independent*, March 1, 8, 1887; *Los Angeles Times* of February 23, 1887; *Visalia Weekly Delta*, February 24, 1887; *Tulare County Times*, March 10, 17, 1887. For the anti-injunction law see *California Statutes*, 1887, 240.

according to the ratio of individual taxes to the district's total tax burden. However, no water rights were absolute. In theory, the water supply belonged to the district, not to individual water users. The questions of when a settler took up his land, or where the land was located in relation to the water supply, did not matter. Nevertheless, the board's authority was strictly limited at critical points. For example, the established courts, not the board or some other administrative-judicial tribunal, would determine compensation in condemnation suits, and the board could not condemn any water rights, dams, or canals, owned by miners or mining companies. Nor could any district's diversions threaten shipping in a navigable stream.[5]

The state had no control or influence over the designation of districts, the construction of irrigation works, the certification or issuance of bonds, the acquisition of water rights, or the distribution of water. The law exalted the principle of local control and served as a monument to sectionalism and the power of special interest groups; it was not a bold reform. Nevertheless, initially much was expected of the Wright Act. The *Tulare County Times*

5. *Cal. Stats.*, 1887, 29. C. C. Wright explained the new law in Richard J. Hinton, *A Report on Irrigation* [1891], 52d Cong., 1st sess., 1893, S. Ex. Doc. 41, (serial 2899), 95–97, and *The Irrigation Age*, 2 (February 1, 1892): 446–448. For a discussion of the Wright Act's basic features, see Frank Adams, *Irrigation Districts in California, 1887–1915* (Sacramento, 1917), 8–9.

Four other irrigation district bills were proposed to the 1887 legislature. A.B. 287 (Vincent), January 26, 1887, *Assembly Bills, 1887*, vol. 3, resembled the Wright Act. However, it restricted taxes to irrigable land and required the approval of two-thirds of all landowners to carry a bond election. A.B. 83 (Butler), January 11, 1887, *Assembly Bills, 1887*, vol. 1, was identical to A.B. 71 and S.B. 38 considered in 1885. It provided for the state engineer to set district boundaries but required the approval of the owners of half the property within the district before bonds could be issued. Perhaps the most "democratic" district bill was S.B. 73 (Langford), *Senate Bills, 1887*, vol. 1. It required a petition from the owners of more than half the acreage within a proposed district to form a district, and restricted the vote in all district elections to landowners, in proportion to the ratio of the size of their landholdings to the total land area of the district. Each of the above bills proposed locally controlled districts. The only state plan was A.B. 64 (Bost), January 11, 1887, *Assembly Bills, 1887*, vol. 1. This bill was virtually identical to Bost's A.B. 464 proposed in 1881. It required the state engineer to set district boundaries, and provided for a state board to supervise the construction of irrigation works, following the approval of plans by a local review board. After the works had been completed, the state would regulate water distribution. Works would be paid for from the proceeds of a uniform tax on all district property.

declared that irrigation districts bolstered "the fundamental principle of American civilization, the sovereignty of the people. . . . The irrigation districts of California will ever be the home of a free, independent, people, true Americans, brave and patriotic in all their instincts." Self-rule and local ownership of water would reaffirm American values by preventing monopoly. In April, 1890, the chairman of the House Committee on Arid Lands echoed a widespread assumption among irrigation district promoters when he declared that the new law "practically overruled" the *Lux* v. *Haggin* decision by recognizing the right of districts "to take water and use it irrespective of riparian rights." Richard J. Hinton of the U.S. Agriculture Department's irrigation office applauded the Wright Act's commitment to public ownership and distribution of water, declaring that it had "unquestionably caused a cessation of litigation over water rights and prior appropriations."[6]

The legislature's unanimous approval of the Wright Act acknowledged that the law offered something to everyone. The district promised to integrate water delivery systems owned by private companies within its boundaries into one coordinated whole, preventing waste through consolidation; the state's water companies had been notoriously inefficient, both in building and operating delivery systems. Even more important, the district would provide irrigators with water at cost, without discriminating among individual users. Large landowners and speculators, long regarded as the bane of California agriculture, would be encouraged to sell their land to small farmers. Both the positive incentive of selling land at great profit to new settlers and the negative incentive of avoiding rapidly increasing taxes on tracts held solely for grain farming or speculation would encourage intensive agriculture and the subdivision of large estates. Wheat would give way to fruit, and baronial ranchos to family farms, not through visionary laws to kill land monopolies, but through a natural, perhaps even inevitable, economic process. Urban as well as rural residents would benefit from the migration of new farmers into the state, as the tax base

6. *Tulare County Times*, September 18, 1890; "Ceding the Arid Lands to the States and Territories," 51st Cong., 2d sess., 1891, H. Rep. 3767 (serial 2888), 189; Hinton, "Irrigation in the United States," 51st Cong., 2d sess., 1891, S. Ex. Doc. 53, (serial 2818), 67.

expanded. Of course, many private land and water companies hoped to profit by selling unprofitable ditches or by constructing new works for fat profits.

Irrigation promoters wasted no time putting the Wright Act to the test. In the eight years following 1887, they organized forty-nine districts covering about 2,000,000 acres of land, 2 percent of the state's area. However, only twenty-four actually issued bonds, for a total debt of $18,000,000. Hinton estimated that among the districts formed from 1887 to 1890, the bonded debt averaged $4.85 an acre in the Sacramento Valley, $5.55 in the San Joaquin Valley, and $32.58 in southern California. The smallest debt was $2.54 an acre in Yuba County's Brown's Valley District; the largest was $83.33 an acre in San Bernardino County's East Riverside District. Four districts were organized in 1887; seven in 1888; six in 1889; eleven in 1890; thirteen in 1891; three in 1892; four in 1893; and the last, fittingly called the Amargoza, in 1895. Southern California spawned thirty districts, thirteen in Los Angeles County, thirteen in San Diego County, and four in San Bernardino. North of the Tehachapis, Tulare and Colusa counties each produced five districts, Fresno four, Stanislaus two, and Yuba, Shasta, and Kern counties one each. The preponderance of districts in southern California reflected more than that section's greater need for irrigation or devotion to citrus farming. Southern California's real estate boom, which lasted from the late 1870s to the late 1880s, had begun to fade at about the same time the Wright Act appeared. Real estate sharks welcomed the new opportunity to sell land, and southern California became the home of the state's most speculative and ephemeral ventures.[7]

7. Adams, *Irrigation Districts in California*; Ray M. Gidney, "The Wright Irrigation Act in California" (M.A. thesis, University of California, Berkeley, 1912); Orson W. Israelsen, "A Discussion of the Irrigation District Movement" (M.S. thesis, University of California, Berkeley, 1914); *Annual Report of the State Board of Horticulture of the State of California for 1892, JCSA*, 30th sess. (Sacramento, 1893) Appendix, vol. 3; *Eleventh Report of the State Mineralogist. Two Years Ending September 15, 1892*, vol. 9; *Fifth Biennial Report of the Department of Engineering of the State of California, December 1, 1914, to November 30, 1916, JCSA*, 42d sess. (Sacramento, 1917), Appendix, 5: 129–130; F. H. Newell, *Report on Agriculture by Irrigation in the Western Part of the United States at the Eleventh Census, 1890*, pt. 20, 1896, H. Mis. Doc. 340 (serial 3021), 37–40; *Thirteenth Annual Report of the U.S. Geological Survey, 1891–1892: Part 3—Irrigation* (Washington, D.C., 1893), 146; and Hinton, *A Report on Irrigation [1891]*, 28. Gidney provides good brief histories of the individual districts.

Irrigation Districts of California, ca. 1892.

At first, many Californians considered the Wright Act a great success. In September 1890, the *Pacific Rural Press* reported that district bonds were selling for 90 to 96 percent of par, and at the end of that year San Francisco's *Chronicle* remarked, "the bonds of a number of districts have been issued and disposed of on favorable terms and work has been commenced in the construction of canals. These bonds are so well thought of as securities that they were taken up by capitalists without the heavy discount too frequent in similar affairs." The *Chronicle* added that while the average cost of irrigation would not exceed $2.50 an acre, irrigation would drive up land values by $50 to $200 an acre. In his address to the legislature at the beginning of 1891, Governor H. H. Markham called the results of the Wright Act "favorable and encouraging."[8]

Even as the governor spoke, however, many of the districts grappled with formidable problems. Colusa County's Central District provides a useful case study. The largest of seven districts organized in the Sacramento Valley, it was the child of Will S. Green, the newspaperman and speculator who had been promoting irrigation in the valley since the drought of 1864. Immediately after the governor signed the Wright Act on March 7, 1887, Green began to drum up support for a district adjoining the Sacramento River.[9] In 1884, as Colusa County surveyor and the Stony Creek Canal Company's leading promoter, Green had surveyed a forty-mile canal between Stony Creek and Cache Creek—essentially the same aqueduct he first proposed in 1864. The newspaperman-booster estimated the cost of the main canal and feeder lines at $600,000, and promised that this network could supply 265,200 acres with water. Even adding $260,000 as a contingency fund, the cost of irrigation would average only $3.24 an acre. At a public meeting held in Maxwell on March 26, a three-man committee was chosen to determine precise district boundaries and secure the

---

8. *San Francisco Chronicle*, December 29, 1889; *Pacific Rural Press* 40 (September 20, 1890): 257; *Inaugural Address of Governor H. H. Markham, Delivered January 8, 1891, JCSA*, 29th sess. (Sacramento, 1891), Appendix, vol. 1.

9. For example, on March 25, 1887, Green wrote to J. De Barth Shorb: "If we get our irrigation schem [sic] started there is a big chance to make some money in land. I know a tract of 1,800 acres on the [Sacramento?] river . . . that can be had at $45 an acre. I want you to come up this spring." Incoming Business Correspondence, Part 2, Box 3, Shorb Collection, Huntington Library.

signatures of fifty landowners necessary to put the organization of a district before the voters. The Sacramento Valley was so thinly populated that most of the sixty-four signatures collected belonged to owners of town lots in the villages of Maxwell and Williams.[10]

In 1887, virtually all the land in the proposed district was planted to grains; the average farm within Green's proposed district contained 870 acres, and the forty largest landowners held estates that averaged 2,225 acres apiece. Aside from residents of Maxwell and Williams, the district contained only 260 voters who lived on the land; most of these people were unmarried farm laborers.[11] The nonresident wheat barons used every conceivable argument to scuttle the district. They complained that irrigation would produce poor fruit and disease and would saturate valley soils with alkali. They complained that the cost of irrigation would kill the wheat and barley industries. They complained that the district had been improperly organized for many reasons, but especially because not all land could be irrigated from a single water source, as required by the Wright Act. And, most of all, they complained that landless voters and town dwellers could carry any district election. On October 19, 1887, "Pioneer" wrote the *Weekly Colusa Sun* expressing the frustration shared by most of the valley's largest farmers:

> That the irrigation law is a blow aimed directly at large land holders is as apparent as a nose on the face. In Colusa county there could not be formed an irrigation district of any considerable dimensions without including one or more small burghs or towns. Now what we want to know is this. Is it right for the many men of small holdings who generally hang

10. *Weekly Colusa Sun*, March 26, April 2, 1887; *Sacramento Daily Union*, March 28, 1887. The best overall surveys of the Central Irrigation District are in Adams, *Irrigation Districts in California*, 11–14; Joseph A. McGowan, *History of the Sacramento Valley* (New York, 1961), 1: 389–401; John P. Ryan, "Notes on Will S. Green, Father of the Glenn-Colusa Irrigation District," unpublished MSS, Bancroft Library.

11. Adams, *Irrigation Districts in California*, 11. The census of 1890 revealed that less than 1 percent of Colusa County's land was irrigated. In Fresno County 7.69 percent of farmland was irrigated, in Los Angeles County 9.23 percent, and in Kern County 21.7 percent. The mountainous counties of Sierra, Mono, Modoc, and Inyo had the highest percentage of land "under ditch." Newell, *Report on Agriculture by Irrigation in the Western Part of the United States at the Eleventh Census, 1890*, 41.

around those little villages and the men with no holdings at all except a cigarette holder, to waltz up to the polls on election day, and cast their vote, and thereby become the dictator to the man with his thousands of acres of land? There is only one way to construe the matter. It places the whole army of men with small holdings, the laborer, the tramps and the paupers on one side and the landlords with their thousands of acres on the other. And the former say to the latter, "we will build an irrigation ditch here or there as we please and we'll make you foot the bills."[12]

A tense atmosphere prevailed in Colusa County as the November district election drew near.

Green conceded that the Wright Act should have restricted participation in district affairs to landholders. But construction of his canal depended on winning the "spite vote" of landless tenants and farm workers, as well as the "pocketbook vote" of small town merchants who hoped to profit from the new farmers a district would attract. Most of Colusa County's land barons did not behave as Green expected them to. Some doubtless hoped to profit from the appreciation of land values, but they also feared that a wave of new settlers would erode their political power and drive up taxes. Then, too, wheat ranches were not solely an economic investment. In nineteenth-century California, they were "badges" worn by the state's landed gentry. Although wheat was a highly speculative crop, the wheat ranch was part of a life-style as well as a source of wealth.[13]

On November 22, 1887, Colusa County residents approved the formation of the Central Irrigation District by a vote of 271 to 51. The district included 160,000 acres on the west side of the

12. *Weekly Colusa Sun,* October 29, 1887. Also see the editorial reprinted from the *Willows Democrat* in the *Sun* of September 8, 1888.

13. The wheat industry in California deserves a book-length study. Historians particularly need to examine the sociology of wheat farming. Where did the wheat ranchers come from and when did they enter the state? What occupations did they hold before taking up their farms? Did they see themselves as a landed aristocracy in the same sense as large farmers in Europe? Were they challenging the model small family farm that prevailed in the Midwest and New England? What did the wheat rancher think of the family farm in California? These and many other questions concerning the wheat industry need to be answered.

Sacramento River in Glenn and Colusa counties, only 60 percent of the acreage initially proposed by Green. The new boundaries were probably drawn to exclude the lands of some disgruntled large wheat farmers. All five of the original directors were residents of Willows, Maxwell, or Colusa; whether they had any practical experience as farmers, let alone as irrigators, is uncertain. In any case, they named C. E. Grunsky, a hydraulic engineer with extensive experience designing canal systems for Miller and Lux as well as for Haggin, Tevis, and Carr, as chief engineer. Following the election, the *Sacramento Daily Union* predicted: "Soon Colusa county will come to the front as the home of thousands of small farmers." Green himself urged the district's opponents to fall into line: "With a five-to-one sentiment against you, you must know that the thing will come, and if you delay it on legal quibbles it is only giving trouble to yourselves and to others. . . . If it is a success you will reap the reward."[14]

The Central District, along with virtually every other district, faced many court challenges. Some suits questioned the constitutionality of the Wright Act; others charged district promoters and officials with violating or ignoring provisions of the new law. Nevertheless, by March 1888, Grunsky completed his engineering surveys. He estimated the average cost of watering district lands at $4.81 an acre—which sum included a liberal allowance for acquiring rights-of-way and fighting legal battles. On April 2, 1888, a $750,000 bond issue carried by a vote of 189 to 36, and the first $100,000 in bonds went on sale in July. Legal obstacles delayed the beginning of construction until November 9, 1889, when Will Green, in a modest ceremony, broke ground for the aqueduct. On November 16, the newspaperman exuberantly predicted that the project would be completed quickly, again assuming that the greed of large landowners would ultimately overcome their objections: "[The work] will go on because everybody can see that the land of Central District has advanced more in value since the letting of the [construction] contract than the entire cost of the work."[15]

14. *Sacramento Daily Union*, November 23, 1887; The *Weekly Colusa Sun*, November 26, 1887. The *Union* reported the results of the election as 310 for and 52 against.

15. *Weekly Colusa Sun*, February 18, March 10, April 7, and June 9, 1888; November 16, 1889; *Alta California*, November 12, 1889.

By February 1890, however, Green's optimism had faded. The drought of 1889 drove many small farmers out of the Sacramento Valley, and most of the abandoned farms were snatched up by the wheat barons. The newspaperman warned that the concentration of wealth in the hands of an elite helped destroy the Roman Empire, and drew a parallel:

> Where, o where are we drifting in America? Look close around you and see what is going on! 180 square miles with but a single school census child! Shall we open [the] Central canal or will the great landlords find means to shut it up? The canal may be stopped up, our rich plains given up to the few, and finally fall into decay and ruin, and those who have promoted it fill unknown graves, but some future Napoleon will read a lesson from it.[16]

Yet criticism of the district continued and intensified. By the early months of 1890, Joseph A. Sutton, a former chairman of the Central Irrigation District's Board of Directors, complained that the district had been badly managed, particularly that construction costs had been padded. Green and Sutton engaged in a running battle during 1890 and 1891. At one point, Sutton described the newspaperman and his "lieutenants" as men "who by their blundering have rendered the business a failure and disastrous." Green responded in kind: "Since Satan sought to rule in Heaven and became the king of hell he has not had a more complete parallel [than Sutton]."[17]

The criticism of district officials was partly warranted. The Wright Act prohibited the sale of district bonds at less than 90¢ on the dollar, discouraging potential investors, bankers, and bond houses, especially given the cloud of litigation that hung over the districts during the late 1880s and early 1890s. Put simply, the bonds were too risky at the price. After failing to sell their securities on the open market, many districts used them almost like depreciated currency to pay off contractors willing to speculate. Accordingly,

16. *Weekly Colusa Sun*, February 8, 1890.

17. *Weekly Colusa Sun*, March 8, 1890; September 26, 1891. The quotes are from the latter issue.

construction companies fattened up bids to include a suitable "discount." Such an arrangement could not last indefinitely, and by 1891 the San Francisco Bridge Company wisely refused to take any more of the Central District bonds. The company suspended work on the canal after completing forty of sixty-one miles. Unfortunately, the forty miles were not continuous and no headgates had been constructed.[18]

Many additional costs were beyond the control of district officials. Of the three major gaps in the completed canal, two were within estates whose owners bitterly opposed the aqueduct; one stretch of 6.5 miles crossed the Glenn ranch, and a 2.5-mile section ran through the Glide ranch. Since the Glenn estate had asked the exorbitant price of $50,000 for the right-of-way, district officials sued to condemn the land. But a sympathetic jury awarded the Glenn trustees $33,000 for the parcel, which Green claimed was four times the prevailing price for comparable land in Colusa County. In the summer of 1891, an association formed by the irrigation districts to help make irrigation district bonds a more attractive investment hired William Hammond Hall to survey the plans of each district and issue an "impartial" report on its chances of success. Hall reported that the Central District had fine soil, sufficient water, and well-designed irrigation works. He predicted that the cost of the district's water system would ultimately reach $950,364.25, but maintained that the difference between this sum and the original $750,000 bond issue was due to the cost of litigation and condemnation rather than fraud or inept management.[19]

Will Green acknowledged the district's failure as early as March 1892, but the final blow came in October 1893. The California Supreme Court discovered that some of the sixty-four petitioners who had appealed to the board of supervisors to form a district owned no land; hence the district had not been legally organized. Al-

18. Adams, *Irrigation Districts in California*, 13; *Weekly Colusa Sun*, September 21, 1891; *Sacramento Daily Union*, August 17, 1891. Central District voters approved an additional $250,000 bond issue in 1891, but the securities did not sell.

19. *Weekly Colusa Sun*, December 6, 1890; August 8, 1891; *The Irrigation Age* 1 (October 15, 1891): 234. William Hammond Hall investigated at least three districts, the Allessandro and Perris as well as the Central. See Wicks and Phillips (irrigation district bond dealers), *Irrigation Bonds: Their Security, Certainty, Desireability* (San Francisco, 1891), a pamphlet in the Bancroft Library.

though the court did not dissolve the district, it prohibited the sale of any more bonds. Of the $574,000 in securities already in circulation, 80 percent had been taken by the San Francisco construction company and the remainder by farmers and land speculators within the district. The bonds had attracted virtually no "outside" investors.[20]

The failure of the Central Irrigation District and the tumbling price of wheat during the 1890s turned Colusa County into a wasteland; the county's already scanty population declined by 50 percent during the 1890s. The San Francisco *Call* commented in 1903:

> Litigation commenced early in the history of the district and has continued until the present time. Owing to the litigation and to the bonded indebtedness incurred for the work already performed on the ditch, sales of lands in the region have been impossible and parties owning large tracts, containing thousands of acres, have found it impossible to divide them into smaller parcels and make sales thereof, so the population has not increased.

Similar conditions prevailed within defunct irrigation districts throughout the state. Of the seven irrigation districts formed within the Sacramento Valley, only one—the Browns Valley District which irrigated land near Marysville from a twenty-six-mile canal completed in February 1893—succeeded. Even that district had not enjoyed complete success. Organized to provide water to 53,000 acres, only 7,000 acres were irrigated from the district's canal in 1915.[21]

In the same month that the *Call* published the editorial quoted above, Willard M. Sheldon of the San Francisco Savings Union organized the Central Canal and Irrigation Company and a companion venture, the Colusa-Glenn-Yolo Land Company. He promised to extend the Central District's aqueduct from the northern

20. *Weekly Colusa Sun,* January 2, March 28, 1892; *California Advocate* 1 (December 1896): 89.

21. *Call,* January 11, 1903; McGowan, *History of the Sacramento Valley,* 1: 396–397. On the continuing litigation over the Central Irrigation District, see the *Call,* January 24, 1901; May 24, 1902; June 26, 1902.

boundary of Glenn County all the way to Woodland, a distance of about eighty-five miles, and to open 300,000 acres to irrigation. The ditch company attracted investors from San Francisco, Fresno, and Los Angeles, as well as from Colusa and Williams. The land company bought 4,600 acres from the Glenn estate for $180,000, hoping settlers and investors would eagerly purchase the land once water was available. The directors of the Central Irrigation District leased the canal and other irrigation works to the water company for fifty years at $25 a year in exchange for Sheldon's promise to finish the canal, install headgates, and provide water to local farmers at rates subject to review by district officials. He also bought up the district's extant bonds at 35¢ on the dollar, relieving residents of their original tax burden as well as the continuing cost of litigation. Between 1903 and 1909, Sheldon and associates spent $330,000 on canals and ditches, even though they never watered more than about 3,000 acres.

The Colusa County section of the canal was finished in 1904, but the anticipated flood of small farmers failed to appear. Like most private irrigation schemes, the project failed to make money, and the company went bankrupt. Still, the Sacramento Valley remained attractive to land speculators because it offered cheap land, rich soil less prone to alkali than the San Joaquin Valley, an excellent transportation artery in the Sacramento River, an absence of litigation over water because the stream's flow was under the direction of the War Department, and ready markets for fruits and vegetables in nearby San Francisco Bay communities. Besides, in the Sacramento Valley many fruits ripened three to six weeks earlier than in southern California. In 1909, James S. and W. S. Kuhn of Pittsburgh, who had extensive experience as sponsors of Carey Act irrigation schemes in Idaho, bought out Sheldon and his backers, formed the Sacramento Valley Irrigation Company, and purchased 150,000 acres of land they expected to sell on easy terms as 20- to 80-acre fruit farms. By 1913, the Kuhns had poured an additional $2,000,000 into the canal system, but they also failed to lure settlers into the valley even though the company promised to turn the water system over to the farmers once it had sold off its land. The company never irrigated more than 16,000 to 20,000 acres and went bankrupt in 1913. Ironically, in March 1920, the new Glenn-Colusa Irrigation District took over the Kuhn properties, and during World

War I the demand for irrigated crops, such as rice, greatly expanded irrigation in the valley. Nevertheless, as in Kern County, old patterns of land use died hard and irrigation failed to create an agricultural eden. As late as 1957, a prominent irrigation district bond trader, J. Rupert Mason, commented in his oral history: "There are 200,000 acres in the Glenn-Colusa District and no small holdings, none. Not even today. It looks like Spain."[22]

Most other irrigation districts shared the fate of the Central District. Only a few well-managed ventures, such as the Modesto and Turlock districts, weathered the financial storms of the 1890s; the remainder died at birth or within a few years.[23] And as the speculative balloons burst, even the soundest districts suffered from the aftershocks. Most potential investors turned against *all* irrigation bonds, and apprehensive farmers refused to settle within *any* district. Examples of wildcat schemes abound. A group of land speculators organized the Manzana District in Los Angeles County on December 5, 1891, even though they knew that the district had no chance of finding a reliable water supply. When the district was created, fewer than a dozen people lived within its borders. The

22. *Call*, January 11, 12, 1903; *San Francisco Chronicle*, April 8, 19, 1903; July 9, 1904; November 26, 1904; "Marketing Irrigation District Bonds," *Transactions of the Commonwealth Club* 6 (December 1911): 531; S. T. Harding, *Water in California* (Palo Alto, California 1960), 89–90; J. Rupert Mason, oral history transcript, Bancroft Library, 32; California State Water Commission, "Engineer's Report on Water Rights of the Glenn-Colusa Irrigation District," June 26, 1920, unpublished report, California Department of Water Resources Archives, Sacramento; "The New California," a pamphlet published by the Kuhns (n.p., 1910), Bancroft Library; and F. H. Griswold, "Home-builders Displace Bonanza Farmers," *Sunset* 26 (May 1911): 578–579. To add to the irony surrounding the fate of the Central District, irrigation promoter Charles F. Lambert suggested that the district would have failed for a more basic reason. Chief Engineer C. E. Grunsky's experience had been restricted to the San Joaquin Valley, where the flow of water in July was much greater than in the Sacramento Valley because of the heavier snowpack in the southern Sierra. Apparently, Grunsky fixed the main canal's grade six feet above the Sacramento's low water mark in July. In the twentieth century, electric pumps lifted irrigation water from the river into the canal, but this technology was not available in the late 1880s and early 1890s. See the Charles Lambert oral history transcript, Bancroft Library, 49–50.

23. Even in the darkest days of the 1890s, the Turlock District remained in relatively good health. For example, the *San Francisco Chronicle*, March 4, 1897, reported that on March 2, the district sold $472,500 in bonds at 90¢ on the dollar. However, the bonds may have been "sold" to a contractor at much more of a discount than reported by the newspaper.

promoters "imported" sufficient residents, by *giving* them land, to secure the fifty signatures necessary to call a district election. Once the bond issue had been approved, bona fide settlers began to buy land. Meanwhile, the land company exchanged its property (worth at most $5 an acre) for $200 an acre in bonds. Apparently, its directors hoped that as real settlers took up the land they might be able to sell the bonds at a price approaching face value. Even more outrageous, the company had not secured title to all the land it sold; part belonged to the Southern Pacific! Many gullible newcomers, some of whom were doubtless speculating on their own, paid dearly for their ignorance.

The Manzana District was land speculation pure and simple. But speculation in water rights also offered great opportunities. The Jamacha District, located only a few miles from San Diego's business district, was organized on November 2, 1891, by the San Miguel Water Company—which hoped to exchange its water rights for a fortune in bonds. Of the $111,000 in bonds issued by the district, the company received $105,000. The remainder paid for a reservoir site and the construction of a small dam. Histories of the Linda Vista District in San Diego County, the Big Rock District in Los Angeles County, and the Rialto District in San Bernardino County, offer similar examples. In the Linda Vista District, a $1,000,000 bond issue was secured by district property worth only $600,000.[24]

Even without the machinations of artful speculators and the opposition of large landowners, the Wright Act faced formidable obstacles. Most district promoters assumed (and their assumption was shared by leaders in the national reclamation movement) that any crop could be raised in any desert soil given a sufficient water supply. They believed that even silt-free water contained nearly miraculous properties as a fertilizer. Unfortunately, the flow of California's streams had not been measured systematically, and the unusually wet period from 1880 to 1887 convinced many Californians that the average yearly water supply was larger than it ac-

24. The material on the Manzana and Jamacha districts is taken from the unpaginated Appendix B of Gidney's "The Wright Irrigation Act in California." Many other districts shared similar experiences. See *Irrigation District Movement in California: A Summary*, Assembly Interim Committee Reports 13, no. 5 (Sacramento, 1955), 18, and Adams, *Irrigation Districts in California*, 28–29, 36–37, 39–41.

tually was. To make matters worse, irrigation boosters routinely underestimated the volume of water needed to irrigate different crops. Most assumed a "duty" of one or two acre-feet of water per acre—about one-third to one-half the amount actually required. Because almost every district overestimated its water supply and underestimated its needs, projects were launched that might not have been undertaken if promoters had had reliable stream flow data or state supervision.

The temptation to choose relatively isolated, sparsely settled areas proved irresistible to speculators. Such locations offered greater opportunities to profit from land sales, and they were also easier to administer as districts. However, since most districts included scattered tracts of land that had been cultivated before 1887, and since many districts simply coordinated and expanded canal networks laid out by private companies, few of the irrigation schemes provided the efficient water distribution system promised by the Wright Act. Moreover, district directors and assessors grappled with many difficult questions of equity. For example, how should land already irrigated be taxed in relation to virgin land? Should established farmers who irrigated land adjoining a stream pay the same district tax as new settlers whose land was located five or ten miles from the water source? And should the water rights of established farmers enjoy some priority over those of new settlers, or should all rights date from the formation of the district? In short, how could the residents of such "mottled" districts be placed on an equal footing?

These thorny problems were compounded by gaps and defects in the Wright Act itself. By placing no limit on the size of bond issues, the law encouraged waste and corruption. Nor did the law require district officials to have any practical experience in irrigation agriculture, let alone in designing or administering water supply systems. This omission contributed to serious errors of judgment. Nor did the statute provide an effective method to condemn existing rights. The burden of quieting the multitude of riparian rights (both within and without district boundaries) in the established courts proved overwhelming, as did arriving at a definition of fair compensation for such confiscated property. Purchasing riparian rights usually proved far safer than suits. The district form did not abolish, or even weaken, these rights, as so many of its proponents had promised. Nor did the law give appropriators within an irrigation district any advantage over those outside district boundaries; the courts observed chronological priorities after 1887 as they had before. If anything, the Wright Act increased litigation over water rights.

Probably the greatest single defect in the law related to the sale of district bonds. The Wright Act restricted interest to 5 percent, even though other types of bonds offered a higher return. Many investors refused to take a chance when traditional securities could be found at 6 percent or better. District farmland and irrigation works provided sound collateral, but the total debt ran much higher in irrigation districts than in other bond-issuing jurisdictions. Though irrigation districts had no debt limit, state law limited school and county bonds to 5 percent of the assessed value of district property and city bonds to 15 percent.[25]

In the end, no law, no matter how carefully drawn, could have prevented the web of litigation spun around most of the irrigation districts. The Wright Act proved as much of a boon to lawyers as it did to land speculators. A mass of suits filed in 1887 and 1888 attempted to undermine individual districts by challenging the procedures followed in their creation, the establishment of boundaries, the issuance of bonds, and the assessment of taxes. Consequently, the 1889 legislature enacted a law to speed up and consolidate the process of judicial review. Had such a law not been adopted, interminable litigation might have prevented any district from issuing bonds. The new law required the state's superior courts, on petition from an irrigation district's board of directors, "to examine and determine the legality and validity of, and approve and confirm, each and all of the proceedings for the organization of said district . . . from and including the petition for the organization of the district, and all other proceedings which may affect the legality or validity of said bonds." The law expedited the review process by requiring the courts to disregard any "error, irregularity, or omission which does not affect the substantial rights of the

25. For discussions of the Wright Act's weaknesses, see Hinton, *Report on Irrigation, 1891*, 29–34; "Marketing Irrigation District Bonds," 515–583; Harding, *Water in California*, 83–84; William E. Smythe, "Irrigation Principles," *Irrigation Age*, 8 (January 1895): 10–12, and *Out West* 18 (May 1903): 656; Edward F. Adams, "California Irrigation District Bonds," *Sunset* 27 (September 1911): 324–327; *Sacramento Daily Union*, July 11, 1891; *San Francisco Chronicle*, January 3, 1897; *Pacific Rural Press* 57 (March 4, 1899): 130.

parties to said proceeding." Moreover, those who wanted to appeal a decision had to do so within ten days.[26]

A certain amount of litigation was inevitable given the wide range of questions posed by the law. For example, was an irrigation district a public or a private entity? Did the legislature have the constitutional authority to delegate the power to organize districts to boards of supervisors? Did the boards have the right to decide what land would be included in, or excluded from, the districts? Did the sale of a parcel of land for unpaid district taxes release that land from future assessments? And could mortgagees sign petitions to organize districts even though the law specified "freeholders"?

Looming over these questions was the broader issue of the constitutionality of the Wright Act. The first court test came in *Turlock Irrigation District* v. *Williams*, decided on May 31, 1888.[27] The plaintiff claimed that irrigation districts were private rather than public corporations and questioned the right of districts to condemn property for a private purpose. The suit also charged that the new law allowed condemnation without due process, apportioned taxes unequally, permitted unlimited taxation even when taxes exceeded benefits, and constituted a usurpation of judicial power both by the legislature and by local boards of supervisors. The California Supreme Court disagreed, upholding the Wright Act on grounds that irrigation districts were "at least *quasi* public corporations" with full powers to condemn property and levy taxes. In no case had individual landowners taken property for their own individual use; the condemnation and tax powers had been reserved to the district. Moreover, taxes for local improvements did not have to be equal or uniform if they were just and reasonable. The court concluded:

> Such a general scheme, by which immigration may be stim-
> ulated, the taxable property of the state increased, the relative
> burdens of taxation upon the whole people decreased, and
> the comfort and advantage of many thriving communities
> subserved, would seem to redound to the common advantage
> of all the people of the state. . . . It is true that incidentally
> private persons and private property may be benefited, but
> the main plan of the legislature, viz., the general welfare of

26. *Cal. Stats.*, 1889, 212.

27. Turlock Irrigation District v. Williams, 76 Cal. 360 (1888).

the whole people, . . . is plain to be seen pervading the whole of the act in question.[28]

The state supreme court unanimously upheld the Wright Act in this and five subsequent cases.[29]

After the *Turlock* v. *Williams* decision, opponents of the Wright Act hired armies of lawyers to look for technical violations of the Wright Act in the organization and operation of individual districts. For example, in one case filed against the Central Irrigation District, unhappy landowners charged that the district's boundaries had been imprecisely described in the organizing petition, that the proper bond had not been filed with the petition, that the petition had not been filed at a "lawful" meeting of the board of supervisors, and that the petition had not been properly publicized in a local newspaper prior to the first district election. Though each of these charges was technically correct, the supreme court did not consider innocent mistakes as significant. The petition had been presented at a highly advertised special meeting of the board and it *had* appeared in the county newspaper (though the misspelled names in the published version constituted a technical violation of the law).[30]

The California Supreme Court's consistent defense of the Wright Act made the Fallbrook case all the more ironic. When the 12,000-acre Fallbrook District was created in San Diego County, it included 40 acres owned by Maria King Bradley, who did not irrigate her land and considered irrigation an unnecessary expense. On July 2, 1892, district voters approved a $6,000 tax assessment to pay for organizing the district. Bradley refused to pay her share ($51.41), so her property was confiscated and sold. Soon thereafter, she filed suit in the U.S. Circuit Court for southern California. Judge Erskine M. Ross agreed to hear the case because Bradley was an English citizen who did not reside in the district, and be-

28. Ibid., p. 369.

29. John E. Bennett, "The District Irrigation Movement in California," *Overland Monthly* 19 (March, 1897): 252–253. Other than *Turlock* v. *Williams,* the most important case concerning the Wright Act's constitutionality was In the Matter of the Bonds of the Madera Irrigation District, 92 Cal. 296 (1891). For a complete list of early court cases concerning irrigation districts see Adams, *Irrigation Districts in California,* 105–112.

30. Central Irrigation District v. De Lappe, 79 Cal. 351 (1889); Crall v. Poso Irrigation District, 87 Cal. 140 (1890).

cause her lawyers claimed that the Wright Act violated the constitutional guarantees of due process and equal protection included in the Fourteenth Amendment. Bradley charged that the $400,000 bond issue approved by district residents exceeded the value of district property, that 242 acres of state and 80 acres of federal land had been improperly included in the district, and that the district did not have an adequate water supply. (District promoters planned to build a catchment basin to capture rain and flood water for irrigation; no stream ran through or near the district.)

Judge Ross issued his controversial ruling on July 22, 1895. He denied that federal courts were bound to observe decisions rendered by state courts when rights protected by the federal constitution were at stake. While streets, highways, and municipal water systems constituted "public uses," irrigation did not because every person in the district could not use the water provided on the same terms as every other; agricultural landowners received the only direct benefits. Ross conceded that all residents received "indirect and collateral" benefits, but they also profited when an individual planted a tree or built a house—benefits had to be direct and available to every member of the community. "No man's property can be constitutionally taken from him without his consent, and transferred to certain other men for their use," the judge declared, "however numerous they may be." Finally, Ross ruled that the Wright Act was "arbitrary, oppressive, and unjust" in that, "from first to last, at no time or place is the owner of land within the district given the opportunity to be heard."[31]

The Los Angeles Times of July 23, 1895, called Ross's reasoning "close" and his conclusions "unavoidable," but the decision paralyzed the state's irrigation districts. One observer noted:

> There were those who had faith in the law and the districts; some had only hopes of both, but when the decision of Judge Ross was announced the heart of everyone failed; confidence was lost; the bonds were valueless upon the market; all activities upon the water systems were shut down; several districts were disorganized; there was heard wailing over lost

31. Bradley v. Fallbrook Irrigation District, 68 Fed. 948 (1895). Also see the *Sacramento Daily Union, San Francisco Chronicle, Call, Los Angeles Times,* July 23, 1895; *Citrograph,* July 27, 1895.

property on one side and about incompleted work on the other; it was as though some giant hand had come from out of the mist and closed upon them, and thousands were squeezed in the vise.[32]

Much damage was done by Ross's decision, but many lawyers—including C. S. Kinney, C. C. Wright, and James A. Waymire—expected the U.S Supreme Court to overturn the decision. *Lux* v. *Haggin* had established irrigation as a "public use," and Kinney noted that "when the legislature has once declared that a certain use is a public one, as it did in the law in question, the courts as a general rule will support it when not satisfied that a great wrong has been committed; and, when there is any doubt as to the purposes, the legislative decision should always stand."[33]

Kinney's prediction proved accurate. On November 16, 1896, the U.S. Supreme Court ruled that irrigation district legislation was similar to swampland reclamation statutes, whose legality had already been upheld by the highest court. In delivering the opinion, Justice Rufus W. Peckham denied that the Wright Act violated the Fourteenth Amendment and urged those injured by the law to press their grievances in local courts. He pointed out that district critics did receive a fair hearing because the petition presented to the board of supervisors recommending the organization of a district had to be publicized for two weeks prior to a regular meeting of the board. Those who disagreed with the boundaries could express their reservations at that meeting. Though the boards had no authority to redraw district boundaries, they could reject petitions, forcing the proponents of a district to accept more modest limits.[34]

-The U.S. Supreme Court verdict stunned opponents of the Wright Act, but they found a new champion in George Hebard Maxwell. Maxwell was a young California attorney who specialized

32. Bennett, "The Irrigation District Movement in California," 256.

33. C. S. Kinney's statement is from the *Rural Californian* 19 (March 1896): 114. For similar opinions, see the *Rural Californian* 18 (September 1895): 434; *San Francisco Chronicle*, April 2, 1896, *Call*, September 24, 1895.

34. Fallbrook Irrigation District v. Bradley, 164 U.S. 112 (1896). Also see *Irrigation Age* 10 (November 1896): 139–143; *Citrograph*, November 21, 1896; *Tulare County Times*, November 19, 1896.

in irrigation district litigation during the 1890s and headed the legal staff of Maria Bradley and company in the Fallbrook case. In October 1896, even before the supreme court had spoken, Maxwell founded the *California Advocate* to publicize the injustice, incompetence, and fraud he saw in the administration of the state's irrigation districts, and to lobby for state and federal reclamation in the Golden State. Subsequently, the lawyer-journalist formed the California State League to rally district critics and further his objectives. The motives of Maxwell and his followers are not entirely clear. Some of his backers, including Miller and Lux, hoped that Maxwell's muckraking would discredit the Wright Act once and for all. However, many district bondholders probably took refuge in the camp of the enemy, hoping that the state or federal government would take over the districts, assume their debts, and pay off the badly depreciated bonds at par; in 1897, a large number of district bonds sold at less than 50¢ on the dollar.

Maxwell filled early issues of the *Advocate* with the hyperbole that characterized his uncompromising personality. In the November 1896 issue, he denounced "the underlying communistic principle of the Wright Act," warning that it would soon spawn a legion of "improvement districts" with the "power to vote unlimited debt." In December, he described the Wright Act as "a menace to every one contemplating settlement in California," warning that "any newcomer to the State may, at any time, against his will, have his property embraced in an irrigation district and taxed even to confiscation to provide irrigation to others." In the same issue he characterized district bonds as "rotten with fraud and all manner of illegality." He noted that payment of the bonds would require taxes far beyond the means of most landholders, taxes that would "involve the confiscation of half or more than half of the lands of every district in California." Maxwell emphasized that most opponents of the district concept favored greater state control over water: "They believe that the state should appoint a competent commission to supervise the construction of works and to distribute the water at the actual cost of such construction and distribution."[35]

35. *California Advocate* 1 (November 1896): 44; (December 1896): 80, 88, 89, 98. Copies of the *Advocate* for 1896–1898 are at the California State Library, California Room, Sacramento. Also see the Frank Adams oral history transcript, Bancroft Library, 80–81; *Tulare County Times*, November 3, 10, 1898. The *San Francisco Chron-*

Thwarted in the U.S. Supreme Court, Maxwell and his fellow district critics turned to the California Legislature in 1897. The Wright Act had been amended many times since 1887, but its essential features remained intact.[36] At the 1891 legislature, the California Association of Irrigation Districts—an organization formed in September 1890 by representatives of about half the state's districts—drafted a bill providing for a state board of irrigation to approve or reject all plans to create new districts. A second bill would have created the State Association of Irrigation Districts, supported by a tax of not more than 1¢ for each $100 of assessed value of property within all the state's districts. This organization would have supervised the construction of irrigation works, as well as the financial operations of each district. Representatives of bona fide irrigation districts hoped that such legislation would discourage the wildcat schemes that had undermined investor confidence in virtually all the districts. Unfortunately, retrenchment became the watchword of the 1891 session. Public hostility toward the many state boards created since the 1870s—including the State Horticulture Commission, the State Fish Commission, the State Viticulture Commission, and the State Forestry Commission—ran high, and in January the governor recommended abolishing several of these boards. Since the irrigation board bill carried a $10,000 appropriation to create yet another commission, it won little support. Nor did the second bill.[37] From 1893 to 1897, many changes were proposed, including state purchases of district bonds, a constitutional amendment limiting the vote in district elections to landowners, and an amendment prohibiting any discount in the price of bonds used to pay for irrigation works, water rights, or rights-of-way.[38] C. C. Wright

---

icle, January 3, 1897, noted that one justification for a state irrigation system was that it would not tax local landowners who did not want to irrigate for the benefit of those who did. The January 3 issue contained many articles discussing conditions in the different irrigation districts.

36. For amendments to the Wright Act see *Cal. Stats.,* 1889, 15–16, 20, 21, 212; 1891, 53, 142–150, 244–246; 1893, 175, 276, 516, 520–524; 1895, 127, 174.

37. *Tulare County Times,* September 18, 1890; January 15, 22, 1891. *Visalia Weekly Delta,* September 18, 1890; January 1, 1891; "Ceding the Arid Lands to the States and Territories," H. Rep. 3767, p. 10; Hinton, "Irrigation in the United States: Progress Report for 1890," 68–70.

38. I. M. Holt, "Our Irrigation Law: Its Defects and Remedies," *Rural Californian* 17 (December 1894): 631–632; 18 (September 1895): 432–433; "Cession of the

favored a state irrigation board to weed out infeasible projects but warned that radical or wholesale changes in the law might renew litigation, "and the battle, which has lasted almost ten years, would have to be fought over again. A conservative administration of the law will cure nearly all the ills complained of."[39]

The legislature that convened in January 1897 met in the shadow of the Fallbrook decision, at a time when the agricultural depression of the 1890s, manifested in the low price of wheat, had reached its nadir. District residents could no longer hope that the courts would cancel their debt, and that debt threatened to increase dramatically. The Wright Act wisely specified that for the first ten years after the issuance of bonds, district taxpayers would pay only the 5 percent annual interest on the securities. District supporters had hoped that by the time farmers began to pay off the principal, the appreciation in crop and property values would render the increased tax burden easier to bear. Consequently, while the Central District's tax burden was only $1 per $100 of assessed valuation in 1897, it threatened to soar in 1898 and 1899 because the law required the retirement of at least 5 percent of the bonds in the eleventh year, 6 percent in the twelfth year, 7 percent in the thirteenth year, and so on until the entire debt had been paid by the twentieth year. The legislature could not erase existing debts, but it could limit the issuance of new bonds and the formation of new districts.

Irrigation was not a hot public issue in the 1897 legislature; a bill to provide state bounties for coyote scalps drew more attention. No district had been created since 1895, and none was likely to be formed while the devastating depression continued. Nevertheless, spokesmen for the irrigation districts and their critics squared off on January 21, 1897, at a joint meeting of the senate and assembly committees on irrigation and water rights. George Maxwell and a legal associate, J. Percy Wright, played a prominent part in the heated debate over the Wright Act. The meeting ended after the

Public Lands, Etc.," 55th Cong., 1st sess., 1897, S. Doc. 130, 31; *Tulare County Times,* December 31, 1896. Delegates to the National Irrigation Congress held in Phoenix, Arizona, December, 1896, spent much of their time discussing possible amendments to the Wright Act. Many features of the Bridgeford Act, adopted by the California legislature in 1897, conformed to those suggestions.

39. *Tulare County Times,* December 10, 1896.

joint committee appointed a nine-man subcommittee to draft new legislation. The subcommittee included strong defenders of the district form, including Assemblyman James A. Waymire, as well as critics.[40]

The group selected E. A. Bridgeford to draft the legislation, and he produced a bill markedly different from the Wright Act. As signed by the governor on March 31, 1897, the new law virtually scuttled the district concept. It required a majority of all landowners, representing a majority of the property values of land susceptible of irrigation, to petition the board of supervisors before an election to form a district could be held. (The Wright Act simply required a petition from fifty freeholders within the proposed district, and they could be town residents as well as farmers.) The new law also required a two-thirds vote to create a district, rather than the simple majority demanded by the 1887 legislation. And while the Wright Act allowed district directors to call bond elections on their own initiative anytime after the first election had been held, the Bridgeford Act required petitions to schedule bond elections as well as to form districts. The directors could dispense with the petition only when the proposed bond issue was for $10,000 or less. Other changes pertaining to bonds provided that the securities be issued for thirty rather than twenty years, with the principal paid in the twenty-first through thirtieth years; that they be exempt from state, county, or municipal taxes; that they could not be sold at less than par; and that they could not be exchanged for land or water rights. The new act did not affect existing district debts, but it did establish a procedure by which disgruntled landowners could petition to withdraw from their district. If the directors approved the petition, the property owner could not be taxed to pay off future bond issues. The law also required the directors of new districts to exclude all land not irrigated, such as pasture, on the demand of the owner or owners. Finally, it provided that special elections could be held to reduce the bonded debt if more bonds had been approved than were necessary to build the irrigation works.[41]

40. *Sacramento Daily Union*, January 19, 22, 28, March 4, 6, 13, 14, 1897; *Sacramento Bee*, January 22, 1897; *Tulare County Times*, January 28; February 4, 1897.

41. *Cal. Stats.*, 1897, 254; Adams, *Irrigation Districts in California*, 47; also see Adams, *Irrigation Districts in California* (Sacramento, 1930), 40. Bridgeford also pushed

The Bridgeford Act achieved its basic objective; no new irrigation district was formed until 1909. In January 1898, George Maxwell scornfully wrote:

No bank in California will loan on an acre of land in any irrigation district. No man who is informed as to existing conditions will purchase property or make investments in an irrigation district unless it be as speculation—gambling on the chances of knocking it out. Home seekers shun the irrigation districts as though they were cursed with the plague. The system hasn't a friend left except those who are bondholders, or a few who are getting cheap water at their neighbor's expense.[42]

In the closing months of the nineteenth century, few Californians could even imagine the remarkable success this maligned institution would enjoy during and after World War I. Of the fifty irrigation districts organized in California before 1893, only twelve were operating in 1910. The fifty original districts contained 2,285,000 acres, but the districts that survived until 1910 included only 606,351 acres and irrigated only 173,793 acres—roughly 5 percent of the state's irrigated land in 1909 or 1910. Only $2,000 of the $7,917,850 in bonds had been paid off in full. Bondholders accepted $5,690,800 in compromised settlements and the courts ruled that over $2,000,000 in bonds had been issued illegally, and hence were void.[43]

The Wright Act failed to transform California's vast wheat ranchos into small, intensively cultivated family farms, as the law's proponents had hoped. In fact, it promoted piecemeal, uncoor-

---

a second law through the legislature providing that any district property which had been sold for delinquent taxes could be redeemed by the former owner within six months of adoption of the law if the new owner had not secured a final patent to the property. In the future, any district landowner whose property was sold for back taxes had one year to pay his taxes and redeem it. *Cal. Stats.*, 1897, 2.

42. *California Advocate*, 3 (January 22, 1898): 4.

43. Ray P. Teele, *Irrigation in the United States* (New York, 1915), 80; Wells A. Hutchins, *Irrigation Districts: Their Organization, Operation, and Financing*, U.S. Department of Agriculture, Technical Bulletin no. 254 (Washington, D.C., 1931), 72–73.

dinated development of the state's water supply, tacitly acknowledging that a bold, comprehensive program of water resource
development was impossible in California. Nevertheless, its influence extended far beyond the state's borders and had a profound
affect on the western reclamation movement of the 1890s. The
district idea won plenty of friends in Washington. Senator William
Morris Stewart's Committee on the Irrigation and Reclamation of
Arid Lands toured the arid states in the summer and fall of 1889.
Its majority report, which reflected Stewart's thinking, concluded
that the Wright Act "bids fair to wisely settle all questions of water
and its use" in the West. Stewart recommended that the federal
government segregate the section's irrigable land, reserve potential
reservoir sites, then let local communities reclaim the land. In 1890,
he introduced a bill to grant irrigation districts the federal pasture
and timber land within their borders.[44] U.S. Geological Survey Director John Wesley Powell also favored local development and control of water but warned that the Wright Act districts "will soon
be in conflict with each other, as there are no means yet provided
for the division of the waters among them." Powell urged that
district boundaries in the West conform to the 150 natural water
basins, to limit conflict between districts sharing the same stream.
He also proposed that the surplus water of each basin be reserved
exclusively for the irrigable lands designated by the U.S.G.S. in
that basin. This marriage of land and water did not threaten established water rights, which Powell wanted to confirm, but it did
challenge state sovereignty. Many of Powell's districts spanned the
borders of two or more states, and western politicians shunned
any plan that implied federal ownership of water. Put simply, they
considered the scientist's plan impractical or unworkable.[45]

In 1891 the first National Irrigation Congress, chaired by
C. C. Wright, unanimously urged Congress to cede the public lands
of the arid region to the states. This touched off a crusade that

44. "Report of the Special Committee of the United States Senate on the
Irrigation and Reclamation of Arid Lands," 51st Cong., 1st sess., 1890, S. Rep. 928,
(serial 2707), 58–59; San Francisco Chronicle, February 20, 1890.

45. See John Wesley Powell's testimony in "Ceding the Arid Lands to the
States and Territories," 1891, H. Rep. 3767, pp. 112, 135, and his discussion of
irrigation districts in the Congressional Record, 52d Cong., 1st sess. (July 21, 1892),
6499–6500. The quote is from p. 112 of the first document.

culminated in passage of the Carey Act in 1894. Support for cession usually went hand in hand with support for federal or state laws modeled on the Wright Act. The Desert Land Act of 1877 had failed to encourage small farming; instead, it provided a bonanza for large land and water companies. The irrigation district was seen as insurance against corporate monopoly. In 1891, Richard J. Hinton sponsored a bill requiring the Secretary of Interior to designate the boundaries of potential districts whose land would then be transferred to the states. No patents to the land would be issued until the state or territory adopted a district law similar to the Wright Act. Proceeds from land sales would help pay for irrigation works. The districts could build their own dams and canals or let private companies do the job. But when private companies did the work, the irrigation system would become community property after twenty-seven years. Hinton promised that his plan insured that the public domain would become "the property and prize of the homestead settler."[46]

In the winter of 1892, Senator Francis E. Warren of Wyoming submitted his own bill providing that after ten years, land ceded to the states would be restored to the public domain if the president decided that district land had not been reclaimed. The bill implied that the states themselves would issue bonds, build the irrigation systems, then sell the land to bona fide settlers to pay for the work. The states would also be allowed to lease land not subject to reclamation to stockmen and sell the district's surplus timber. William Ellsworth Smythe, then editor of the *Irrigation Age* and a prominent leader in the reclamation movement, opposed Warren's plan, charging that irrigation districts could not work in thinly settled parts of the West such as Wyoming and Arizona. There land had little value, and the cost of dams and canals would far exceed the value of irrigated crops. Warren's bill died in committee.[47]

46. Richard J. Hinton's bill is reprinted in "Ceding the Arid Lands to the States and Territories," 1891, 7–8. Also see his "Irrigation in the United States: Progress Report for 1890," 32–33.

47. Francis E. Warren's bill was reprinted in the *Congressional Record*, 52d Cong., 1st sess. (July 21, 1892), 6485–6486. For William E. Smythe's comments see *The Irrigation Age* 2 (November 1, 1891): 264; (February 1, 1892): 442; (March 15, 1892): 522.

Much of the support for cession came from land and water companies whose directors hoped to sell out to the districts or build canals and ditches, as they had done in California; the depression of 1893 cripped these companies, and the cries for cession coupled with autonomous districts faded. Nevertheless, many states copied the Wright and Bridgeford acts, including Washington in 1890, Kansas and Nevada in 1891, Idaho, Nebraska, and Oregon in 1895, Colorado in 1901, Texas in 1905, Montana and Wyoming in 1907, New Mexico and Utah in 1909, Arizona in 1912, Oklahoma in 1915, and North and South Dakota in 1917. Some states, such as Nevada, copied the Wright Act verbatim. But others learned from California's mistakes. For example, by 1915, half the states that had enacted district legislation required consent from the owners of a majority of acreage and a majority of assessed property to form a district—not just a majority of voters, as in the Wright Act. Other states required larger majorities to organize districts or issue bonds. In all states, the bonds were a lien on district property, but not all states taxed the same property: some excluded town lots, some taxed farmland but not improvements on that land, and a few states excluded all nonirrigable land from assessments. Some states allowed the sale of bonds at less than face value, others did not. Still, all the states and territories honored the two basic principles contained in the Wright Act: democratic home rule and bonding land to pay for irrigation.[48]

The irrigation district was a significant institutional innovation. Its emphasis on planning, cooperation, community control, and interdependence represented a sharp break with nineteenth-century agrarian values. But the irrigation district form had serious weaknesses, as did private and mutual water companies. Private companies had proven inefficient, unpopular, and unprofitable;

48. C. S. Kinney, *A Treatise on the Law of Irrigation and Water Rights and the Arid Region Doctrine of Appropriation* (San Francisco, 1912), 3: 2518–2581; Wells A. Hutchins, *Summary of Irrigation-District Statutes of Western States,* U.S.D.A. Miscellaneous Publications no. 103 (Washington, D.C., 1931); Teele, *Irrigation in the United States,* 105–123; A. E. Chandler, *Elements of Western Water Law* (San Francisco, 1913), 140–142; Roy E. Huffman, *Irrigation Development and Public Water Policy* (New York, 1953), 75. There were two phases in the development of irrigation districts before World War I, spanning the years 1890–1895 and 1907–1917. In the first phase only California and Washington actually formed any districts, and of the seven districts created in Washington before 1895, only two issued bonds.

the depression of 1893 dried up investment capital and left most companies bankrupt. The popularity of mutual water companies increased dramatically during the 1880s and 1890s in parts of the San Joaquin Valley and, particularly, in the Los Angeles basin. The Anaheim, Santa Ana Valley, San Antonio, Etiwanda, Riverside, Bear Valley, and Temescal water companies all date from this period. They were community-owned operations which reduced conflict by attaching water rights to specific parcels of land. The mutual company had two distinct advantages over the irrigation district: it was much easier to organize, and since stock ownership corresponded to landownership, (usually one share per acre), the largest water users had the greatest say in determining company policies. However, the mutual company was not an effective tool to raise the money needed to build new irrigation works.[49] The district provided a mechanism to regulate the use of water and raise money, but the bonds did not sell. Consequently, by the end of the 1890s irrigation boosters turned to Sacramento and Washington for help. Government aid was embraced as a cure for the agricultural stagnation of the "terrible nineties."

49. On the nature and organization of mutual water companies see Kinney, *A Treatise on the Law*, 3: 2659–2678; Alexander, *The Life of George Chaffee* (Melbourne, 1928), 51–52; Luther A. Ingersoll, *Ingersoll's Century Annals of San Bernardino County* (Los Angeles, 1904), 227–228; Beatrice Lee, "The History and Development of the Ontario Colony" (M.A. thesis, University of Southern California, 1929), 27, 38–39, 54, 65.

10

The Beginnings of Federal Reclamation in California

By 1900, California stood as the leading agricultural state in the arid West. Already it exhibited many of its distinctly twentieth-century characteristics: it was highly mechanized; increasingly devoted to high-value fruits, fibers, nuts, and vegetables; dependent on a large force of migrant workers; and dominated by those large farms Carey McWilliams called "factories in the field." The agricultural revolution that began in the late 1870s, as citrus orchards began to spread across the Los Angeles basin, reached full tide by 1900. Forage crops and cereals constituted 96.1 percent of California's agricultural production in 1879 but only 56.7 percent in 1899. In 1880, California's fruit growers held their first annual convention, and three years later the state established a board of horticulture, mainly to gather information on the host of voracious insects that plagued orchards. The number of bearing orange trees increased from 41,000 in 1870, to 280,000 in 1880, to 1.2 million in 1890. The nation's expanding rail network and the advent of refrigerated boxcars—along with urban growth in southern California—created new markets. The first full freight car of oranges left Los Angeles as part of a special "orange train" on February 14, 1886, and in June 1888, a carload of refrigerated apricots and cherries survived the trip from Suisun, on the upper arm of San Francisco Bay, to New York without reicing. By 1900, California held a virtual monopoly on the production of such fruits as plums and apricots, and 25 percent of all fruits and vegetables canned in the United States were packed in the Golden State. California farmers profited from changes in the American diet as the standard fare of meat, bread, and starchy vegetables such as potatoes, gave way to a more diversified diet.[1]

The new century held great agricultural promise, but crop diversification depended directly on irrigation, and the expansion of irrigation stalled in the middle and late 1890s. The depression of 1893 killed investor interest in the land and water companies responsible for the growth of irrigation from 1868 to 1887, and the irrigation district stood discredited. To make matters worse, the cost of irrigation rapidly increased during the 1890s. Because normal stream flow had long since been appropriated, new supplies depended on building reservoirs to store flood water or on pumps to tap underground aquifers. Almost by default, state and national aid emerged as the likely answer to the problem of financing new irrigation projects.[2]

In 1878, California's surveyor-general estimated that about 200,000 acres of the state's farmland were irrigated. Yet by 1890, this area had swelled to 1,004,233 acres—an increase of 500 percent. In that year, the Golden State contained the largest number of irrigated acres of any state in the arid West, over 100,000 more than its nearest rival, Colorado. California produced an average annual crop yield of $19 per acre, far ahead of Colorado's $13.12, though the latter state led in both total land area irrigated (1.34 percent to 1.01 percent) and in the percentage of farms irrigated (31.09 percent to 17.86 percent). That California was a state of extremes was evident in that the average size of its irrigated farms under 160 acres was 30 acres (only Utah and New Mexico had a smaller number)

---

1. Earl Pomeroy, *The Pacific Slope* (New York, 1965), 102, 110–111; Ralph J. Roske, *Everyman's Eden: A History of California* (New York, 1968), 399–400, 405; Imre E. Quastler, "American Images of California Agriculture, 1800–1890" (Ph.D. diss., University of Kansas, 1971), 205–206.

2. The issue of who should build reservoirs was more than financial. As mentioned earlier, state courts had not yet decided whether the flood water of late spring and early summer was distinct from the normal flow. If it became necessary to condemn existing water rights before reservoirs could be built, then the job was best left to the state or nation. Many westerners believed that only the state or nation could produce a harmonious, efficient statewide system, and they feared that flimsy reservoirs constructed by private companies or irrigation districts posed a great danger to public safety.

while the average size of its irrigated tracts larger than 160 acres was 547 acres (even larger than the irrigated ranches in grazing states such as Nevada and Wyoming). The value of California's irrigation network was three times greater than Colorado's and four times greater than Utah's. At $150 an acre, the average price of California's irrigated land was nearly double the figure for its closest rival, Utah.[3]

During the 1880s, the greatest growth in irrigation occurred in the San Joaquin Valley and in southern California. In 1890, Tulare County led the state with 168,455 irrigated acres, Kern followed with 154,665 acres, and Fresno ran third with 105,665 acres. The two leading irrigation counties south of the Tehachapis, Los Angeles and San Bernardino, counted 70,164 and 37,907 irrigated acres, respectively. Reliable statistics comparing the growth of irrigation county by county during the 1880s are not available, but population statistics and property values reflect the impact of irrigation. In 1870, the combined population of the seven counties where irrigation was practiced on the largest scale (Los Angeles, San Diego, San Bernardino, Kern, Tulare, Fresno, and Merced) was 40,849; by 1890, their population had increased to 296,719. While the population of the state as a whole roughly doubled from 1870 to 1890, the population of the above counties increased by more than 700 percent. In the same twenty-year span the value of property in these counties increased from $22,513,820 to $198,356,127. By 1890, land that sold for $5 to $25 an acre in Los Angeles and San Bernardino counties in 1870 fetched $100 to $1,100 an acre; land around Fresno that had sold for $3 to $20 an acre twenty years earlier sold readily for $75 to $750 an acre. Of course, irrigation was not the only reason for soaring property values. New rail lines, the flood of invalids and health-seekers, and crafty land speculators all did their part to push prices up. Still, with the notable exception of baronial Kern County, population growth went hand in hand with the expansion of irrigation.[4]

3. F. H. Newell, *Report on Agriculture by Irrigation in the Western Part of the United States at the Eleventh Census, 1890* (Washington, D.C., 1894), vii, 2, 3, 6, 10, 33, 90; *Abstract of the Twelfth Census of the United States, 1900* (Washington, D.C., 1904), 234.

4. *Report on the Statistics of Agriculture in the United States at the Eleventh Census, 1890* (Washington, D.C., 1895), vii, 41, 116, 117; *Report of the Special [Stewart] Committee of the United States Senate on the Irrigation and Reclamation of Arid Lands*, 51st Cong.,

The irrigation boom of the 1880s all but disappeared in the following decade, though pockets of prosperity persisted in southern California. The state's irrigated land increased by more than 500 percent during the eighties, but by less than 44 percent in the next decade—and most of that increase came from 1890–1893. While the value of Colorado's farm property increased by 37 percent and Utah's by 90 percent during the 1890s, California's increased by a scant 3 percent. And although the Golden State could not compare with Georgia, Louisiana, or Alabama, it led the arid West in tenant farmers; 23.1 percent of its cultivators did not own the land they worked. Worst of all, sharp *declines* in the amount of irrigated land occurred in many Central Valley counties during the 1890s: Colusa County suffered a 41.8 percent loss, Amador County 62.8 percent, El Dorado County 21.6 percent, Kern County 27.2 percent, Lake County 45.4 percent, and Plumas County 16.9 percent.[5]

In part, these statistics reflected the demise of the wheat industry. By the middle 1880s, California ranked as the second largest wheat producing state in the nation. In 1885, 3,750,000 acres were planted to that crop, perhaps two-thirds of the state's cultivated land. But prices fell dramatically during the decade from 1885 to 1895. Wheat that sold for $1 a bushel in 1883 or 1884 fetched only 53¢ in 1893, and the price remained at that level for the rest of the 1890s. In 1900, 2,683,405 acres of wheat yielded 36,534,407 bushels worth 55¢ each, and the Central Valley's exhausted soils produced the fewest bushels per acre of any arid or semiarid state. To compound problems, stiff competition from Australia, Argentina, Canada, and Russia cut into California's international markets. The

1st sess. 1890, S. Rep. 928, 1: 54; *Report of Irrigation Investigations in California*, U.S.D.A. Office of Experiment Stations, Bulletin no. 100, 57th Cong., 1st sess. 1902, S. Doc. 356, 217; *Annual Report of the State Board of Horticulture of the State of California for 1892* (Sacramento, 1893), 44-45; Richard J. Hinton, *Irrigation in the United States: Progress Report for 1890*, 51st Cong., 2d sess., 1891, S. Ex. Doc. 53, (serial 2818) 54; *San Francisco Chronicle*, January 1, 1887; *Pacific Rural Press* 38 (August 10, 1889): 115–116. Of course, statistics can be deceptive. Most irrigation in Tulare and Kern counties, for example, produced forage. So while these counties contained far more irrigated land than Fresno County, in 1889 property in the latter county was worth about three times the assessed value of Kern County's property and $10,000,000 more than property in Tulare County.

5. *Twelfth Census of the United States Taken in the Year 1900: Agriculture, Part 2, Crops and Irrigation* (Washington, D.C., 1902), 227, 291, 826.

Irrigated Land in California, 1900. *First Annual Report of the Reclamation Service, 1902* (Washington, D.C., 1903).

state's wheat farmers had never been able to compete with Midwestern growers for national markets, so they exported 40 to 50 percent of their crop to Europe. In the nineties, their competitive position also suffered from the declining quality of their product. Club wheat, the most common variety, stayed in the shell well when left in the field after harvest; it was highly resistant to mold and rot. But exhausted wheat lands produced a flour with a low gluten content, so the grain had to be mixed with wheat raised out of state. By 1910, the industry had all but disappeared, with 478,217 acres returning a crop of only 6,203,206 bushels. In 1899, a traveler observed the dry lands west of the Kings River:

> The plains are given up to desolation. Eight or ten years ago large crops of wheat were raised on this land . . . and farm-houses built on nearly every quarter section. . . . But not a spear of anything green grows on the place this year. . . . The houses of former inhabitants are empty, the doors swing open or shut with the wind. Drifting sand is piled to the top of many fences. The windmills, with their broken arms, swing idly in the breeze. Like a veritable city of the dead, vacant residences on every side greet the traveller by horse team as he pursues his weary way across these seemingly endless plains.

Similar scenes greeted travelers through the Sacramento and upper San Joaquin valleys.[6]

The wheat boom of the 1880s left a painful legacy. Declining crop yields due to soil exhaustion, along with the high price of farm machinery, encouraged large growers to expand the size of their estates, driving many small farmers out of business. By 1890, the champions of wheat could no longer tout it as the "poor man's crop," as they had in 1870. In 1880, Yolo, Colusa, Butte, and Tehama counties included 71 estates of more than 5,000 acres; in all, these princely domains totaled 797,761 acres. Ten years later the number of landowners holding more than 5,000 acres had swelled to 106 who owned 1,479,104 acres. Not surprisingly, the population of

6. Crop statistics are from the annual reports of the Commissioner and Secretary of Agriculture during the 1880s and 1890s. Also see the *Twelfth Census of the United States, 1900: Agriculture, Part 2, Crops and Irrigation*, 28, 29, 92–93; *Thirteenth Census of the United States, 1910*, vol. 6 (Washington, D.C., 1913), 143, 160; and *Water from Sacramento River, California, for Irrigation Purposes*, 59th Cong., 1st sess., 1906, S. Rep. 2900, (serial 4905), 4. The quote is as reprinted in William L. Preston, *Vanishing Landscapes: Land and Life in the Tulare Lake Basin* (Berkeley, Los Angeles, London, 1981), 134–135.

these four counties increased only 4 percent during the 1880s. In Fresno County the number of landowners who held 5,000 acres or more declined from 44 to 41 in the years from 1875 to 1890, but those 41 landowners held 943,557 acres—about 100,000 more than the 44 largest landowners in 1875. By 1900, many of these large wheat farms were on the market at prices ranging from $10 to $40 an acre.[7]

From 1888 to 1897, neither the governor nor the legislature paid much attention to the irrigation question. For example, four issues dominated the 1894 gubernatorial campaign: the depression, government spending, the "currency question," and railroad rate regulation. All three political parties, the Populists no less than the Democrats and Republicans, called for sharp reductions in state spending, though each supported a government-owned transcontinental railroad.[8] Retrenchment dominated the administration of Governor James Budd. He abolished state aid to aged indigents, cut fifteen companies from the state militia, slashed appropriations for state insane asylums, abolished the state viticulture commission, and eliminated the state bounty for coyote scalps. The governor, who in the 1870s had favored a state irrigation system, did not oppose all state public works. For example, he asked the legislature to draft a plan to improve navigation on the state's rivers and protect against floods. While he wanted the federal government to pay for the work, he argued that the state should do it if Washington refused. But Budd paid no attention to the financial

7. The statistics are from W. H. Mills, "California Land-Holdings," a speech delivered to the Chit-Chat Club of San Francisco in December, 1891. A copy is contained in *Pamphlets on California Lands,* vol. 2, Bancroft Library, University of California, Berkeley. For other examples of Mills's statements on the evil effects of land monopoly, see the *Call,* January 13, 1896; December 19, 1897. For an excellent brief survey of Mills's role in promoting irrigation in California, see Richard J. Orsi, "*The Octopus* Reconsidered: The Southern Pacific and Agricultural Modernization in California, 1865–1915," *California Historical Quarterly* 54 (Fall 1975): 197–220. For a sampling of editorial statements against land monopoly see the *San Francisco Chronicle,* January 1, 1887; January 3, 1897; *Weekly Colusa Sun,* March 26, 1887; December 15, 1891; February 20, 1892; *Pacific Rural Press* 48 (July 14, 1894): 18; the *Country Gentleman* 64 (June 29, 1899): 505. Of course, not all large farmers raised wheat; many were cattlemen, especially in Fresno County where Miller and Lux owned 239,486 acres in 1890.

8. Erik Falk Peterson, "The End of an Era: California's Gubernatorial Election of 1894," *Pacific Historical Review* 38 (May 1969): 141–156.

plight of the state's irrigation districts, nor did he mention irrigation in any of his major speeches.[9]

During the period from 1893 to 1897, the irrigation crusade all but evaporated. But it revived rapidly during the drought of 1898 to 1900. The drought's effects were not uniform. In 1898–1899, San Francisco and Sacramento enjoyed 77 percent of their normal rainfall while Los Angeles received only 35 percent; in 1899–1900, Sacramento received 105 percent of normal percipitation, San Francisco 84 percent, and Los Angeles only 48 percent. Nevertheless, the 1898 statistics are misleading because most of northern California's rain fell during a two-week period in late March and early April; this limited the value of the precipitation to farmers. Overall, 1898 was California's driest year in two decades.[10]

During the drought, George Maxwell's career blossomed. In the early months of 1898, he continued to ponder the irrigation district problem. In March, he recommended that the state issue its own district bonds to raise the money needed to complete and own all the district water systems. His plan called for the state to reimburse bondholders for the amount they actually paid for the securities in money or services—75 percent of par in the Modesto, Turlock, and Escondido districts, 50 percent of par in the Central and Tulare districts, and so forth. The new bonds would be issued for a longer period, to reduce district taxes, and no land would be included in districts without the approval of the landowner and a competent, impartial state engineer who would determine whether the local water supply was adequate to irrigate all the land within the district. Moreover, no towns would be included in the districts, and each landowner would be able to avoid district taxes by paying for his water right in full upon completion of the irrigation works. Maxwell noted: "There can be no doubt that a State irrigation system could be devised which would do away with all the defects

9. *First Biennial Message of Governor James H. Budd to the Legislature of the State of California, 1897, JCSA,* 32d sess. (Sacramento, 1897), Appendix, vol. 1; *Second Biennial Message . . . , JCSA* 33d sess. (Sacramento, 1899), Appendix, vol 1. Also see the *Citrograph,* July 22, 1899.

10. "Drought in California," *Transactions of the Commonwealth Club* 21 (December 28, 1926): 473–526. For specific reports on the drought, see the *Pacific Rural Press* 56 (December 31, 1898): 426; 57 (February 11, 1899): 82; (February 18, 1899): 98; (April 29, 1899): 258; 60 (September 22, 1900): 178; (October 13, 1900): 226.

of the present district system, and under which the disasters in the districts could be relieved."[11]

However, in late 1898 or early 1899, Maxwell dropped state reclamation from his program. Apparently, the young attorney first recognized the national significance of irrigation in 1896, when he attended the Irrigation Congress meeting in Phoenix. In the following year, after a meeting of the Trans-Mississippi Industrial Congress in Wichita, Kansas, Maxwell formed the National Irrigation Association. Western railroads heavily subsidized the new booster organization in anticipation that a national reclamation program would drive up the value of their land and increase freight and passenger traffic. So would state reclamation, but most of the arid West was sparsely populated, which increased the cost of irrigation, and the western states enjoyed only limited financial resources. By 1898, Maxwell's program had been endorsed by the National Board of Trade, the National Business Men's League, and the National Association of Manufacturers.[12] Meanwhile, the crusade for federal reclamation won new respectability from the efforts of Captain Hiram Martin Chittenden of the U.S. Army Corps of

11. George H. Maxwell, "The Irrigation District: The Inherent Defects Which Have Caused Its Failure Can Only Be Remedied by a State System," *Irrigation Age* 12 (June, 1898): 250–253; *Citrograph*, July 2, 30; August 20, 1898; *Call*, March 15, 1898; *Pacific Rural Press* 56 (August 6, 1898): 84-85; *California Advocate* 3 (July 9, 1898). Maxwell's program included creation of a state board of irrigation and public works to supervise construction of state reclamation projects; state control over water and water rights; a "just, equitable, and comprehensive" set of irrigation laws for California; opposition to the cession of public lands to the states; formation of a national arid land commission to survey the public lands capable of irrigation; and reservation of forest land and watersheds. Clearly, in 1898 Maxwell considered state and federal reclamation compatible, though he did not define the responsibilities of each.

12. William Ellsworth Smythe, *The Conquest of Arid America*, 2d ed. (New York, 1905), 272–273. Samuel P. Hays, *Conservation and the Gospel of Efficiency: The Progressive Conservation Movement, 1890–1920* (Cambridge, Mass., 1959), 9–11, claims that Maxwell did not form the National Irrigation Association until 1899. But the irrigation publicist had begun to publish the *National Advocate*, a journal probably subsidized by the same group that bankrolled the association, in 1897.

Maxwell soon lost interest in the *California Advocate*, though not in the politics of California water, and the journal ceased publication in late 1898 or early 1899. The very small circulation of the *California Advocate*, *National Advocate*, *Homemaker*, and *Talisman*, each of which Maxwell edited, suggests that they were intended primarily for circulation among businessmen, politicians, and other "decision-makers" rather than designed to win broad public support for federal reclamation.

Engineers. In his famous report, "Preliminary Examination of Reservoir Sites in Wyoming and Colorado," Chittenden recommended that the national government build and operate reservoirs in the arid West, and furnish free water to farmers (distributed under state laws).[13] This report inspired Maxwell, and the California drought of 1898–1899 may well have driven home the point that reservoir construction was too big a job for the states. He may also have decided that Congress would authorize the reclamation of private as well as public land, eliminating the need for state aid. And given the balance of power in California, with its strong sectional rivalries, perhaps he expected southern California to be neglected or overlooked in a state irrigation program.

In any case, in 1899 Maxwell emerged as the nation's chief publicist for federal reclamation. For example, in February he testified before the Senate Committee on Commerce in favor of a bill to use $5,000,000 from the Rivers and Harbors appropriation to build reservoirs in the arid West. Maxwell maintained that irrigation was a national, rather than a sectional, issue. Arid land reclamation was as important as navigation and flood control, and simple justice dictated that money set aside for internal improvements, such as that contained in the Rivers and Harbors fund, be dispersed impartially, in the West as well as in the East and South. In testimony before the commerce committee, Maxwell promised that eastern businessmen would profit mightily from western development: "Every new home in the West would make an increased market for the Eastern manufacturers and the farmers would feed the workers in the Eastern factories." But the fear of domestic turmoil also figured prominently in Maxwell's thinking. In a piece published in *Irrigation Age* in September 1899, the lobbyist warned:

> We are passing through a period of prosperity when there is work for all who want it. But hard times are sure to come again when men will be thrown out of employment. Labor-saving machinery is constantly lessening the need of human labor. Our wage-earning population is increasing at an enormous rate. Year by year occupation must be found for the new workers who are growing to youth and manhood. Labor

13. Hiram Martin Chittenden, *Preliminary Report on Examination of Reservoir Sites in Wyoming and Colorado*, 55th Cong., 2d sess., 1897, H. Ex. Doc. 141.

organizations have worked wonders in dignifying labor and maintaining fair wages. But they can not create work where there is none. They should use all their influence to open a channel through which all surplus labor can constantly return to the land, and Arid America beckons to them with open arms.

In addition to offering a solution to the labor problem, Maxwell considered arid land reclamation a tool to preserve and restore Republican institutions and values in the face of industrialization, urbanization, and the flood of immigrants who entered the United States during the 1890s. The dreary, crowded, stultifying tenements of Boston and New York, he claimed, planted and nourished seeds of revolution in a potentially explosive landless proletariat.[14]

The drought of the late 1890s was not confined to California, and the California legislature of 1899 reflected growing support for federal reclamation throughout the arid West. In a Senate Joint Resolution sent to Congress, the lawmakers bemoaned the state's agricultural stagnation and declared that "the building of storage reservoirs is far beyond the means of the state." As a first step, the resolution called for federal reservoir and canal surveys. An Assembly Joint Resolution specifically asked for a survey of reservoirs to "confine and husband" the waters of the Stanislaus, Tuolumne, Merced, Fresno, San Joaquin, Kings, and Kern rivers, to "provide for the sufficient irrigation of the whole valley of the San Joaquin." Both resolutions urged federal construction and operation of the proposed irrigation works. Since Maxwell was in Washington for at least part of the California legislative session, his part in the adoption of these resolutions is unclear.[15]

14. *Pacific Rural Press* 57 (March 4, 1899): 133; George H. Maxwell, "Reclamation of Arid America," *Irrigation Age* 13 (September 1899): 407–409. In an article published in the *Citrograph*, October 28, 1899, Maxwell argued that the state should not pay for work that was exclusively a federal responsibility. He also warned of a potential conflict between the state and the nation over the use of navigable waters such as the Sacramento River. At this time he favored leasing public grazing lands to create a fund to provide additional water for private land already irrigated and appropriations under the rivers and harbors acts to construct reservoirs and canals to serve the unsettled public lands.

15. Senate Joint Resolution no. 8, February 9, 1899, *California Statutes*, 1899, 444; Assembly Joint Resolution no. 7, March 6, 1899, *Cal. Stats.*, 1899, 497; *Tulare County Times*, January 19, 26, 1899; *Visalia Weekly Delta*, February 23, 1899.

Not all Californians were willing to jump on the bandwagon for federal reclamation. In the late spring of 1899, a group of San Francisco bankers and businessmen issued a statement that questioned the wisdom of relying on federal aid:

> The drought of 1898 cost the State of California over $40,000,000. For twenty-five years we have applied to the National Legislature without relief. It refuses to regard the irrigation question as a national one. No interstate questions are involved in the sources of water supply of our State, as such sources are almost all within our geographical boundaries. . . . Of late years the attention of the National Legislature has been turned toward the irrigation by storage reservoirs of lands still owned by the United States. We have no such lands of any appreciable amount in this State, therefore our chances of relief from this quarter are less, in our opinion, than they were years ago. . . . Private capital will not invest in storage reservoirs, as the return must be small or the public oppressed. The Wright Irrigation Act has been a practical failure. Moreover, the control exercised by the Boards of Supervisors over such investments frightens private capital.[16]

This left the alternative of direct state action. The state party organized in April 1899, when it became clear that 1899 would be as dry as 1898. Its leaders included George Davidson, a professor of astronomy at the University of California who had served on the Alexander Commission in 1873; William Thomas, a prominent San Francisco attorney; W. H. Mills, land agent for the Southern Pacific; I. W. Hellmann, president of the Nevada Bank; E. B. Pond, president of the San Francisco Savings Union; Philip L. Lilienthal, manager of the Anglo-Californian Bank Limited; Hugh Craig, president of the San Francisco Chamber of Commerce; and F. W. Dohrmann, president of the Merchant's Association. They unanimously agreed that the federal government would not pay for reclamation in California, but that the state could issue bonds and retire them using

16. As reprinted from the *Citrograph* in the *Irrigation Age* 13 (July 1899): 356–357.

proceeds from the sale of irrigation water and electrical power generated at the reservoirs. The state party also urged San Francisco businessmen to raise the money to pay for surveys of California reservoir sites because the city's "very existence depends upon the success of the country." The group called for a convention to discuss the state's water problems and coordinate the efforts of conservationists dedicated to "storing the floods" with those interested in preserving the state's forests.[17]

The state party drew many of its supporters from San Francisco's business community. Those banks that held or owned land in the Central Valley recognized the value of state reclamation, even if they had not invested in land or water companies or irrigation district bonds. The collapse of the wheat industry and the failure of most irrigation districts left San Francisco financial institutions holding hundreds of mortgages on unsalable land or land rapidly declining in value. In addition, the dramatic growth of southern California's population and citrus industry convinced many businessmen that Los Angeles would soon eclipse San Francisco as the commercial and economic heartland of the Pacific Coast, unless the tide of immigration could be turned northward. While the population of southern California's seven counties did not increase as rapidly during the 1890s as during the preceding decade, those counties absorbed 90 percent of the state's rural population growth during the decade and most of the overall increase of 22.4 percent as well. Northern Californians turned to such organizations as the California Board of Trade for help. Formed in 1887, by the turn of the century it joined the Sacramento and San Joaquin Development Associations in advertising the lands of the Central Valley and in calling for a subdivision of the moribund wheat ranches.[18]

17. William Thomas to J. M. Gleaves (president of the California Water and Forest Society), April 24, 1899; George Davidson to F. H. Newell (chief hydrographer, U.S. Geological Survey), May 1, 1899, George Davidson Collection, Bancroft Library; *Citrograph*, June 24, 1899.

18. See the annual reports of the California State Board of Trade in the Bancroft Library. On the competition betweeen northern and southern California for immigrants see the *Twelfth Annual Report, 1901* (San Francisco, 1902), 14. On the subdivision of the large wheat farms see the *Fifteenth Annual Report, 1904–1905* (San Francisco, 1905), 4. On the Sacramento Valley Development Association see the

The need to stimulate small farming reflected the desire of businessmen for a more rational, orderly economy. True, San Francisco's financial leaders were mainly concerned with how the immediate problems of the "terrible nineties"—drought, the collapse of the wheat industry, rural depopulation, and declining property values—affected their pocketbooks. But they also recognized that by coordinating their efforts, and by recognizing that their economic prosperity depended upon the health of farming in the Central Valley, they could alleviate the boom-and-bust cycles characteristic of the California economy in the nineteenth century. In this sense, irrigation offered the foundation for planned, comprehensive economic growth.[19]

As the proposal for a statewide irrigation convention gained support, George Maxwell and his growing legion of supporters opened fire on the state party. In a form letter dated October 20, 1899, bearing the letterhead of the National Irrigation Association, Maxwell charged: "The vast possibilities of mismanagement and corruption in the future which lurk in the movement started by Mr. William Thomas, of San Francisco, attorney for the Irrigation District Bondholders looking to the issuance of millions upon millions of state bonds to store and distribute the flood waters of California, should arouse the active interest of every citizen and property owner in California." In October 1899, Maxwell spoke throughout southern California against state reclamation. Several newspapers echoed his suspicions concerning the state party's motivations. For example, the *San Jose Herald* editorialized:

Fresno Daily Republican, January 16, 1900; "Five Years Review: Being a Report of [the] Executive Committee of the Sacramento Valley Development Association," January 21, 1905, Sacramento County Pamphlets, Bancroft Library. On formation of the San Joaquin Valley Development Association, see the *Tulare County Times*, February 22, May 3, 1900.

19. The keen interest of San Francisco businessmen in reclamation can be seen in the annual reports of the city's chamber of commerce. See *Forty-Ninth Annual Report of the Chamber of Commerce of San Francisco, 1899* (San Francisco, 1899), 9, 10; *Fiftieth Annual Report* (San Francisco, 1900), 42, 53; *Fifty-First Annual Report* (San Francisco, 1901), 40, 44, 51, 54, 58, 70–71; *Fifty-Second Annual Report* (San Francisco, 1902), 34, 66. For an excellent analysis of the changing nature of business in California built on the "organizational synthesis," see Mansel Blackford, *The Politics of Business in California, 1890–1920* (Columbus, Ohio, 1977). Gerald D. Nash's *State Government and Economic Development, A History of Administrative Policies in California, 1849-1933* (Berkeley, 1964), also deserves close reading.

It is probable that a proposition will be sprung to bond the State to construct reservoirs and other works for irrigation. . . . Who are behind it? It is hard to say, but we may guess. The Wright law irrigation bondholders might find a way to repair the value of their securities in connection with an issue of State irrigation bonds. The money lenders of San Francisco will see in it a chance for driving a thriving business in buying and speculating in bonds. What other interests may be in the background remains to be seen.

The *Los Angeles Evening Express* claimed that the scheme would primarily benefit the private contractors selected to build dams and canals.[20]

Among the groups that worked hard to arrange a water convention was the California State Association for the Storage of Flood Waters, another group spawned by the drought. This association, whose leaders included such prominent members of the state party as William Thomas, George Davidson, and W. H. Mills, pitched its appeal to a wide constituency. It promised that cheap hydroelectric power would revive mining, stimulate industrial growth by providing an alternative to imported coal, and provide a means to tap hitherto inaccessible underground water for irrigation. In addition, storage reservoirs would facilitate the drainage of swamplands and aid navigation by increasing stream flow during summer months.[21]

Four hundred and eighty-one delegates attended the irrigation convention, held at San Francisco's Palace Hotel on November

20. A copy of the October 20, 1899, letter is in the *Water Supply of California* collection of pamphlets, vol. 3, Bancroft Library. The *San Jose Herald* editorial was reprinted in the *Citrograph*, October 28, 1899, an issue that contained many articles on irrigation and the storage question. For other examples of newspaper suspicions concerning state reclamation, see the editorials reprinted in the *Citrograph*, October 21 and November 4, 1899. The *Los Angeles Evening Express* attacked state reclamation in its issues of October 16 and 21, 1899. The National Irrigation Association never contained more than a handful of active members. It had been formed largely to lend respectability to Maxwell's lobbying efforts.

21. See the leaflet entitled, "The California State Association for the Storage of Flood Waters," in the folder "California Water and Forest Association," Carton 16, George Davidson Collection, Bancroft Library. Also see the *Pacific Rural Press* 56 (December 31, 1898): 431.

14, 1899. The Association for the Storage of Flood Waters issued formal invitations to the governor, lieutenant governor, chief justice of the supreme court, two members of each county board of supervisors, three delegates from each assembly district, the California Press Association, fifty delegates at large appointed by the governor, the president or a representative of each private water company or irrigation district, the mayors of the state's major towns and cities, and assorted representatives from boards of trade, chambers of commerce, and other civic and business associations.[22]

San Francisco Mayor James D. Phelan, a charter member of the California State Association for the Storage of Flood Waters, delivered the water convention's opening address, and remarked: "I am informed that already there are two parties—one contending that it is the duty of the State to provide for the storage of flood waters, and the other that it is the duty of the Federal Government." The "nationalists," with their glowing promise of free storage works, won out. Maxwell's appointment to the critical legislative committee of the new California Water and Forest Association, formed as a result of the meeting, symbolized that victory. Although the convention adopted a platform endorsing federal reservoirs paid for by proceeds from federal grazing leases, it also endorsed joint federal-state river improvement projects, hydrographic surveys, and the construction of reservoirs designed to generate electrical power. In most respects, the platform echoed the program of the National Irrigation Congress. For example, it recommended that all irrigation water rights be attached permanently to individual parcels of land; that beneficial use should be the first test of any water right; and that a national commission should resolve conflicts over interstate streams. The convention charged the state with the responsibility to collect stream flow data, reform its outdated water laws, and create an irrigation tribunal to prepare a comprehensive record of valid water claims and resolve conflicts over water rights out of court. The *Pacific Rural Press* noted that the platform planks did not provoke serious debate:

A glance at them will show why the meeting was quiet. As Mayor Phelan premised, everyone was in favor of Govern-

22. George Davidson to Henry T. Gage (governor of California), September 15, 1899, Davidson Collection; the *Citrograph*, October 28, 1899.

ment work, and the allusions to Uncle Sam even by those who trusted rather to State initiative were respectful in the extreme. The sentiments of the assembly, as signified by the applause, were clearly along national lines, and the proposition of bonding and taxation [as in irrigation districts] to promote reservoir building was hardly heard of. Those who have faith in such a measure as most expeditious and practicable were content to hold their views in abeyance. For this reason no issue was joined during the first day of the convention, and approval of the plans of the National Irrigation Congress grew more and more emphatic.

The state party conceded the first round, and demands for state reclamation abated during the first decade of the twentieth century. But the battle was far from over.[23]

In 1900, the future growth of irrigation agriculture in California remained uncertain. The 1900 census revealed that a pattern set in the 1870s and 1880s had continued during the 1890s. Eight of the arid West's eleven states and territories grew faster than California from 1870 to 1890, and during the 1890s nine did. At the turn of the century, census officials classified 40 percent of California's inhabitants as "urban," about the same percentage as in

23. The quotes are from the *Pacific Rural Press* 58 (November 18, 1899): 322. Also see the *Press* 58 (November 25, 1899): 338, 341–342; (December 9, 1899): 370; (December 23, 1899): 406; *San Francisco Chronicle*, November 12, 15, 16, 17, 1899; *Call*, November 16, 1899; *Visalia Weekly Delta*, November 16, 1899; *Tulare County Times*, November 16, 1899; *Citrograph*, November 25, 1899. The first issue of *Water and Forest*, the publication of the new Water and Forest Association, noted that the convention had "indorsed [sic] the platform of the National Irrigation Association, syllable for syllable, as presented and urged by its representative, George H. Maxwell." See *Water and Forest* 1 (September 1900): 7. For a brief survey of the Water and Forest Association's early history, see *Forestry and Irrigation* 11 (August 1905): 364–367.

Though the state and federal factions disagreed about many issues, both sides feared that the destruction of California's forests by fire, overgrazing, clear-cutting, and erosion would restrict irrigation in the Central Valley and coastal plains of southern California, or make it impossible. To protect the headwaters of California's major streams, as well as to ameliorate flood damage and improve navigation, a handful of Californians, led by the fledgling Sierra Club, favored the creation of one massive forest reserve stretching from the Tehachapis to the Cascade Range in far northern California. Watershed protection was at least as important as the fear of a timber famine, the desire to preserve scenic beauty and animal habitats, or the resistance to corporate monopoly of the public domain and its resources.

Pennsylvania and Maryland.[24] Moreover, the drought had already lasted two years, and the immediate prospect for federal reservoirs appeared slim. True, federal resource agencies had been active in the Golden State since the late 1880s. In 1888 and 1889, Congress set aside $350,000 to pay for John Wesley Powell's survey of reservoir sites in the arid West. William Hammond Hall directed the U.S. Geological Survey's work in California. Beginning in July 1889, Hall and his staff surveyed and mapped storage sites at Clear Lake, 100 miles north of San Francisco, and in the Stanislaus, Tuolumne, and Merced watersheds. In all they inspected over thirty dam sites and reserved 21,000 acres adjoining those sites from private entry. They also measured the flow of the Kern, San Joaquin, Tule, Merced, Tuolumne, Mokelumne, and several other streams. Congress cut off funds for the Powell survey in 1890, but the Geological Survey's Hydrographic Branch, headed by Frederick Haynes Newell, began systematic stream flow measurements in California in the mid-1890s. Then, in 1899 and 1900, the California Legislature established a pattern of state-federal cooperation in resource surveys when it agreed to match federal appropriations for the U.S.G.S., Bureau of Forestry, and Office of Irrigation Investigations dollar for dollar.[25]

24. *Water and Forest* commented on the 1900 census statistics in 1 (September 1900): 7. In 1900, California contained only two large irrigation reservoirs, the Sweetwater Dam in San Diego County and the Bear Valley Dam in San Bernardino County.

25. Through 1901, in addition to the $350,000 set aside for surveys of dam sites, Congress had granted the U.S.G.S. $357,000 for gauging western streams and $700,000 for hydrographic surveys and the construction of irrigation works on Indian reservations. An additional $198,000 had been appropriated for the irrigation investigations conducted by the Department of Agriculture, first under Richard Hinton and later under Elwood Mead. See Alfred R. Golze, *Reclamation in the United States* (New York, 1952), 23. For examples of U.S.G.S. survey work in California prior to passage of the Newlands Act, see C. E. Grunsky, *Irrigation Near Bakersfield, California*, Water Supply and Irrigation Papers of the United States Geological Survey, no. 17 (Washington, D.C., 1898); *Irrigation Near Fresno, California*, Water Supply and Irrigation Papers, no. 18 (Washington, D.C., 1898); *Irrigation Near Merced, California*, Water Supply and Irrigation Papers, no. 19 (Washington, D.C., 1899); Alfred E. Chandler, *Water Storage on Cache Creek, California*, Water Supply and Irrigation Papers, no. 45 (Washington, D.C., 1901); J. B. Lippincott, *Storage of Water on Kings River, California*, Water Supply and Irrigation Papers, no. 58 (Washington, D.C., 1902); *Development and Application of Water Near San Bernardino, Colton, and Riverside, California*, Water Supply and Irrigation Papers, nos. 59, 60 (Washington, D.C., 1902). The activities of Elwood Mead's Office of Irrigation Investigations can be traced in the reports of his office for 1900–1902, published as U.S.D.A. Office of Experiment Stations Bulletins no. 104 (Washington, D.C., 1902), 119 (Washington, D.C., 1902):

George Maxwell considered California's support for federal reclamation essential. The Golden State offered many attractions, including an extraordinarily mild climate, a long growing season, rich soil, an abundance of potential reservoir sites, a well-developed transportation network, and plenty of engineers skilled at building waterworks. However, from 1900 to 1902, Maxwell spent most of his time trying to win support for his program from eastern merchants, businessmen, civic leaders, and labor organizations. Meanwhile, the Water and Forest Association devoted most of its attention to water law reform. Some of its members considered new water laws an indispensable foundation for federal work. Others hoped those laws could protect *against* federal confiscation of established rights.

When Theodore Roosevelt signed the Newlands Reclamation Act into law on June 17, 1902,[26] little had been done to draft construction plans for irrigation projects in California. As a result, early Reclamation Bureau files at the National Archives contain hundreds of appeals for aid from business and civic groups as well as from farmers. By 1907, the new Reclamation Service had launched twenty-four projects throughout the West, but finding suitable sites in California was no easy job.[27] For example, though the vast Los Angeles basin offered a marvelous climate, the San Gabriel and San Bernardino Mountains provided few canyons suitable for res-

---

no. 133 (Washington, D.C., 1903). Also see Mead's *Irrigation Investigations in California*, 56th Cong., 2d sess., 1901, S. Doc. 180 (serial 4033); *Report of Irrigation Investigations in California*, Office of Experiment Stations Bulletin no. 100, 57th Cong., 1st sess. 1902, S. Doc. 356.

26. The best survey of federal reclamation is Paul Wallace Gates, *History of Public Land Law Development* (Washington, D.C., 1968), 635–698. Also see the excellent introductory chapter in Donald C. Swain's *Federal Conservation Policy, 1921–1933* (Berkeley, 1963); E. Louise Peffer, *The Closing of the Public Domain: Disposal and Reservation Policies, 1900–1950* (Stanford, Calif. 1951), 32–62; William E. Warne, *The Bureau of Reclamation* (New York, 1973); Michael C. Robinson, *Water for the West: The Bureau of Reclamation, 1902–1977* (Chicago, 1979).

27. For example, see "(930), California; Surveys and Investigations thru 1910," Records of the Bureau of Reclamation, R.G. 115, National Archives, as well as the files for 1911–1914; 1915–1918; and January–June 1919. Each of these files contains separate folders for different parts of California such as the Honey Lake Valley, Victor Valley, Kings River Basin, Santa Barbara, San Diego, Palo Verde, and Chuckawalla valleys.

ervoir sites and a completely inadequate surface water supply. In 1902, at least two-thirds of the basin's orchards depended on underground water for irrigation. The Reclamation Service and U.S. Geological Survey carefully examined underground supplies within the region but could do little to augment the supply.[28] On the other hand, the Sacramento Valley offered an abundant water supply and plenty of good reservoir sites, but most streams emptied into the Sacramento River, a navigable stream under the jurisdiction of the Army Corps of Engineers. Reservoirs in the Sacramento Valley might affect navigation and touch off a battle with the politically powerful, well-entrenched Army agency.[29] The Owens Valley, in the eastern Sierra, offered 60,000 acres of irrigable public land and an excellent reservoir site. However, the valley was 4,000 feet above sea level, the cold climate restricted the growing season, and the soil was heavily alkaline. Owens Valley farmers could be expected to raise only low-value forage crops, as they had in the past. In addition, many landowners there refused to cooperate with the Reclamation Service because they did not stand to benefit directly from a federal project.[30]

In October 1903, a Reclamation Service board of engineers recommended that the most practical irrigation projects from an engineering standpoint were at Clear Lake, in the Coast Range 100 miles north of San Franciso; on the Kings River near Fresno; and on the Colorado River at Yuma, just north of the U.S.-Mexico border. However, it warned: "As measured by the ease with which agreement could be entered into with irrigators in dealing with the

28. J. B. Lippincott to Hydrographer, U.S. Geological Survey, August 6, 1903, "(930), California: Surveys and Investigations thru 1910," RG 115; Walter C. Mendenhall, "The Underground Waters of Southern California," *Forestry and Irrigation* 10 (October 1904): 448–455; idem, "Studies of California Ground Waters," *Forestry and Irrigation* 11 (August 1905): 382–384; and his address to the Twelfth National Irrigation Congress (1904), *Proceedings of the International Irrigation Congress* (Galveston, Tex. 1905), 150–157.

29. J. B. Lippincott to F. H. Newell, November 28, 1903, "(930–8), California: Surveys and Investigations, Kings River Storage," RG 115.

30. *Forestry and Irrigation* 9 (November 1903): 558–559; *Second Annual Report of the Reclamation Service, 1902–1903* (Washington, D.C., 1904), 54–55, 93–96; F. H. Newell, "The Reclamation Service and the Owens Valley," *Out West* 23 (October 1905): 454–461; and A. P. Davis to the Secretary of the Interior, September 1, 1905, "(527), Owens Valley: Preliminary Reports on General Plans," RG 115.

land and water questions, the order is just reversed."[31] Half of California remained part of the public domain, but most of that land was in the Mojave or Colorado deserts, or in the virtually inaccessible northeastern corner of the state. From the beginning, the Reclamation Service's first director, Frederick Haynes Newell, recognized that federal projects in California "will primarily benefit lands in private ownership."[32]

The fate of the Clear Lake and Kings River schemes illustrated many of the obstacles faced by the Reclamation Service in California and the entire arid West. After Lake Tahoe, which the Service expected to acquire for its Truckee-Carson Project in western Nevada, Clear Lake was the largest natural storage reservoir in the state. It covered 40,000 acres, and J. B. Lippincott, the Reclamation Service's chief officer in California, predicted that a dam at the lake's outlet to the Cache Creek Valley could raise the water level six feet, providing a supply adequate to irrigate 250,000 acres. The Geological Survey had already discussed such a project in its *Water Supply and Irrigation Paper Number 45*. Because the project would utilize a natural storage reservoir, the cost would be much lower than most reclamation projects in California—about $500,000. Then, too, the Reclamation Service needed at least a few projects in the West that provided quick returns to farmers. Its political support could disappear overnight if construction took too long.

Most of all, the Reclamation Service hoped to win a monopoly over the lake's water, and no other site in California offered such an opportunity. In March 1878, the California Legislature had designated Clear Lake as navigable. Morris Bien, the Reclamation Service's chief legal officer, suggested that "the legislature regarded this body of water as subject to its own control, and not open to use or occupation by private parties, as in the case of non-navigable bodies of water." In 1904, California Attorney-General U. S. Webb concurred with this interpretation. He reasoned that since the lake's

31. J. B. Lippincott and W. H. Sanders to F. H. Newell, October 15, 1903, "(153–11), Yuma Project: Reports of the Engineers, Estimates of Construction, etc., thru 1906," RG 115. The four-man board of engineers included Lippincott, the Reclamation Service's chief engineer in California and the southwest, and Morris Bien, head of the Service's Legal and Lands Division in Washington.

32. *First Annual Report of the Reclamation Service from June 17 to December 1, 1902* (Washington, D.C., 1903), 105.

navigability might be affected by diversions for irrigation, no grant could be made without express approval of the legislature. In effect, state laws regulating the acquisition of water rights did not apply. Hence, in theory, the state could give the nation all the water in the lake, or at least that which had not been claimed and put to beneficial use prior to 1878.

The superb soil and seventeen inches of average annual rainfall enjoyed in the Clear Lake region, added to the attractions already mentioned, put it at the top of the list. However, in the opening years of the century, a group of investors spent $100,000 to acquire water rights on Cache Creek and another $14,000 to buy land at the outlet of the lake. Then they built twenty-five miles of canals in the expectation of selling 16,000 acres or more to farmers. In October 1903, Lippincott noted: "It was suggested that they submit a proposition to us, based on the guaranteeing to them of a water supply for their lands as a recognition of their existing rights, leaving the project to us for complete development. They decline to deal with us on any basis." Later, the irrigation promoters offered to sell out to the government for $1,500,000, but even under intense pressure from California Governor George Pardee, they refused to drop their price below $750,000. Payment of such an exorbitant price would have encouraged other speculators who were busy buying up land at the lake in the hope they could sell to the government at a fat profit. In addition, in 1904, the Reclamation Service had only $2,000,000 to spend on its California projects. Had it paid the price, not much money would have been left for other work in the Golden State.[33]

The Reclamation Service might have tried to force the issue by having the water rights of these developers invalidated in the courts. But the courts moved slowly, and the contest could have seriously damaged the Reclamation Service's image. Already, critics

---

33. Lippincott and Sanders to Newell, October 15, 1903, "(153–11), Yuma Project," RG 115; Lippincott to Governor George Pardee, May 14, June 4, 1904; Morris Bien to Pardee, May 27, July 26, 1904; Bien to A. B. Nye (Pardee's private secretary), May 27, 1904, all in the Pardee Collection, Bancroft Library; U. S. Webb (California attorney general) to Arthur C. Huston, October 7, 1913, Hiram Johnson Collection, Bancroft Library; *Pacific Rural Press* 66 (October 17, 1903): 246; J. B. Lippincott, "The Reclamation Service in California," *Forestry and Irrigation* 10 (April 1904): 162–169.

of federal reclamation grumbled that the initial projects contained mainly private land, which they thought violated the homemaking spirit of the Newlands Act. A decision to *force* federal reclamation on private landowners would have compounded the problem. In any case, the federal government stood both to gain and to lose from supporting state control over Clear Lake. Morris Bien often argued that the West's surplus water belonged to the nation, not to the states, a position inconsistent with state sovereignty at Clear Lake.[34]

The Reclamation Service also encountered recalcitrant water users on the Kings River. As at Clear Lake, the Geological Survey had already surveyed storage sites along the river and published its findings in *Water Supply and Irrigation Paper Number 58*. J. B. Lippincott conducted the survey and estimated that a reservoir at Clark's Valley could store 217,196 acre-feet of water, at an average cost of $10.15 per acre-foot, and a reservoir at Pine Flat could store an additional 78,197 acre-feet. Lippincott found the appeal of integrating the half-dozen major canal companies, and dozens of smaller ditches, overpowering. Here was an opportunity to promote efficiency and eliminate conflicts among water users. Even though all the land around Fresno had long since passed into private ownership, the town served as the "raisin capital" of the San Joaquin Valley, and irrigated crops from that region returned much more per acre than land in the Cache Creek Valley, where wheat remained the dominant crop. Moreover, in 1900 a majority of water users on the Kings River had organized the Kings River Storage Association. They had several motives. First, they wanted a federally financed dam. Farmers along the Kings River usually had an adequate supply of water for irrigation in May and June, but not during the summer or fall. Moreover, after years of litigation, the largest water companies had buried the hatchet. In 1897, the Fresno Canal and Irrigation Company, Peoples Ditch Company, Last Chance Water Ditch Company, and Lemoore Canal and Irrigation Company signed a voluntary agreement allocating water among the four companies. Subsequently, several mutual water

---

34. On the issue of state versus federal sovereignty over the West's water, see Donald J. Pisani, "State vs. Nation: Federal Reclamation and Water Rights in the Progressive Era," *Pacific Historical Review* 51 (August 1982): 265–282.

companies also ratified the agreement. At least in theory, Kings River irrigators agreed that the $40,000 average annual expenditure for litigation over the normal flow of the river could be better spent building dams to store floodwater to augment the supply. A new supply of water would reduce conflict, at least if the recipients could agree how to apportion it.

Apparently, J. B. Lippincott won the support of L. A. Nares of Liverpool, who managed the Fresno Canal Company and Consolidated Canal Company for the English capitalists who owned the ditches; Frank Short, who owned or managed the 60,000-acre Laguna de Tache ranch, which held extensive riparian rights; and the Alta Irrigation District, whose water supply had been cut off by the courts every year on July 1. While these interests wanted reservoirs and a coordinated irrigation system, however, they balked at repaying the government or turning their water rights over to the Reclamation Service. To make matters worse, Newell heard rumors that Lippincott was speculating in land near the river. The Reclamation Service's chief discounted the charge. Nevertheless, he wrote Lippincott on November 27, 1903:

> the more I consider the matter the less desirable it seems to me it is for us to go into the King [sic] River country at present, or until we have reclaimed public land somewhere in California. We are the subject of scrutiny and attack from various sources [nationally] and the most serious charge which can or may be made against us is that we are diverting the work from the reclamation of public land to assist speculators in disposing of their holdings. For the present at least we must be extremely cautious not to take up enterprises which give any foundation for this charge. In all of the other States public land reclamation is under consideration although the area of public land may not be very large. Even in Arizona [the Salt River Project] we have reserved large areas of public land around Phoenix and are in a position to reclaim this if water is not taken by owners of small tracts. In the case of the King [sic] River, there is no public land to reclaim. . . . There has been no popular movement to induce us to take up work and if we begin operations here it will be charged that we have been induced to do so by the speculative element. In short,

the men who are clamoring for examination and construction of works in the part of California where there are public lands, will make it extremely uncomfortable for us and perhaps destructive if we voluntarily neglect them and take up the work on King River. In fact I do not think we could recommend this to the Secretary of the Interior in the face of the strong opposition and statement that we are interfering with private enterprise.

Ten days later Newell reminded Lippincott that 1904 was an election year and that the enemies of federal reclamation would exploit any opportunity to discredit the Roosevelt Administration's policies. The Reclamation Service's director reported an "active, persistent and definite" *popular*, as opposed to corporate, demand for federal reclamation in the Honey Lake Basin, sixty miles north of Lake Tahoe. However, Lippincott quickly responded that the public land left in that valley was of poor quality and was located too far from water.[35]

The Reclamation Service refused to give up on the San Joaquin Valley, even in the absence of attractive projects. An engineering board investigated 7,880 square miles of land from Bakersfield to Madera County, north of Fresno, in the fall of 1905. This tract still contained scattered blocks of government land, though they were far removed from surface water sources. The board, which included Newell's second-in-command, Arthur P. Davis, reported: "Within this area it is found that the amount of underground water, which lies near the surface, and is available for irrigation purposes is enormous and that its value is but little recognized." Cheap power was needed to raise the water to the surface, so the board recommended putting the entire Tuolumne River watershed perma-

35. For a superb discussion of Kings River water conflicts, see Maass and Anderson, . . . *and the Desert Shall Rejoice*, chap. 6, especially pp. 232–237. Also see the *Second Annual Report of the Reclamation Service, 1902-1903* (Washington, D.C., 1904), 53–54; Lippincott and Sanders to Newell, October 15, 1903, "(153–11), Yuma Project;" RG 115; J. B. Lippincott to Hydrographer, U.S.G.S., July 26, 1904, "(930–8), California: Surveys and Investigations, Kings River Project"; Newell to Lippincott, November 27, December 7, 1903, in the same file. That file also contains accounts of meetings of the Kings River Storage Association held on November 28, December 10, 1903. Finally, see Lippincott to George Pardee, December 1, 1903, April 20, 1904, Pardee Collection, Bancroft Library.

nently off-limits to private developers to protect potential power plant sites. But the plan ran into two snags. By the end of 1905, most of the reclamation money earmarked for California had been spent or committed; there was no money left to build hydroelectric plants. In addition, Newell hesitated to reserve such a large watershed. Throughout the West, private developers claimed that the Reclamation Service had blocked legitimate projects by reserving land and reservoir sites it could not put to immediate use. Newell treated many of his western critics with thinly veiled contempt, but he was more cautious in California, particularly since the city of San Francisco had applied to the federal government for the right to dam the Tuolumne and to flood Hetch-Hetchy Valley. Other states, such as Nevada, had a much greater need for federal reclamation and a small delegation to Congress; the howls of critics there were not likely to undermine the federal program. The board's recommendations went unheeded.[36]

Elimination of the Clear Lake and Kings River projects in 1903 and early 1904 left the Reclamation Service with only one major irrigation project. Vast tracts of rich, alluvial soil bordered the Colorado River in California and Arizona, and most of it still belonged to the nation. Because frosts and freezes were all but unknown in the warm, dry desert climate, farmers could raise abundant crops year round; pioneer irrigators around Blythe and Needles boasted of five or six alfalfa crops a year and wheat harvests of sixty bushels an acre. The Colorado contained an abundant water supply and also offered many potential reservoir sites. Residents along the river seemed eager to cooperate with the Reclamation Service, whose officials hoped that a new mining boom in Arizona would create ready markets for farm products raised on a federal project. Lippincott crowed: "here is an opportunity to 'Build the State.' Here is a sleeping empire at our doors awaiting the touch of some Siegfried to awaken it."[37]

36. J. B. Lippincott, A. P. Davis, W. C. Mendenhall, and O. H. Ensign to F. H. Newell, November 22, 1905; and Newell to Davis, December 2, 1905, "(989), Sacramento Valley; General Reports thru 1908," RG 115; *Fifth Annual Report of the Reclamation Service, 1906* (Washington, D.C., 1907), 99.

37. J. B. Lippincott to Hydrographer, U.S.G.S., August 6, 1903, "(930), California: Surveys and Investigations thru 1910," RG 115; Lippincott and Sanders to

The most important attraction, at least to Lippincott, was the 500,000 acres of land in the Imperial Valley. The Colorado River Basin offered plenty of good public land, but most of it was too far above the level of the stream to irrigate using gravity-fed canals—even with the help of diversion dams to raise the water. Throughout the West, the Reclamation Service competed with land-grant railroads and land development companies for settlers, so it had a strong incentive to keep the cost of reclamation as low as possible. Since the Imperial Valley was below the level of the river and contained many thousands of acres of government land, it offered a chance to expand the project and reduce its per-acre cost. The valley had been opened to irrigation in June 1901 by the California Development Company, which by 1904 provided water to 70,000 acres. However, the company's canal followed an old flood channel through Mexico before it reentered the United States and farmers complained that they were receiving less water than promised. In April 1904, Lippincott reported to Governor George Pardee:

> The farmers at that locality [in the Imperial Valley] informed me that their losses, due to insufficient water supply, this spring will aggregate $600,000. This is because the canals are not adequate to carry the water. Water rights have been sold for 271,000 acres of land in California, and the capacity of the canal is not sufficient in my judgment for the irrigation of more than say 60,000 or 70,000 acres. In addition to this I understand the Company has numerous contracts in Mexico [to provide water to private lands]. We are keeping our hands off the situation, but I think something will break down there pretty soon and when the break comes we hope to be in a position to stand in the breach.

Lippincott acknowledged that an "All-American Canal" directly from the Colorado River into the valley would reduce the flow of

---

Newell, October 15, 1903, "(153–11) Yuma Project"; *First Annual Report of the Reclamation Service from June 17 to December 1, 1902* (Washington, D.C., 1903), 121; "Possible Irrigation Projects [in California]," *Forestry and Irrigation* 9 (March 1903): 142–145; Sharlot M. Hall, "The Problem of the Colorado River," *Out West* 25 (October 1906): 305–322.

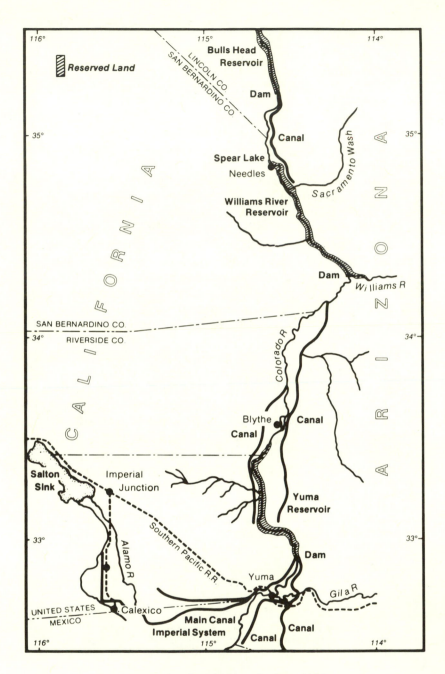

Proposed Colorado River Project, 1903.

water into Mexico and might precipitate international conflict. Nevertheless, he urged that "we certainly should not give up a very large amount to them."[38]

Despite Lippincott's enthusiasm for building the Colorado River Projects around the Imperial Valley, other Reclamation Service officials, including A. P. Davis and George Wisner, opposed his recommendation. Norris Hundley has suggested that the Reclamation Service backed off from buying out the California Development Company because the United States could not own property in Mexico. He also implies that the company, by opening a new stretch of canal through Mexico in May 1904, increased the water flow into the Imperial Valley and reduced the dissatisfaction of the farmers.[39] Lippincott's critics also thought that adding the Imperial Valley was financially infeasible. The $20 an acre cost of the All-American Canal, added to the debt incurred by purchasing the company's property, would have dramatically increased the cost of irrigation in the valley. George Wisner commented:

> It seems to me that the present is not the proper time for the Government to purchase the rights of the California Development Co. The Company claims that it has valuable assets in the way of lands and water rights and that if left alone they can successfully reclaim the lands of the Imperial Valley. They have expended a large amount of money to inaugurate this

38. The long quote is from J. B. Lippincott to George Pardee, April 20, 1904, the shorter quote from Lippincott to Pardee, December 1, 1903, both in the Pardee Collection, Bancroft Library; Lippincott to Newell, September 23, 1904, "(187–B), Colorado River Project, Board and Engineering Reports and General Correspondence, Imperial Valley thru 1909," RG 115; *Irrigation Age* 20 (February 1905): 103. Lippincott strongly opposed the California Development Company's claim to 10,000 second-feet of water from the Colorado River, insisting that such a grant would allow the company to monopolize the flow of the river four out of every five years. See his letter to F. H. Newell, February 29, 1904, "Irrigation, 4: Diverting the Waters of the Colorado, Imperial," Box 9–B, Thomas R. Bard Collection, Huntington Library. Also see the file entitled "Irrigation, 5: Colorado, Imperial, Rio Grande, Mexico," in the same collection. For background to the history and water conflicts of the Colorado River, see Norris Hundley, *Dividing the Waters: A Century of Controversy Between the United States and Mexico* (Berkeley, 1966); *Water and the West: The Colorado River Compact and the Politics of Water in the American West* (Berkeley, 1975).

39. Norris Hundley, *Water and the West*, 26. Many farmers in the Imperial Valley also opposed federal reclamation because of the 160-acre limitation. See Lippincott to Newell, January 14, 1908, "(187–B), Colorado River Project," RG 115.

project, and should be given a fair opportunity to demonstrate
what they can do. If they are successful they should be given
the right to use such amount of water of the Colorado River
as they actually put to beneficial use, and if they are not
successful their property rights will assume their true value,
and the Government may then be able to take up the project
on such a financial basis as to render it a feasible project to
undertake.

In 1904 to 1907, a massive flood all but destroyed the California
Development Company's Imperial Valley empire and led to renewed calls for an All-American Canal. However, the Reclamation
Service was unable to raise the money to build the expensive aqueduct until 1928.[40]

Davis's opposition to Lippincott's All-American Canal underscored a sharp disagreement over how the Colorado River should
be used. Lippincott pushed for building diversion canals as quickly
as possible, even before a thorough topographic map of the river
basin had been completed. In March 1903, he warned Newell that
the Reclamation Service would face severe public criticism unless
it began work immediately. The construction of large dams on the
Colorado, he claimed, would tax the Reclamation Service's limited
budget and delay the construction of other western projects. Davis
took a more comprehensive view, arguing that the river's tremendous load of silt could be turned to advantage. He proposed four
dams along the lower Colorado, the first at the mouth of the Williams River above Parker, Arizona. This "high dam" would capture
enough water to irrigate 400,000 acres and generate electricity besides. And as it filled with silt, thousands of acres behind the dam,
particularly flood plains, alkali flats, and other low-lying areas,

40. George J. Wisner to F. H. Newell, November 29, 1904, "(187–B), Colorado
River Project: Board and Engineering Reports and General Correspondence, Imperial Valley thru 1909," RG 115. For similar sentiments see A. P. Davis to Wisner,
December 5, 1904, in the same file. Congress finally authorized the All-American
Canal when it passed the Boulder Canyon Act in 1928, though the ditch was not
completed until the early 1940s. The hostility of officials of the California Development Company toward the Reclamation Service reflected the resentment of private land and ditch companies whose plans were challenged or thwarted by federal
reclamation. See L. M. Holt, "How the Reclamation Service is Robbing the Settler,"
Overland Monthly 50 (November 1907): 510–512.

would gradually be reclaimed or "produced." After the silt destroyed the reservoir's storage capacity, water could be released from the dam's lower outlets, exposing the new farmland. Meanwhile, the Reclamation Service would have constructed another dam farther upstream. Eventually, Davis hoped that 1,200,000 acres could be reclaimed or created along the river at a cost of $22,000,000.[41]

Pending thorough surveys, the Reclamation Service initially withdrew from entry all the arable public land along the Colorado River from the Grand Canyon to the Mexican line, as well as several major reservoir sites. However, the Service's limited budget did not permit launching Davis's bold project, and government engineers had great difficulty finding good reservoir sites on the lower Colorado. There, the river's gradient was so small that the accumulated silt from 225,049 square miles of land upstream settled out. Engineers found the closest bedrock at Bulls-Head, thirty miles upstream from Needles, at a depth of 130 feet.

By the summer or fall of 1903, the Reclamation Service decided to build the first unit in the Colorado River Project at Yuma, on the California-Arizona border. The Southern Pacific Railroad ran through the town, providing the transportation needed to build the irrigation works and later carry crops to market. Moreover, a large majority of local farmers supported federal reclamation. Farming around Yuma began in 1897, but the Colorado often flooded alluvial farmland and local ditch companies built poorly constructed, uncoordinated canals. Irrigators formed the Yuma County Water Users Association in November 1903 and appealed to the federal government for assistance. Equally important, the Yuma Indian Reservation offered thousands of acres of easily irrigable land adjoining the river in California.[42]

In 1904, the Reclamation Service persuaded Congress to split

41. A. P. Davis to J. B. Lippincott, October 10, 1902, "(187), Colorado River Project, thru 1905," RG 115. Also see Lippincott to Davis, October 18, 1902; Davis, George Wisner, and W. H. Sanders to Newell, September 26, 1904 in the same file; Lippincott to Newell, March 23, 1903, "(930–8), California: Surveys and Investigations, Kings River Storage"; *Irrigation Age* 18 (December 1902): 53–54; *Second Annual Report of the Reclamation Service* (Washington, D.C., 1904), 123–161.

42. J. B. Lippincott, "The Reclamation Service in California," *Forestry and Irrigation* 10 (April 1904): 162–169; Hall, "The Problem of the Colorado River," 312; U.S. Reclamation Service (Portland, Ore. Office) to George Pardee, August 25, 1905, Pardee Collection, Bancroft Library.

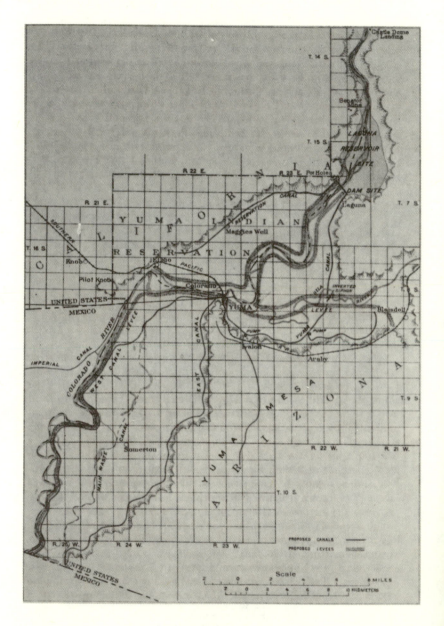

Yuma Project, 1903. *Third Annual Report of the Reclamation Service,
1903–1904* (Washington, D.C., 1905).

up the Indian reservation and allot each Indian five acres. Apparently, Congress bought the allotment scheme as a way to promote the family farm—a policy consistent with the earlier Dawes Act. After allotment, about half the remaining land, or 6,500 acres, was thrown open to settlement under the Newlands Act. The Reclamation Service's motive was clear. The Indian reservation included the largest block of public land within the Yuma Project. Although the project contained 97,000 acres in 1906, only 27 percent of the land belonged to the nation. Once Congress provided for allotment, a board of engineers approved the project on April 8, 1904, and the secretary of the interior promised $3,000,000 to pay for construction on May 10.[43]

The project contained four major features: a settling dam, irrigation canals in both California and Arizona, flood control levees, and drainage ditches to prevent the buildup of alkali and remove seepage water that collected behind the levees. The Laguna Dam, constructed twelve miles upstream from Yuma, served as the capstone of the project. It was erected at a site where low granite hills came nearly to the edge of the river, offering substantial abutments. These, in themselves, were not sufficient to anchor the dam. Its design repeated many features of the British "floating dams" on the Ganges and Nile rivers; the structure's 600,000-ton weight helped keep it in place even though it could not be anchored to bedrock. The 226-foot-wide, nineteen-foot-high dam was not designed primarily to store water; instead, it provided a ten-mile-long settling basin. The headgates could be raised to skim off the upper foot of water from the river into canals, leaving the silt behind. At the base of the dam, sluice gates provided a way to flush the dam of debris as it accumulated.[44]

Unfortunately, like most other government irrigation schemes,

43. Ethan Allen Hitchcock, Secretary of the Interior, to Director of the U.S.G.S., May 10, 1904, "(153–11), Yuma Project: Reports of the Engineers, Estimates of Construction, Etc. thru 1906," RG 115; *Los Angeles Times*, May 10, 1904; *Ninth Annual Report of the Reclamation Service, 1909–1910* (Washington, D.C., 1911), 74. The Reclamation Service exhibited little interest in the welfare of American Indians, but plenty of interest in their land. It persuaded Congress, as a logical extension of the Dawes Act, to allot land in very small parcels, a step made possible by irrigation. Of course, this opened new land to white farmers. The fascinating relationship between federal reclamation and the allotment policy remains to be studied by historians.

44. *Official Proceedings of the Thirteenth International Irrigation Congress, 1905* (Portland, Ore. 1905), 209–210; "National Irrigation: Work and progress under the

the Yuma Project failed to live up to expectations. The Reclamation Service suffered from a "love-hate," or perhaps a "dependence-avoidance," relationship with private landowners and ditch companies. It faced a staggering dilemma: federal reclamation had been sold to Congress largely as a plan to provide homes for landless urban workers, tenant farmers, and other disgruntled residents of the humid East; but these people lacked the experience to build successful desert irrigation projects. Even though Newell and his lieutenants stubbornly, and foolishly, refused to exploit the skills and experience of the Department of Agriculture, they did appreciate the need for experienced, seasoned farmers to insure the stability of projects and help teach and "acclimate" new residents. Nevertheless, established farmers, whose water rights frequently antedated the Newlands Act by many years, often perceived national reclamation as a threat rather than an opportunity. They expected some compensation for forfeiting the advantage time had given them to enter repayment contracts with the federal government. Invariably, Reclamation Service officials promised established water users reliable canals and a water supply equal to that which they had enjoyed in the past. In some cases they promised even more water. But many old farmers joined up for a more immediate reason: to make money selling land to new settlers.

Ironically, the Newlands Act's famous 160-acre limitation encouraged land speculation. The law did not require farmers to divest themselves of *all* land over 160 acres immediately, especially if they left excess holdings unirrigated. Moreover, the law provided no specific procedure to regulate the disposal of surplus land. Since 1902, the process has varied from time to time and place to place; but everywhere farmers have played important roles in determining Reclamation Bureau policy. Speculation was not confined to large landowners. In most projects, the 160-acre limitation was a moot restriction, because the secretary of the interior confined the size of farms to as little as forty acres; on the Yuma project, farms varied from forty to one hundred acres, depending on the location and quality of land. This limitation was designed to promote in-

Reclamation Act," *Maxwell's Talisman* 6 (March 1906): 27; "The Yuma Reclamation Project," *Forestry and Irrigation* 12 (March 1906): 143–144; Hall, "The Problem of the Colorado River," 324–325. For technical features of the Yuma Project, also see George Wharton James, *Reclaiming the Arid West* (New York, 1917), 86–102; *Fund for Reclamation of Arid Lands*, 61st Cong., 3d sess., 1911, H. Doc. 1262 (serial 6022), 29–33.

tensive farming and community life, as well as reduce the cost of construction by keeping projects compact, thus reducing the length of canals. However, this sop to scientific farming encouraged even more speculation, and Reclamation Service officials often winked at outrageously inflated land prices within the projects, especially during the boom of World War I.[45]

By 1907, the Reclamation Service had signed up 91 percent of the private landowners within the Yuma Project; it had also bought out the four or five major canal companies within project boundaries. Here again, the need to pacify local vested interests left a legacy of suspicion among new settlers. The Service paid considerably more than the canal systems were worth to avoid ugly condemnation suits. As mentioned earlier, in many parts of the arid West private companies persistently assaulted the Reclamation Service as arbitrary and capricious in its treatment of vested rights. The nation had the power to condemn all property necessary to build irrigation projects, but the Newlands Act failed to provide a legal process to speed up the resolution of suits. And the cost of litigation swelled the debt of farmers as did the cost of "accommodation" out of court. Usually the Service favored the latter alternative to protect its reputation and speed up construction. But such arrangements angered many farmers. Not surprisingly, by 1909, when the Laguna Dam was finished, the per-acre cost of reclamation within the Yuma Project had ballooned from the initial estimate of $35 an acre to $55 an acre. Two years later, it stood at $65 an acre and by 1917 reached $75 an acre.[46]

45. For early examples of Reclamation Service concern with land speculation, the 160-acre limitation, and winning the support of established farmers, see J. B. Lippincott to George Maxwell, Executive Chairman of the National Irrigation Association, May 2, 1903, "(153–11), Yuma Project: Reports of the Engineers, Estimates of Construction, Etc., thru 1906," RG 115. Davis et al. to Newell, April 8, 1904, in the same file. Also see the *Fifth Annual Report of the Reclamation Service, 1906* (Washington, D.C., 1907), 100. For an excellent discussion of the 160-acre limitation, see Clayton R. Koppes, "Public Water, Private Land: Origins of the Acreage Limitation Controversy, 1933–1953," *Pacific Historical Review* 47 (November 1978): 607–636.

46. *Ninth Annual Report of the Reclamation Service, 1909-1910* (Washington, D.C., 1911), 78. The miscalculations of Reclamation Service officials contributed to the fact that project construction costs usually ran far higher than initial estimates. For example, contrary to expectations, Yuma Project officials found that citrus fruits would not grow well on the project's lowlands. Subsequently, water was pumped up to more fertile mesa lands which produced lush citrus crops and demonstrated the feasibility of diversified farming in the desert (at an enormous cost ultimately borne by the settlers).

Such problems simply scratched the surface. The Colorado River had not been systematically measured during the 1890s—a wet decade save for its closing years. In its enthusiasm and inexperience, the Reclamation Service overestimated the water supply for irrigation and underestimated the thirstiness of desert soils. This error was particularly dangerous on the lower Colorado where many farmers practiced irrigation year round. In 1912, the Service reported that from 1902 to 1911, the stream's flow varied from a maximum of 25,900,000 acre-feet to a minimum of 7,960,000 acre-feet with a mean of 16,500,000. This was much less water than anticipated, and the erratic volume underscored the need for storage, which A. P. Davis had perceived from the beginning. Without upstream storage, which the Reclamation Service began to search for once again in 1914, no new Colorado River units were possible.[47]

As of June 30, 1912, the Reclamation Service claimed that the Yuma Project was 73.7 percent complete, but only 10,500 acres were under cultivation. Much of this land had been watered in 1902, and instead of high-value citrus crops, farmers raised mainly alfalfa and occasionally barley or corn. Canals silted up much faster than expected, and drainage ditches did not effectively solve the problem of alkali buildup in the ground water and topsoil. Both problems added to the expense faced by project residents. The Reclamation Service continued to dazzle the nation with its technological ingenuity. For example, on the Yuma Project a 930-foot-long inverted siphon carried water under the Colorado River into the Yuma Valley. But this work could not hide the rate at which settlers abandoned their farms. Within the Yuma Indian Reservation, 173 homesteads were opened to settlers beginning in March, 1910. Of these, 68 were forfeited in 1910; 17 in 1911; 10 in 1912; and 4 in 1913. When Californian Franklin K. Lane became secretary of the interior in 1913, he launched an investigation of the much-criticized federal reclamation program. The head of a special committee appointed to study conditions on the Yuma Project concluded:

The local water users on the Reservation Unit in the Yuma

47. *Tenth Annual Report of the Reclamation Service, 1910–1911* (Washington, D.C., 1912), 72; *Fourteenth Annual Report of the Reclamation Service, 1914-1915* (Washington, D.C., 1915), 323–324; Report of J. B. Lippincott to Arizona and California River Regulation Committee, October 1, 1913, "(187), Colorado River Project thru 1913," RG 115.

Project, some one hundred and seventy-five in number, have been deceived, mistreated and burdened to such an extent that unless relief is afforded they must abandon their farms and seek homes elsewhere. Large sums of money have been expended in the construction of an irrigation system, the Legune [sic] Dam, the carrying and lateral systems, drainage systems, which is a failure, and the general physical condition of the works and the installation of the structures is evidence at the present time of the incapacity, wastefulness and improper administration of the affairs.

Since the purpose of such complaints was to reduce the debt of project farmers, these pleas exaggerated problems on the project. Many conditions, including the string of dry years from 1909 to 1911, were beyond the government's control. Still, the sheer number of complaints in Reclamation Bureau files help explain Lane's decision to fire Frederick Haynes Newell in 1915. Fair or not, on the nation's irrigation projects, Newell and his lieutenants were portrayed frequently as heartless, bungling, petty tyrants, totally unresponsive to the criticism of settlers.[48]

Had the pace of settlement lived up to expectations, the Reclamation Service would have escaped much of this criticism. But to government engineers, half a project was no project at all, and the canal network expanded much faster than the rate of settlement. Land served by ditches increased from 16,000 acres in 1912 to 71,200 acres in 1915. However, in the latter year only 27,000 acres were irrigated. The Reclamation Service's report for 1915–1916 listed the project's size as 128,000 acres, about 30 percent more land than the original Yuma Project approved in 1904.[49]

48. Charles Johnson to John D. Works, U.S. senator from California, August 14, 1915, John Works Collection, Bancroft Library. Also see Everett Teasdale to Works, August 3, 1915, in the same collection. The "Yuma Irrigation Project" files, RG 115 for 1913–1915, contain many similar complaints. Partly as a result of this discontent, the Reclamation Service allowed for graduated repayment of individual debts to the government. This had been authorized by Congress on February 13, 1911, but applied only to those irrigation projects designated by the Reclamation Service for special consideration. See *Eleventh Annual Report of the Reclamation Service, 1911–1912* (Washington, D.C., 1913), 204–205.

49. *Fourteenth Annual Report of the Reclamation Service, 1914-1915* (Washington, D.C., 1915), 59–63; *Fifteenth Annual Report of the Reclamation Service, 1915–1916* (Washington, D.C., 1916), 5–6, 13, 68–83. The Reclamation Service frequently changed

That southern California had been chosen as the home of the state's first federal reclamation project angered many northern Californians. The population of southern California grew at a much faster rate than that of the northern section, and the Yuma Project offered one more inducement for immigrants to settle south of the Tehachapis. Politics dictated that the second project would belong to the north, and after considerable searching, the Reclamation Service found a potential project on the California-Oregon border. In its quest to irrigate public land, the Service discovered that by draining Tule and lower Klamath lakes, over 100,000 acres of prime farmland would be exposed. Nevertheless, the federal government faced a thorny legal problem. Though the lakes were navigable interstate bodies of water, and hence under federal jurisdiction, once drained the lake beds belonged to the states. Federal officials used gentle pressure to persuade Oregon and California to deed the land to the nation. For example, Lippincott advised California Governor George Pardee in September 1904: "We feel very much inclined to vigorously push a large project for Northern California, and the adjustment of this matter with reference to these lake beds we consider a vital matter in the case."[50]

The Klamath Project, as it came to be called, held many attractions besides a large block of public land. Government officials and settlers alike initially assumed that drained land within the Klamath Project would be as rich and productive as reclaimed marshland in the Sacramento Valley. The soil was volcanic, and volcanic soils usually yielded abundant crops. In addition, the lake beds were level, and required little preparation for irrigation. The Klamath River provided an abundant water supply, and the re-

---

the size of projects. On the Yuma Project, the project's size was listed at from 86,000 acres to over twice that area, largely depending on whether future additions were included. Usually, the project was listed as between 90,000 and 110,000 acres.

50. J. B. Lippincott to George Pardee, September 29, 1904, Pardee Collection, Bancroft Library. Also see Lippincott to Pardee, March 9, 1905. The Reclamation Service drafted the state act transferring ownership of the lake beds to the nation. See Morris Bien to Pardee, December 23, 1904; January 27, 1905, Pardee Collection. Copies of the California law, and a similar Oregon statute, are included in "(281), Klamath: Engineering Reports thru 1905," RG 115. The federal law was approved on February 9, 1905. It was reprinted in the *Fifth Annual Report of the Reclamation Service, 1906* (Washington, D.C., 1907), 20.

gion's mountain lakes promised cheap storage facilities. Moreover, such an interstate project, where water stored in one state was used in another, raised legal issues private ditch companies could not solve; so the nation could not be considered an interloper. Finally, the scheme appealed to Reclamation Service engineers because it involved swampland reclamation and flood control, as well as irrigation.

As usual, before the government could begin work, peace had to be made with established interests. As soon as word of the project leaked to the public, land speculators tried to file on the land under Tule and lower Klamath lakes in state land offices; the Reclamation Service enlisted the aid of Governor Pardee to block these entries.[51] Most of the region's private ditch companies welcomed the opportunity to sell out to the federal government, but the Klamath Canal Company proved more stubborn. This company had been formed in January 1904 to construct a canal from upper Klamath Lake to Tule Lake. It wanted to reclaim 60,000 acres and offered irrigators and potential irrigators water at $10 an acre— a price that persuaded the owners of 20,000 acres to enter contracts with the firm. Nevertheless, the promise of cheaper water from the government undermined the company's plans. As a matter of routine policy, the Reclamation Service withdrew all the public land, reservoir sites, and hydroelectric power sites in the region, and Lippincott urged Newell to use the Service's withdrawal power, and power to deny rights-of-way, to checkmate the company. Although the promoters promised to begin construction as soon as possible, Lippincott smelled speculation: "Personally I believe it is the same old proposition over again of a Civil Engineer seeing some natural opportunities, endeavoring to seize them, work them up into shape, and dispose of the proposition to some other outfit he may find who will build." The War and Justice departments supported the Reclamation Service's cause on grounds that the company's diversion from a navigable interstate body of water was illegal without federal approval. Neither Klamath nor Tule Lake

51. J. B. Lippincott to George Pardee, January 29, 1905, Pardee Collection, Bancroft Library. James, in *Reclaiming the Arid West*, 304–305, estimated that land under ditch that sold for $20–$30 an acre before the Klamath Project opened sold for $50–$100 an acre by 1917. Land not served by canals appreciated in value from $5–$6 an acre to $10–$20 an acre during the same period.

was used extensively for transportation, but early in 1905 the Justice Department secured a temporary injunction blocking the use of upper Klamath Lake for irrigation.[52]

Originally, the Klamath Canal Company offered to sell its property and water rights to the federal government for $250,000, but once the Justice Department filed suit to prevent diversions from Klamath Lake, the company cut its price, ultimately settling for $150,000.[53] In 1906, the Reclamation Service spent another $170,000 to purchase the property of the Jesse D. Carr Land and Livestock Company, which included an important reservoir site at Clear Lake in Modoc County, just south of the California-Oregon border. The Klamath Project Water Users Association later charged that the Carr property had been up for sale at $35,000 before the government moved into the region and that, in its haste to cut out private interests and get the project started, the government sharply increased the project's cost, thus imposing an added debt on farmers.[54]

As approved in 1905, the Klamath Project contained 236,000 acres, and government engineers estimated the project's cost at $4,500,000, or about $18 an acre. Because the Reclamation Service decided to use natural reservoirs at Clear Lake in California and Horsefly Lake in Oregon, the project promised the lowest per-acre

52. The quote is from J. B. Lippincott to the Chief Engineer (F. H. Newell), July 29, 1904, "(281), Klamath: Engineering Reports thru 1906," RG 115. Also see Lippincott, et al., to the Chief Engineer, April 14, 1905 in the same file; Lippincott's speech, "The Klamath Project—Status of Investigations, 1904," *Proceedings of the Twelfth International Irrigation Congress of 1904* (Galveston, Tex. 1905), 190–198; N. C. Briggs to E. J. Mitchell; Briggs to Thomas R. Bard, July 23, 1904; J. B. Lippincott to Bard, August 15, 1904, "Irrigation," Box 7-D, File 3, Thomas Bard Collection, Huntington Library. Bard to Lippincott, August 3, 1904; Bard to Frank Flint, July 28, 1904, "Irrigation: National, 1," Box 9–A, Thomas Bard Collection, Huntington Library.

53. Memorandum of Proposed Recommendations to be made by the Board of Engineers of the United States Reclamation Service to the Secretary of the Interior Relative to the Purchase of the Property of the Klamath Canal Co., May 3, 1905, "(281), Klamath: Engineering Reports thru 1905," RG 115; *Fifth Annual Report of the Reclamation Service, 1906* (Washington, D.C., 1907), 29.

54. *Forestry and Irrigation* 12 (November 1906): 518; *Irrigation Age* 22 (December 1906): 53. For settlers complaints about the Reclamation Service's acquisition of private property, see *Report of the Committee on Irrigation and Reclamation of Arid Lands on the Investigation of Irrigation Projects*, 61st Cong., 3d sess., 1911, S. Rep. 1281 (serial 5846), 718–731. The Service's director, F. H. Newell, denied that the government had paid too much for reservoir sites. In each case, he argued, it made a "good bargain." Project officials responded to the charges on pp. 747-762.

reclamation cost of any federal scheme. However, the Service decided that only about 50,000 acres under lower Klamath and Tule lakes could be easily reclaimed, so the percentage of public land within the project shrank to about 23 percent by 1906. Not surprisingly, given cattle ranching's dominance in the region's economy, 40,000 acres within the proposed project were held in tracts larger than 160 acres.[55]

The Klamath Project received intense criticism from the farmers it served. As on the Yuma Project, settlement lagged. This helps explain why the average $30 per acre construction debt in 1911 was nearly double the cost estimate made by Reclamation Service officials in 1904 and 1905. In addition, much of the project's marshland turned out to be worthless, and the per-acre debt jumped sharply when government engineers reduced the project's total acreage from 236,000 to 160,000 acres. The Reclamation Service spent $300,000 trying to reclaim the Klamath Lake bed only to find that the soil was choked with lime and black alkali. In many parts of the arid West, alkali could be leached or washed out of the soil. But this required an extensive, expensive drainage system, and most of the project soil was not porous enough to flush clean. The cost of buying up riparian rights on Lower Klamath and Tule lakes, along with the cost of the Carr property and four private irrigation systems, also inflated project costs. Not surprisingly, when the U.S. Senate Committee on Arid Lands met with the Klamath Project Water Users Association during an inspection tour of federal reclamation projects in 1910, the farmers complained about poor construction and inefficient service and called for "practical men in charge of the various construction and engineering work; not boys, but men with some practical knowledge of irrigation and farming."[56]

55. *Forestry and Irrigation* 11 (June 1905): 276–278; 12 (March 1906): 115–118; *The Irrigation Age* 22 (November 1906): 8-12; *Fifth Annual Report of the Reclamation Service, 1906* (Washington, D.C., 1907). Because of the sale of excess holdings, the average size farm within the Klamath Project dropped from 104 acres in 1907 to 69 acres in 1910 (*Tenth Annual Report of the Reclamation Service, 1910–1911* [Washington, D.C., 1912], 202–207).

56. *Report of the Committee on Irrigation and Reclamation of Arid Lands on the Investigation of Irrigation Projects*, 717. In part because of mounting criticism from farmers, President William Howard Taft ordered the Army Corps of Engineers to inspect the government reclamation projects in 1910. The Army's report on the Klamath project is in *Fund for Reclamation of Arid Lands*, 61st Cong., 3d sess., 1911,

For all the above problems, the government's biggest mistake was in choosing a location so far removed from large agricultural markets, especially a location where the average fifteen inches of rainfall per year allowed many farmers and stockmen to get by without irrigation. The project was 451 miles from San Francisco and 443 miles from Portland. And while the Southern Pacific's rail line ran directly through Yuma, farmers on the Klamath Project were 36 miles or more from the Klamath Lake Railroad's closest station. In 1916, when the Klamath Project was 62 percent complete, 30,123 acres of public land remained open to settlement and only about 27,000 acres were under irrigation. Project farmers raised forage crops which returned a low $13.85 per acre in that year. Moreover, the number of project farms declined from 405 in 1912 to 352 in 1915, and the number of tenant farmers increased from 40 to 105 during the same period. As on the other federal projects, prosperity came only with World War I.[57]

Forced to confine their efforts to remote sections of California, Reclamation Service officials hoped one day to begin a larger reclamation project that would win greater public recognition and political support. As mentioned earlier, in 1902 and 1903 the Sacramento Valley, especially Clear Lake, seemed a happy choice. The valley's rich soil largely eliminated the alkali and drainage problems suffered by farmers on the Yuma and Klamath projects. Moreover, the Sacramento River carried an enormous volume of water, and the rim of the valley contained many suitable reservoir sites. Since 1896, the U.S. Geological Survey had measured the stream's volume at a station in Iron Canyon, above Red Bluff. Finally, the wheat

H. Doc. 1262 (serial 6022), 119–127. Klamath Project farmers wanted graduated payments and an extension of the repayment schedule from ten to twenty years. They also wanted the government to write off project costs incurred through the ignorance or bad judgment of Reclamation Service officials. Such mistakes, they claimed, doubled the cost of irrigation works as compared with those provided by the Klamath region's private ditch companies.

57. *Fourteenth Annual Report of the Reclamation Service, 1914–1915* (Washington, D.C., 1915), 249–250; *Fifteenth Annual Report of the Reclamation Service, 1915–1916* (Washington, 1916), 373; *Nineteenth Annual Report of the Reclamation Service, 1919–1920* (Washington, D.C., 1920), 320–330. Even during the boom days of World War I, the Klamath Project remained too isolated to raise high value cash crops. Alfalfa and timothy remained the project's staples.

industry's rapid decline made the valley a perfect laboratory to promote diversified agriculture. Irrigation would encourage crop rotation and help restore soil fertility. It would also boost the per-acre economic return and help lure new settlers into the Sacramento Valley.

Nevertheless, federal reclamation encountered a mixed reception in the valley. Many large landowners, such as Will S. Green, opposed federal efforts because of the 160-acre limitation and federal restrictions on established water rights. (The Reclamation Service tried to put all project water rights on an equal footing; those who had irrigated their land for 20 years had no stronger claims than new settlers.)[58] However, the Sacramento Valley Development Association, formed in 1900 to arrest the migration of people out of the valley, pushed for a comprehensive $40,000,000 Sacramento Valley Project as early as 1905. At the time, the entire Reclamation Fund contained only $28,000,000, so the proposal had no chance of winning approval in Washington.[59]

During the years from 1903 to 1906, the Reclamation Service contented itself with survey work in the Sacramento Valley. In cooperation with the U.S. Geological Survey, it set up stream measurement stations on the Sacramento River's major tributaries; prepared extensive topographic maps of the valley's west side; and inspected twenty-four potential reservoir sites.[60] Apparently, the overall Sacramento Valley Project remained as Newell had outlined it in an early report:

58. F. H. Newell reported that at the annual convention of the California Water and Forest Association held in 1903, he met many Sacramento Valley residents who opposed federal reclamation. See Newell to George Maxwell, December 13, 1903, "(848), Orland Project: Preliminary Reports and General Plans, etc. to December 31, 1909," RG 115.

59. For the Sacramento Valley Development Association's activities, see the *San Francisco Chronicle*, March 6, 8, 11, 12, 13, 15, 17, 29, 1902; *Forestry and Irrigation* 11 (June 1905): 283-284; *Pacific Rural Press* 70 (November 11, 1905): 308; Mary Montgomery and Marion Clawson, *History of Legislation and Policy Formation of the Central Valley Project* (Berkeley, 1946), 14–16.

60. *Second Annual Report of the Reclamation Service, 1902-1903* (Washington, D.C., 1904), 43; *Forestry and Irrigation* 10 (May 1904): 202; *Fifth Annual Report of the Reclamation Service, 1906* (Washington, D.C., 1907), 5, 94–98; H. E. Green, U.S.G.S. Engineer, to George Pardee, April 20, 1904; Morris Bien to Pardee, March 31, 1904; F. H. Newell to Pardee, December 30, 1904, January 18, 1905, February 8, 1905, Pardee Collection, Bancroft Library.

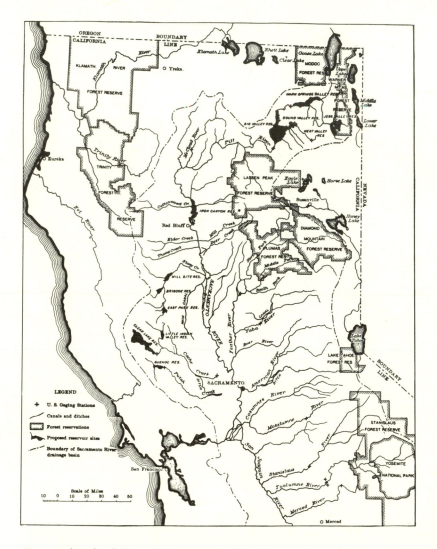

Proposed Federal Dams in Sacramento River Basin, 1905. *Fourth Annual Report of the Reclamation Service, 1904–1905* (Washington, D.C., 1906).

The general idea for the extension of irrigation in [the] Sacramento Valley is to ultimately construct a large diversion canal from the head of the valley down its western side. During high stages of [the] Sacramento River this canal would be supplied with water from the river itself, and during the low stages from numerous reservoirs in the Coast Range. The system would be operated in such a way that navigation of [the] Sacramento River would not be interfered with by irrigation diversions.

As mentioned earlier, entrenched interests prevented the Reclamation Service from beginning the Sacramento Valley Project at Clear Lake, as originally intended.[61]

In addition to the recalcitrance of many farmers and large landowners, the Reclamation Service faced formidable opposition from another federal agency, the much older Army Corps of Engineers. The Corps had participated informally in flood control planning in the Sacramento Valley since the 1870s, when the controversy over hydraulic mining debris erupted. However, the Corps' basic responsibility was maintaining the river's navigability. Soon after passage of the Newlands Act, the Corps' ranking officers in California warned the Reclamation Service that shipping required a flow of at least 10,000 cubic feet per second of water. The stream carried this amount only a few weeks a year, and the Reclamation Service bitterly complained that irrigation should take precedence over navigation during summer months. Had the restriction been applied to private irrigation companies, the Reclamation Service might have swallowed the edict more easily. But the War Department made no protest against illegal diversions made through the canal originally constructed by Will Green's Central Irrigation District. The Corps had taken heavy fire from the California Miners Association and other groups that questioned the design of debris-restraining dams constructed by the Army on the Yuba River during the 1890s. J. B. Lippincott hoped that such protests might result

61. *Second Annual Report of the Reclamation Service, 1902–1903* (Washington, D.C., 1904), 53.

in Congress transferring the responsibilities of the Corps to the Reclamation Service.[62]

Both the Corps and the Reclamation Service acknowledged the value of reservoirs to impound water for irrigation, but they disagreed as to the effectiveness of reservoirs in flood control. Frederick H. Newell, who would serve on the famous Inland Waterways Commission in 1907, recognized that watershed protection, flood control, and irrigation were interrelated:

> The two problems [of flood control and irrigation] are in fact one and inseparable. It is believed that it will be difficult to reclaim the Lower Sacramento Valley except by building storage reservoirs, to be operated in connection with levees, and in such way as to offer the greatest protection to the latter. The safety of the reclaimed areas can be further insured by protecting and extending the forested area near the headwaters of the streams. The investigations previously made on the Lower Sacramento Valley [by the Army Corps of Engineers and State of California], having for their purpose the reclaiming of the flood basins, have been confined to the problems of channel enlargement and levee protection, the great object being to pass the flood flow to tide water without utilization.[63]

Newell and Lippincott both believed that construction of the major reservoirs surveyed by the Reclamation Service would eliminate flooding in the valley. Lippincott suggested that a dam at Iron Canyon above Red Bluff would reduce the Sacramento River's "flood wave" by 50 percent.[64] However, the Corps argued that reservoirs

62. J. B. Lippincott to F. H. Newell, July 18, 1905, "(930–8), California: Surveys and Investigations, Kings River Project," RG 115; Newell to E. G. Hopson, October 12, 1909, and Hopson to Newell, October 22, 1909, "(989), Sacramento Valley: General Reports thru 1908," RG 115; J. B. Lippincott to George Pardee, February 14, 15, 1905, Pardee Collection, Bancroft Library.

63. *Fifth Annual Report of the Reclamation Service, 1906* (Washington, D.C., 1907), 98.

64. J. B. Lippincott, "General Outlook for Reclamation Work in California," *Forestry and Irrigation* 11 (August 1905): 353; also see Lippincott's "The Sacramento Valley Irrigation Project" in the volume of pamphlets entitled *Sacramento County* at the Bancroft Library.

could not capture enough water to provide flood protection, especially when used primarily to store water for farmers, and that reservoirs, by reducing the rate of stream flow, limited a stream's ability to scour out its channel and carry debris in suspension.[65]

Even though the Yuma and Klamath projects took economic precedence, Lippincott urged that construction begin as soon as possible on a small Sacramento Valley irrigation project. In April 1906, he wrote Newell: "The Sacramento Valley offers the greatest opportunity for irrigation development at the least cost, and with the least complications of anything that I am familiar with in the State."[66] By the summer of 1906, Newell approved construction of a very small project at Orland, twenty-five miles west of Chico, if the cost could be kept under $1,000,000. He noted: "By [building a small unit] it may be practicable to secure an allotment for work wholly within the State of California and to keep with us the sentiment of the people in the Sacramento Valley."[67] Just as the Reclamation Service considered the Yuma Project as the first step toward comprehensive development of the Colorado River, the Orland Project was expected to herald full development of the Sacramento.

The Orland Project included 14,000 acres of land, less than 1 percent of the Sacramento Valley's irrigable soil. Nine landowners held about 9,000 acres, the project contained no government land, and the fields were impoverished from years of ruthless wheat and barley farming. The *Irrigation Age* clearly perceived the Reclamation Service's objectives: "When . . . these . . . acres now producing a small amount of wheat shall become highly productive, when oranges and lemons and walnuts and almonds are being shipped from this small area by the hundreds of car loads, when the land shall have increased in value from less than $10 to many hundreds

65. On the Army Corps of Engineers opposition to reservoirs for flood control, see Samuel P. Hays, *Conservation and the Gospel of Efficiency: The Progressive Conservation Movement, 1890–1920* (Cambridge, Mass., 1959), 200–218, Arthur E. Morgan, *Dams and Other Disasters* (Boston, 1971), 252–309.

66. J. B. Lippincott to F. H. Newell, April 17, 1906, "(930) California: Surveys and Investigations thru 1910," RG 115. Also see Lippincott to Newell, May 3, November 14, 1905, "(989), Sacramento Valley: General Reports thru 1908."

67. F. H. Newell to D. C. Henny, July 26, 1906, "(989) Sacramento Valley: General Reports thru 1908," RG 115.

of dollars per acre, then it is hoped that other parts of the valley will desire to be similarly improved, and that the large landowners will consider it to their benefit to encourage similar work elsewhere."[68]

The Orland Project did not divert water directly from the Sacramento River. Instead, it tapped a large, relatively unused tributary, Stony Creek, and stored the water at the East Park Reservoir, forty miles southwest of Orland. Orland was also attractive because its water users association eagerly assisted the Reclamation Service in persuading landowners to sell off their excess holdings and sign repayment contracts; project farms were restricted to forty acres. The association also helped arrange amicable negotiations by which several small ditch and power companies sold their property to the nation.[69]

The Reclamation Service completed the East Park Reservoir in June 1911, though the project was not finished until 1916, at which time it contained over 20,000 acres. In 1915 the average irrigated project farm contained only twenty-five acres, but a handful of landowners persistently refused to break up their estates or utilize irrigation. Diversified agriculture had not yet arrived in the Sacramento Valley. Well over three-fourths of the project land produced alfalfa, and dairying quickly became the project's basic industry. The number of milk cows increased from 125 in 1911 to 4,000 by 1917, when two creameries served dairy farmers. In 1915, only ninety acres were planted to nuts, eighty-eight to deciduous fruit, and eighty-seven to citrus fruit. The value of crops per acre

68. *Irrigation Age*, 22 (May 1907): 213; 23 (June 1908): 235. In its May, 1908, issue (p. 203) the *Age* commented: "[the Orland Project's] chief value will . . . be as a positive demonstration to the landowners of the Valley of the unparalleled advantages of the Government system, and with that end in view it is planned to make the Orland Project the model irrigation system, or rather, the Orland Unit was selected because it has all the essential elements of such a model system."

69. On early planning for the Orland Project, see A. P. Davis to D. C. Henny, January 2, 1907; Davis, *et al.* to F. H. Newell, August 5, 1907; Henny to Newell, February 20, 1907; Newell to Henny, February 28, 1907, "(848) Orland Project: Preliminary Reports and General Plans, etc. to December 31, 1909," R.G. 115. Also see *Sixth Annual Report of the Reclamation Service, 1906-1907* (Washington, D.C., 1907), 70–72; *Seventh Annual Report of the Reclamation Service, 1907–1908* (Washington, D.C., 1908), 63–65. A board of engineers formally recommended construction of the Orland Project on November 13, 1906, and the secretary of interior approved preliminary project plans in December.

was very low, $26.99. This sum soared to $71.90 during World War I, but the value of crops still remained considerably less than on most federal projects. Moreover, land speculation posed a problem as it had on the Yuma and Klamath projects. One professional speculator later estimated that 75 percent of the project's early residents joined the game. He purchased project land for $10 an acre, sold it for $40, bought it back at $75 an acre, and resold it for $125 an acre. "Everyone was happy along the line. Everybody made money." Nevertheless, high land prices slowed the rate of settlement.[70]

The limited success of the Yuma, Klamath, and Orland projects did not discourage Sacramento Valley boosters, many of whom thought that the valley's future hinged on the construction of a dam at Iron Canyon at the north end of the valley. In 1907, the Sacramento Valley Development Association created the Iron Canyon Association to lobby for a reservoir. The year was particularly significant because it witnessed a massive flood on the Sacramento River and the creation of the federal Inland Waterways Commission, which was charged by President Theodore Roosevelt with multiple-purpose water planning. Both Frederick Haynes Newell and Francis G. Newlands took seats on the commission.[71]

In July 1907, Secretary of the Interior James R. Garfield visited the Iron Canyon site. The Association linked the construction of a 226,900 acre-foot reservoir at Iron Canyon to construction of a 1,000,000 acre-foot structure at Big Valley on the Pit River, a tributary of the Sacramento. The group promised that the dam would

70. For a fascinating account of land speculation within the Orland Project, see Charles F. Lambert's oral history transcript, Bancroft Library. The quote is from p. 26 of the transcript. George Wharton James noted in his 1917 history of the Reclamation Service: "As soon as the project was assured, things began to move; prospective settlers began to make inquiries, the price of land went up with a bound from $15 and $20 per acre to $50, $60, $75, and in the course of three years to $125 and $150" (*Reclaiming the Arid West*, 112). Also see the *Pacific Rural Press* 79 (June 14, 1911): 489, 498; *Eleventh Annual Report of the Reclamation Service, 1911–1912* (Washington, D.C., 1913), 59–62; *Thirteenth Annual Report of the Reclamation Service, 1913–1914* (Washington, D.C., 1915), 67–72; *Nineteenth Annual Report of the Reclamation Service, 1919–1920* (Washington, D.C., 1920), 101.

71. For the genesis of multiple-purpose water planning and the work of the Inland Waterways Commission, see Hays, *Conservation and the Gospel of Efficiency*, 91–114.

irrigate more than 100,000 acres adjoining the river and reduce flood damage. At the time, California's flood control works consisted largely of levees and bypass channels—no impoundment dams. Sacramento Valley landowners paid for flood protection by taxing their land through swampland reclamation districts. Members of the Iron Canyon Association argued that the Reclamation Service should build the dam because the stream was navigable and co-ordinated river basin planning was beyond the means of private enterprise, irrigation districts, or even the state. They also knew that the Service did not charge interest on its irrigation works and that sales of hydroelectric power could help pay for the dam.

The Sacramento River offered the Reclamation Service a chance to use the multiple-use concept of water development to win a larger constituency in California; nowhere in the West was a stream used for more different purposes. Nevertheless, when the Senate Committee on Arid Lands held a hearing at Red Bluff on November 10, 1910, the Service's enthusiasm for the project had visibly cooled. Preliminary surveys raised doubts whether the Iron Canyon could provide a suitable foundation for the dam, and government en-gineers estimated the cost of reclaiming the 100,000 acres at from $111 to $185 an acre—far above the cost of reclamation on the Yuma, Klamath, or Orland projects. Moreover, the War Department still insisted on a minimum flow of 10,000 cubic feet per second at Red Bluff to protect navigation; the Iron Canyon reservoir would violate that requirement, at least part of the year. The legislative commit-tee's report acknowledged that power sales could help pay for the project but warned that the federal government would have to compete with private electrical companies, so the revenue could not be predicted. The lawmakers concluded that the Iron Canyon Project was best left to private enterprise: "The engineering features . . . are difficult, largely indeterminate, and indicating a high acre cost of reclaimed lands unless navigation interests be abandoned or at least limited to a very considerable extent."[72]

Undaunted, officials in the Iron Canyon Project Association continued to push the project and helped pay for additional federal surveys. Promised benefits changed with the times. In 1914, with

72. *Report of the Committee on Irrigation and Reclamation of Arid Lands on the Investigation of Irrigation Projects*, 769–773.

the great flood of 1907 still fresh in the minds of northern Californians, representatives of the association met with Secretary of the Interior Franklin Lane and assured him that the Iron Canyon reservoir alone could store 20 to 25 percent of the river's volume during a five-to eight-day flood and, if used in conjunction with a reservoir on the Pit River, the two dams would have reduced the Sacramento River's flood flow in 1907 by 50 percent. Association officials were also quick to recognize the project's broader implications. W. A. Beard, vice-president of the association, wrote to the secretary of the interior in February 1914: "We regard the Iron Canyon project as a step in the direction of comprehensive co-ordinate development and control of the river systems of the Great Interior Basin of California and, in a broader sense, an initial step in the direction of a new national policy with reference to the flowing waters of the country."[73]

By the middle 1920s, a persistent drought, and a massive increase in private irrigation diversions from the Sacramento River during World War I, forced the Corps of Engineers to acknowledge that other uses of the river should take precedence over navigation—at least in the upper valley.[74] But the Reclamation Service had no money to launch the project and even if it had, the expenditure of $10,000,000 to $20,000,000 to reclaim private land would have been highly controversial. In 1927, the Service's "Walker Young report" revived interest in the Iron Canyon reservoir. That survey, according to W. A. Beard, promised that about half the anticipated 800,000 acre-feet of water released from an Iron Canyon reservoir yearly would be sufficient to protect delta lands from saltwater incursions (a problem discussed in chapter 12). However, by the end of the 1920s, new questions were raised about the soundness of the dam site, and an increasing number of State of California water planners claimed that a more efficient, larger reservoir could

73. W. A. Beard to the Secretary of the Interior, February 25, 1914, attached to Beard to Senator John D. Works, March 24, 1914, and John Ellison (president of the Iron Canyon Association) to Works, February 5, 1914, John Works Collection, Bancroft Library.

74. Homer J. Gault (Reclamation Service engineer) and W. F. McClure (California State engineer), *Report on Iron Canyon Project, California* (Washington, D.C., 1921).

be constructed upstream from Redding at the present site of Shasta Dam.[75]

In conclusion, federal reclamation had little effect on the development of California agriculture or the state's water policies until the nation assumed the responsibility for building the Central Valley Project during the mid-1930s. The federal government had been no more successful at stimulating irrigation, or providing efficient water resource management, than the state. In California, government engineers quickly dropped the pretext that their basic goal was to reclaim public land and provide new homes in the desert. In 1917, the Reclamation Service irrigated less than 100,000 acres in California, much of which had been farmed before 1902. Nevertheless, the Reclamation Service had planted the seeds of three ideas which won widespread acceptance in the arid West during the 1930s and after: the cost of irrigation could no longer be carried exclusively by farmers; power revenue could be used to subsidize reclamation; and different uses of water could be coordinated to promote efficiency and win the bureau a new constituency in the cities of the West. Meanwhile, because the Reclamation Service had little money to pay for new irrigation projects, because many California farmers feared the Newlands Act's 160-acre limitation, and because most Californians resented natural resource policies conceived and directed from Washington, attention shifted back to Sacramento. There, during the first decades of the twentieth century, the state began to exercise greater control over its water supply, though it would not seriously consider comprehensive state reclamation until the 1920s.

75. See "Comments of [California] State Engineer and Consultants on Report of Walker Young of U.S. Bureau of Reclamation on Iron Canyon Project, 1927," unpublished document, California Department of Water Resources Archives (hereafter DWR). Also see the wide variety of correspondence, reports, and other documents in the file entitled "Iron Canyon Project Correspondence, April 19, 1919 to Dec. 5, 1929," microfilm reel no. 1185, DWR Archives.

11

The State Asserts Itself: Irrigation and the Law in the Progressive Period

In 1900, the California state government exercised little control over the acquisition of water rights, the distribution of water, or the resolution of conflicts among users. It had done little to stimulate irrigation and repeatedly ducked or sidestepped the need for water law reform. Nor did it play any part in regulating the rates charged by municipal and rural water companies. The lawmakers allowed the courts to define the nature and extent of water rights, and in 1887 they all but deeded any state claim to California's unappropriated water to the autonomous, locally controlled irrigation districts created under the Wright Act. However, at the end of the 1890s and during the first two decades of the twentieth century, an alliance of civic and business leaders, reform-minded politicians, economic boosters, and water resource specialists, launched a crusade which culminated in important new laws. By modern standards, the accomplishments of these reformers were modest. But the new statutes represented a sharp break with the past. They stimulated irrigation agriculture and laid the legal foundation for both the Central Valley Project of the 1930s and the State Water Project of the 1960s and 1970s.

During the drought of 1898–1899, water law reform won support from the proponents of both state and federal reclamation. In July 1899, a committee of prominent San Franciscans wrote A. C. True, director of the U.S. Department of Agriculture's Office of

Experiment Stations, asking him to appoint Elwood Mead, head of the Office of Irrigation Investigations under True, to survey irrigation and water rights in California. Those who signed the letter included the manager of the State Board of Trade; the presidents of Stanford University, the German Savings and Loan Society, and the French Savings Bank of California; and the vice-president of the Crocker-Woolworth National Bank. They noted that nowhere in the United States were irrigation problems "more important, more intricate, or more pressing" than in California. "We can offer . . . examples of every form of evil which can be found in Anglo-Saxon dealings with water in arid and semiarid districts. Great sums have been lost in irrigation enterprises. Still greater sums are endangered. Water titles are uncertain. The litigation is appalling."[1] Subsequently, the California Water and Forest Association, which held its first meeting in November, raised $10,000 to help pay for Mead's study and a survey of reservoir sites on the Kings, Salinas, Yuba, and other streams by the U.S. Geological Survey's Hydrographic Branch. Frederick Haynes Newell headed the Hydrographic Branch. Like Mead, he recognized that the survey work in California could enhance the reputation of his office, as well as his chance to direct a federal reclamation program in the future. The Water and Forest Association's officers expected Mead and Newell's agencies to match or exceed the amount collected by the association.[2]

1. *Report of Irrigation Investigations in California*, U.S.D.A. Office of Experiment Stations (hereafter OES) Bulletin no. 100, 57th Cong., 1st sess. 1902, S. Doc. 356, 22. Elwood Mead, a civil engineer by profession, became a leading expert on western water law by 1900. His office published many surveys of water rights conflicts including *Water Rights on the Missouri River and Its Tributaries*, U.S.D.A., OES Bulletin no. 58 (Washington, 1899); *Abstract of Laws for Acquiring Titles to Water from the Missouri River and its Tributaries with the Legal Forms in Use*, Bulletin no. 60 (Washington, 1899); *Water-Right Problems of the Bear River*, Bulletin no. 70 (Washington, 1899); and *Irrigation Laws of the Northwest Territories of Canada and Wyoming*, Bulletin no. 96 (Washington, 1901). The best survey of Mead's career is J. R. Kluger's "Elwood Mead: Irrigation Engineer and Social Planner" (Ph.D. diss., University of Arizona, 1970). Also see Paul K. Conkin, "The Vision of Elwood Mead," *Agricultural History* 34 (April 1960): 88–97.

2. In 1893, the California legislature appropriated $250,000 for river improvements on condition that the federal government match that sum, which was done in 1896. This set the precedent for extensive cooperation in various water surveys conducted during the Progressive era and after. See the *Pacific Rural Press* 64 (August

By the summer of 1900, Mead had assigned a half-dozen assistants the job of surveying irrigation and water rights controversies on an equal number of streams. In his annual report for 1900, the Secretary of Agriculture proudly noted: "This is the largest and most comprehensive inquiry regarding irrigation laws, customs, and conditions which has been undertaken in this country."[3]

Mead offered many reasons to explain the "retarded" growth of irrigation in California. The state's farmers had long feared becoming the serfs of private ditch companies, and the Wright Act's failure reinforced the traditional prejudices against irrigation discussed in chapter 3. In the Sacramento Valley, farmers still believed that bumper wheat and fruit crops could be produced without irrigation and that irrigation would touch off malaria epidemics. Moreover, the state contained more competing groups of water users, who had invested far more money in their water projects, than any other western state. What other western state contained two stream systems like the Sacramento and San Joaquin rivers, which were used by miners, shippers, and towns as well as by irrigators? And California's water conflicts threatened to become even more complicated and acute as the demands of fledgling hydroelectric power companies expanded.[4]

Mead concluded that conflicts over water rights had been the main reason for the languishing state of California agriculture during the 1890s:

There are few places in the world where rural life has the attractions or possibilities which go with the irrigated home in California, yet immigration is almost at a standstill and population in some of the farmed districts has decreased in

30, 1902): 129. Also see *Water and Forest* 1 (September 1900): 7; (November 1900): 4–5; George Davidson to F. H. Newell, April 10, 1900, George Davidson Collection, Bancroft Library: Elwood Mead to A. C. True, November 18, 23, 1899, Bureau of Agricultural Engineering Records, Irrigation Investigations Division, General Correspondence, 1898–1902, RG 8, Federal Records Center, Suitland, Md.

3. *Annual Reports of the Department of Agriculture for the Fiscal Year Ended June 30, 1900* (Washington, D.C., 1900), 202.

4. *Report of Irrigation Investigations in California*, 18, 31–32, 34.

the past ten years. It is certain that some potent but not natural cause is responsible for this, and this cause seems to be a lack of certainty or stability in water rights which has given an added hazard to ditch building and been a prolific source of litigation and neighborhood ill feeling. Farmers who desire to avoid the courts and live on terms of peace and concord with their neighbors avoid districts where these conditions prevail.

"Floating" water rights had been one major source of conflict. California's water companies had been allowed to claim huge quantities of water apart from the land. They routinely filed for far more water than needed to prevent competition from rival companies and convince potential investors that they enjoyed a water monopoly in a particular region. Most companies promised water users a certain quantity of water per acre, but since water rights did not attach to individual tracts of land or farms, the security of a company's water supply ultimately depended on its financial ability to resist the expensive legal challenges of other companies. Litigation killed many water projects and crippled even more.

Despite the enormous cost of litigation, an expense which bankrupted many small farmers, the legal system resolved few water rights conflicts. The courts acted only on the appeal of one or more contestants, usually after rival water users had failed to settle their differences informally. In the absence of a state engineering office, the courts relied almost entirely on biased witnesses for hydrographic information. Nor was the leisurely pace of proceedings likely to cool overheated tempers. Interminable stays, injunctions, and appeals increased the possibility that once a decision had been reached, the conditions which prompted the suit in the first place might no longer exist. In any case, court tests rarely included all interested parties, so decisions were inevitably incomplete. Then, too, enforcing a court decree was no easy matter; contempt proceedings were expensive and subject to the same delays as water rights suits. Worst of all, the courts simply defined rights of private property; they rarely considered the "public interest." Many other water rights problems discovered by Mead have been discussed in earlier chapters. These included the in-

definite nature of riparian rights, the multitude of purely speculative claims, and the slow process of quieting contested claims.[5]

Mead's individual investigators provided an abundance of
illustrations to reinforce his conclusions. William Ellsworth Smythe
found an incredible array of water claims when he surveyed agricultural conditions in California's Honey Lake Valley, some sixty
miles north of Lake Tahoe. For example, in 1873 W. B. Sargeant
filed on all the surplus water in the Susan River, the largest stream
in the valley, "over and above the 2,000 inches claimed by A. A.
Smith." Ten years later, J. H. Slater claimed "the waters in Caribou
and Silver lakes and tributaries" without estimating either the total
water in the lakes or the amount he actually needed. And in 1887,
D. W. Ridenour and Charles Lawson demanded "all water here
flowing in Gold Run Creek" despite thirteen older claims to the
same stream. Smythe estimated that the Honey Lake Basin's water
supply could irrigate about 100,000 acres of land. However, under
the law of 1872 claims had been filed to a water supply sufficient
to irrigate nearly 230,000,000 acres, which is over twice the total
land area of California and many times the irrigable acreage in the
entire arid West!

The basin's first water case was settled in 1864, and by 1900
fifty-three suits had been thrashed out in the courts. Most court
tests fell into three categories: conflicts between rival appropriators;
conflicts produced by uncertain or ambiguous court decisions; and
conflicts between irrigation companies and consumers. Smythe
charged that the courts had done as much to promote conflict as
to resolve it. They frequently granted water to the extent "here-
tofore used," or granted a specific quantity of water without considering the number of acres a farmer cultivated, or the crops he
raised. Moreover, since the courts usually overestimated stream
flow in adjudicating water cases, their judgments did not provide
much help in dry years. Smythe bitterly concluded:

---

5. *Report of Irrigation Investigations in California,* 19, 35, 36, 41, 43. The quote
is from p. 19. In a May 31, 1900, letter to Elon H. Hooker in the Benjamin Ide
Wheeler Collection, University of California Archives, Mead claimed that litigation
over water rights had "practically put an end to [irrigation] development [in Cali-
fornia] and threatens to bankrupt canal companies. A continuation of this situation
will be destructive to the state's agriculture."

> The fault lies not with the people, not even with the lawyers, though the latter inevitably fatten upon the misfortunes of the community. The fault lies with the irrigation laws of California, which are notable alike for what they contain and what they omit. If deliberately devised to plague the people, no system which man's evil genius could invent would effect the result more surely than the system which invites them to make such reckless claims as we have seen in the case of Honey Lake Valley, and then leaves them to fight it out to the bitter end.

Put simply, irrigation could not expand in the Honey Lake Valley until the state's water laws had been revised.[6]

Conditions in the Honey Lake Basin were hardly unique. In 1856, James Moore filed the first appropriation on Cache Creek, in Yolo County. Theoretically, appropriative rights were limited to beneficial use. However, the state's courts had neither the experience, revenue, or inclination to gather hydrographic data in the field. Consequently, water cases usually hinged on the eloquence or deceit of high-priced attorneys, or the relative political power and social standing of contestants in their local communities. In any case, ultimately the state supreme court confirmed Moore's right to 432 cubic feet of water per second, more than twice the capacity of his ditch. Many residents of the Capay Valley, according to Mead's investigator, had been *forced* to raise wheat because the Moore and Capay ditches monopolized the valley's water supply. In Monterey County, 70 claims had been filed against the Salinas River. Only 10 of the claimants actually diverted water, but many of the remainder filed and refiled to keep their paper claims alive. Finally, of the 316 claims filed against the San Joaquin River and its tributaries, 6 were for the entire stream, and the remainder constituted 8 times its greatest flood flow and 172 times its average volume![7]

6. *Report of Irrigation Investigations in California*, 84, 85, 88, 91, 94–95. The quote is from p. 91.

7. *Report of Irrigation Investigations in California*, 170, 190, 195, 232. For an excellent discussion of water rights litigation in Yolo County, see Rosemary McDonald More, "The Influence of Water-Rights Litigation upon Irrigation Farming

Each of Mead's five investigators ended his report with suggestions for reforming California's archaic water laws.[8] Once their
work had been completed, they joined Mead in Berkeley to draft
a comprehensive list of reform proposals. They agreed that the
state should declare all unappropriated water public property and
create a "board of control" to determine and record existing rights
and regulate future appropriations. They recommended that the
board consist of an attorney, a businessman, and a civil engineer
selected by the state supreme court. It would complete and maintain the record of water rights, determine the volume of unused
water in the state, and fix the water rates charged by private ditch
companies. A state hydraulic engineer would serve as the board's
executive officer and, through his lieutenants, supervise the distribution of water among irrigators. To facilitate this task, the state
would be divided into water districts whose boundaries conformed
to natural watersheds. All future water rights would be attached
to the land and limited to the amount of water actually needed, as
determined by the state engineer. Moreover, beneficial use would
limit riparian as well as appropriative rights. The group suggested
that the federal government build storage reservoirs in California,
but only to serve public lands. The report did not define the part
the state government should play in promoting reclamation, though
it did recommend the establishment of state administrative control
over the creation of irrigation districts and the construction of their
dams and canals. Mead and his aides concluded by suggesting that
California's governor appoint a special commission to frame new
water laws. The panel's recommendations closely resembled the
"Wyoming Idea," the legal and administrative system Mead had
drafted in 1889 while he served as Wyoming state engineer.[9]

---

in Yolo County, California" (M.A. thesis, University of California, Berkeley, 1960).
More claims that Moore spent nearly $250,000 on litigation in the years before 1870
alone (p. 11).

8. *Report of Irrigation Investigations in California,* 112–113, 255–258, 322–325,
346–361, 392–395.

9. *Report of Irrigation Investigations in California,* 397–400. Elwood Mead described the Wyoming code of irrigation laws in his *Irrigation Institutions* (New York,
1903), 247–274.

Mead warned that both state and federal reclamation would be impossible without a dramatic increase in state administrative control over water. For example, some way had to be found to protect the water stored in state or federal reservoirs from being claimed or pilfered by established appropriators or riparian owners upstream from land selected for reclamation.[10]

Bulletin number 100 contributed to Elwood Mead's reputation as the West's preeminent expert on water law and arid land reclamation. It prompted University of California President Benjamin Ide Wheeler to ask Mead to organize a Department of Irrigation at the university, and insured that the Office of Irrigation Investigations would participate in future state-federal hydrographic studies. Mead hoped that the publicity his report received would help persuade Congress to establish a separate Bureau of Irrigation under his control in the Department of Agriculture. Bureau status promised larger appropriations. And the leadership of such a bureau would make Mead the likely choice to head any federal reclamation program approved by Congress. Nevertheless, Bulletin number 100 proved also to be a liability because it put Mead squarely in the "state party" camp. Mead's faith in state sovereignty over water, and his demand that federal reclamation be restricted to the public lands, won the wrath of powerful enemies including George Maxwell and Frederick Haynes Newell.[11]

10. *Report of the Secretary of Agriculture, 1901* (Washington, D.C., 1901), 85–86. Mead recognized the link between water law reform and other Progressive objectives. He hoped that Progressives who feared the corporate monopolization of oil, copper, coal and iron, might join in the crusade against water monopoly. Expanding state control over water also went hand in hand with the Progressive demand for public ownership of utilities. See *Report of Irrigation Investigations in California*, 64.

11. *Pacific Rural Press*, February 23, and June 22, 1901; January 11, 1902; Francis E. Warren to Elwood Mead, June 1, 1901; J. M. Wilson to Mead, August 22, 1901; William E. Smythe to Mead, October 28, 1901; California Water and Forest Association to James Wilson, Secretary of Agriculture, November 22, 1901, Bureau of Agricultural Engineering Records, Irrigation Investigations Division, General Correspondence, 1898–1902, RG 8, Federal Records Center, Suitland, Md. For reports of other work done by the Office of Irrigation Investigations in California see *Report of Irrigation Investigations, 1900*, OES Bulletin no. 104 (Washington, D.C., 1902), 137–146, for a discussion of measurements of the Gage Canal in Riverside; *Report of Irrigation Investigations, 1901*, OES Bulletin no. 119 (Washington, D.C., 1902), 103–189, for a study of the duty of water under the Gage Canal and in the Tule River

Nevertheless, Bulletin number 100 encouraged reformers in the California Water and Forest Association. William Ellsworth Smythe, the association's first vice-president, stumped the state in favor of Mead's recommendations beginning in the summer of 1900. In August 1901, he promised Mead to work for the election of a governor and legislature pledged to reform. If he failed, he vowed to lead an independent movement "which, at least, may give [us] the balance of power in the Legislature and enable us to demand the reform at the end of a club." In the following month, Smythe described Bulletin number 100 as "a thirteen-inch gun directed against Fort Water Monopoly. . . . I shall make every effort in my power, from now until the election of the next legislature, to see that the views expressed in the report, together with the overwhelming evidence on which they are based, are brought to the attention of the people."[12]

Smythe was a valuable publicist. He had almost single-handedly launched the national reclamation crusade in April 1891, when he published the first issue of *Irrigation Age*, which he edited until 1895.[13] In 1901, Smythe became a regular staff writer for Charles F.

Basin as well as a study of the San Bernardino Valley's underground water supply; and *Report of Irrigation Investigations, 1902*, OES Bulletin no. 133 (Washington, D.C., 1903), 151–165, for a survey of irrigation systems on Stony Creek, a tributary of the Sacramento River. For the rivalry between Mead and Newell over the federal reclamation program, see Professor Lawrence B. Lee (San Jose State University), "Elwood Mead and the Beginnings of National Reclamation," typescript manuscript.

12. William Ellsworth Smythe to Elwood Mead, August 29, September 14, 1901, Irrigation Investigations Division, General Correspondence, 1898–1902, RG 8. On Smythe's early lobbying efforts on behalf of water law reform, see *Pacific Rural Press* 60 (August 18, 1900): 98; (November 17, 1900): 306.

13. There is no full-length biography of William Ellsworth Smythe. However, see Lawrence B. Lee's introduction to a recent reprint (Seattle, 1969) of Smythe's classic *Conquest of Arid America* (New York, 1900); idem, "William Ellsworth Smythe and the Irrigation Movement: A Reconsideration," *Pacific Historical Review* 41 (August 1972): 289–311; idem, "William E. Smythe and San Diego, 1901–1908," *Journal of San Diego History* 19 (Winter 1973): 10–24; Martin E. Carlson, "William E. Smythe: Irrigation Crusader," *Journal of the West* 7 (January 1968): 41–47; Stanley R. Davison, "The Leadership of the Reclamation Movement, 1875–1902" (Ph.D. diss., University of California, Berkeley, 1951).

Smythe believed that while the nineteenth century had been an age of individualism, the twentieth would be an age of cooperation and organization, in agriculture as well as industry. Planned "scientific colonization" would replace the haphazard pattern of agricultural settlement that prevailed during the nineteenth

Lummis's *Land of Sunshine*, which became *Out West* in the following year. During the first decade of the twentieth century, *Out West* was California's second most popular magazine after the *Overland Monthly*. Smythe used his regular column, "The 20th Century West," to sell the ideas of Mead and the California Water and Forest Association. He proposed that the state reclaim private land using taxes levied against the land directly benefited. As an alternative, he suggested that California follow the lead of New Zealand by purchasing or condemning the state's largest farms, subdividing them into family plots with planned villages at strategic locations, and leasing the new homesteads to small farmers for an annual fee of 5 percent of the state's cost of buying and improving the land. The state would build roads, bridges, and other public works, in addition to canals, to make the rural communities attractive places to live. A lease system, Smythe hoped, would dramatically increase immigration into California by providing protection against the twin evils of land monopoly and speculation.[14]

The Water and Forest Association's meeting of December 20, 1901, was sparsely attended. J. M. Wilson, Mead's chief lieutenant in California, reported to his boss: "Our report [Bulletin number

---

century. It would also develop the full potential of democratic economic institutions such as marketing cooperatives, consumer-owned banks and insurance companies, and utility companies. Smythe neatly summarized his philosophy in *Out West* 19 (October 1903), when he wrote: "After private ownership, mutual ownership! After destructive competition, constructive organization and association! After combination of the few for the exploitation of the many, combination of the many for the benefit of all" (p. 439).

14. Smythe discussed the New Zealand experience at length in *Out West* (February, 1902): 202–209. The assumption that the federal government would refuse to build irrigation works in any state that had not compiled a full record of water claims, established control over future grants, provided for adjudication of water conflicts and for state distribution of water was very common in 1901 and 1902. For example, William Thomas, president of the Water and Forest Association, wrote Mead on December 31, 1901, noting that at its annual meeting, members of the association had recognized that "unless we could have a code of laws which would secure to the public any flood waters which the National Government might choose to store and permit to flow down the natural channel of the stream, we could expect no assistance from the Federal Government." Irrigation Investigations Division, "General Correspondence, 1898–1902," RG 8. For Smythe's view of the proper roles of the state and federal government in promoting reclamation in California, see *Land of Sunshine* 15 (July 1901): 65–72; (November, 1901): 382; (December, 1901): 491; *Out West* 16 (January 1902): 79; (April, 1902): 437.

100] seemed to please everybody very much and will I think be very satisfactory to all those who would like to see the reformation of the laws, but there are others who will probably not find it so satisfactory." The group agreed to appoint a special commission to draft a new water code for submission to the legislature which would convene in January 1903. The blue-ribbon panel subsequently appointed included Mead; Frederick Haynes Newell of the Geological Survey's Hydrographic Branch; Judge John D. Works of Los Angeles, a lawyer experienced in water rights litigation and a former justice of the California Supreme Court; Presidents David Starr Jordan of Stanford and Benjamin Ide Wheeler of the University of California; Professors C. D. Marx of Stanford and Frank Soule of U.C., both prominent engineers; and Chief Justice of the California Supreme Court William H. Beatty. Beatty, who had little sympathy for reform, served as an ex officio member. Works, Jordan, Wheeler, Marx, and Soule had all been active members of the Water and Forest Association (whose membership numbered around 5,000 at the end of 1902). In March 1902, the *San Francisco Chronicle* editorialized: "All the members of the Commission, it is believed, wish that at the beginning of our irrigation development California had enacted such laws as are now in force in Wyoming. The problem which they have to deal with is what approach to that system we can now make with due regard to vested rights."[15]

By October 1902, the model water law was ready. Judge Works drafted the legislation, which echoed the recommendations contained in Bulletin number 100. The proposed law was not as "radical" as it might have been. Some members of the Water and Forest Association favored state ownership of *all* water and irrigation works. To appease them, Works included a provision allowing the state to buy up or condemn *all* private water rights, even though no member of the code commission favored wholesale condemnation or purchase. The bill created a four-member board of engineers whose members had to be hydraulic or civil engineers from different parts of the state. The board was responsible for granting

---

15. J. M. Wilson to Elwood Mead, December 24, 1901, Irrigation Investigations Divison, "General Correspondence, 1898–1902," RG 8; *San Francisco Chronicle*, March 1, 1902. For a concise summary of the Water and Forest Association's water law reform objectives, see *Water and Forest* 1 (October 1901): 4.

future water rights, determining the amount of water needed to raise different crops under different conditions, scaling down inflated water claims, distributing the water supply to prevent waste and conflict, and adjudicating disputed claims. Moreover, on the appeal of 25 percent of those served by a private water company, the board could also set water rates. This provision reflected the common assumption that the county boards of supervisors did not have the expert knowledge needed to set rates and were too susceptible to political pressure from local water companies to make equitable decisions. To facilitate the board's work, the law required all water companies to provide the state with a detailed financial statement each year. The act specified that no new water rights would be granted until the board had completed a thorough inventory of the state's water supply and had compiled a full record of existing claims. Riparian owners and appropriators alike would receive only the water they needed, not the amount they claimed or even the amount they actually used. Limiting riparian rights to beneficial use was vital; otherwise, the state's surplus water supply could not be determined. Finally, the bill promised the federal government full use of the state's flood waters and granted it the right to purchase or condemn any water rights needed to insure the success of an irrigation project. Thus, the "Works bill" regulated virtually every step in the process of recording old claims, issuing new rights, doling out the state's water, and settling legal conflicts among irrigators. The bill had two serious gaps. It failed to provide state control over municipal water supplies and completely ignored underground water.[16]

Intense criticism of the Works bill surfaced almost immediately in southern California. Critics, led by spokesmen for the region's water companies, irrigation districts, and chambers of commerce, charged that the bill violated the concept of "home rule" by giving the proposed Board of Engineers "unlimited" and "dictatorial" powers, both judicial and administrative. For example,

16. The Works bill was reprinted in full in a supplement to *Water and Forest* 2 (October, 1902). It was amended in December, 1902, after the Water and Forest Association's annual meeting. The provisions mentioned are from the revised version. For discussions of the bill, see the *Call*, October 14, 1902; *San Francisco Chronicle*, November 23, December 6, 1902; *Sacramento Daily Union*, December 7, 1902.

though the law allowed water users to appeal board decisions to the courts, it did not require the board to provide claimants any hearing at all; therefore, the process of determining rights would be arbitrary. Moreover, some critics feared that the board's power to tamper with existing rights in the name of preventing waste would reopen water conflicts already settled by the courts. They also worried about losing control over the distribution of water to an expensive and "unresponsive" new bureaucracy based in Sacramento. And if, as many suspected, the Works bill was designed to serve as the foundation for a comprehensive, state-controlled reservoir and canal system, the financial burden might increase dramatically in the future. The *Riverside Daily Press* recognized the most immediate threats to southern California irrigators when it warned against renewed litigation and editorialized: "We prefer to be are [sic] own judges of the amount of water our land needs. If Riverside, by her enterprise and her wealth, has acquired a good supply, as she has, that is no reason why we should be called upon to 'divy' with some of our less fortunate neighbors. That is socialism gone to seed."[17]

Canal companies, riparian owners, and other opponents of water law reform found strong allies in George Maxwell and, surprisingly, William Ellsworth Smythe. Maxwell argued that water law reform was "nothing more than a proposition to defeat the whole national irrigation movement by interminable delays." In the late 1890s, Maxwell did everything possible to destroy California's irrigation districts. But in 1902 he became a champion of home rule:

> In all matters relating to the adjudication of rights on streams and the division of the flow between irrigators, it has always seemed to me that it would be far from beneficial in many parts of California to create a State political machine at Sacramento, with power to appoint local officers throughout the

17. The quote is from the *Riverside Daily Press*, December 30, 1902. Also see the *Press* of December 29, 1902; *Los Angeles Times*, December 30, 1902; *San Diego Union*, December 27, 29, 30, 1902. Most of these issues reported on a convention staged at Riverside on December 29 to protest the Works Bill.

State to distribute the water. . . . Could not the desired result be reached in a much simpler way by some plan of local control and self-government on the part of the irrigators themselves? In other words, could not each stream or hydrographic basin, where it was desired by the irrigators, be organized into a local district for administrative purposes only, leaving it to the irrigators themselves to determine by vote as to whether such a system should be inaugurated?

In January, 1903, Maxwell wrote to Governor George Pardee explaining that "as the laws of California now stand, the national government can come right in and build irrigation works to utilize any of the unused or unappropriated waters of the State . . . without in any way interfering with vested rights, or being involved in any complications with state officials or rights claimed by the State. Such complications would inevitably follow the inauguration of such a system as that proposed by the Works Bill or any similar code of irrigation laws."[18]

Smythe's motivation is harder to understand because his thinking changed so abruptly. Lawrence B. Lee has suggested that the irrigation crusader's unsuccessful bid for a congressional seat in the fall of 1902 left him deeply in debt and forced him to turn to the state's water companies for financial assistance. Certainly,

18. See George Maxwell's statements in *Out West* 16 (May 1902): 546–555; *Forestry and Irrigation* 8 (November 1902): 444–447. The block quote is from p. 553 of the *Out West* article. Maxwell's letter to George Pardee of January 16, 1903, is in the Pardee Collection, Bancroft Library. Also see William Thomas to Pardee, March 14, September 5, 11, 1903, in the same collection; J. M. Wilson to Elwood Mead, December 19, 1902; Frank Adams to Mead, December 20, 1902, Irrigation Investigations Division, "General Correspondence, 1898–1902," RG 8. Wilson reported that opposition to the Works Bill had surfaced at the December meeting of the Water and Forest Association. Adams reported on a meeting of the irrigation code commission held on December 19. He revealed that "Judge Works doubts very much if this one [bill] will pass."

On December 5, 1902, C. B. Booth, chairman of Maxwell's National Irrigation Association, warned U.S. Senator Thomas Bard: "The evident purpose [of the Works bill] to form a political bureau, by which shall be appointed an army of political ditch tenders to take charge of the interests of the people connected with the distribution of water, is something that is most pernicious from every point of view." See "Irrigation: National, 1," Box 9–A, Thomas Bard Collection, Huntington Library. Also see George H. Maxwell to Bard, November 26, 1902, in the same file.

he served as their unofficial spokesman during the debate over the Works bill.[19] In any case, Smythe claimed that the proposed law was unsympathetic to federal reclamation; failed to establish public ownership of water; did not safeguard the principle of local control; and contained no provision for adjudicating disputes between holders of established rights. He charged that the bill ought to be called "an act for the protection and encouragement of private speculation in the Water supply of California."[20]

Judge Works responded to Maxwell's and Smythe's criticisms by noting that although he had drafted the bill, every member of the code commission had approved it. According to the judge, members of that group disagreed on only one major issue: whether the legislature had the constitutional power to define the total acreage of riparian land in California and limit riparian owners to beneficial use. He traced all opposition to the bill back to private companies intent on preserving hard-won monopolies.[21]

The death of the Works bill did not satisfy George Maxwell. He also wanted to kill the state appropriation earmarked for the Department of Agriculture to continue the Office of Irrigation In-

19. See Lawrence B. Lee's brief biography of Smythe in *The Reader's Encyclopedia of the American West,* ed. Howard Lamar (New York, 1977), 1125–1126, and his typescript "Elwood Mead and the Beginnings of National Reclamation." The *Los Angeles Times,* February 24, 1903, identified Smythe as "the authorized representative of a number of irrigation companies throughout the southern part of the State." This was particularly revealing because the *Times* opposed the Works Bill.

20. William E. Smythe, "The Failure of the Water and Forest Commission," *Out West* 17 (December 1902): 751–757; ibid, 18 (March 1903): 381–389; Smythe to State Irrigation Convention, October 7, 1904, Smythe Correspondence folder, George Pardee Collection, Bancroft Library.

21. John D. Works, "Answers to Objections Made to the Irrigation Bill Proposed by the Water and Forest Association," and "Should the Irrigation Bill Pass?", undated pamphlets in a volume entitled *Irrigation in California* at the Bancroft Library. Also see Works to Pardee, January 9 and 14, in the Pardee Collection at the Bancroft and his letter reprinted in the *San Francisco Call,* January 28, 1903.

The *San Francisco Chronicle's* editor agreed with Works. See, for example, the issue of February 27, 1903. However, the *Pacific Rural Press* noted: "Opposition [to the Works Bill] proceeded from two sources, one which desired to be let alone in present possessions secured through the courts or otherwise, and the other which desired a nationalization of the whole question of property in water. Each side freely used the help of the other in condemning the proposed enactment, and both opponents used weapons they should not have liked to slay the measure." See the *Press* 65 (February 28, 1903): 130.

vestigation's hydrographic work in California. Nevertheless, even though Maxwell enlisted the vast political power of the railroad in his struggle with Mead, Governor Pardee refused to approve any appropriation for joint study unless $10,000 went to Mead's office.[22]

The fate of the Works bill served as a reminder that the sectionalism that helped block a state irrigation system and water law reform in the 1870s and 1880s still prevailed. Recognizing that opposition to the bill had been centered in southern California, several leaders in the Water and Forest Association proposed that the measure be redrawn to apply exclusively to that part of California north of the Tehachapis. However, a deep rift appeared in the association. Early in 1903, the organization's vice-president and former president, William Thomas, resigned in protest over the tactics used by Smythe, Maxwell, and other members of the association to defeat the water bill.[23] Thomas, and many other charter members of the association, believed that the conservation society had been taken over by the enemies of reform. Consequently, they banded together

22. See J. B. Lippincott to Frederick Haynes Newell, March 31, 1903, and the attached correspondence pertaining to A.B. 75 and cooperative investigations in California in "Correspondence—J. B. Lippincott, 1905," Frederick Haynes Newell Collection, Box 6, Library of Congress. In particular see Maxwell to Lippincott, February 8, 1903. This collection of correspondence shows Maxwell's enormous influence on the California legislature. Also see Elwood Mead to Benjamin Ide Wheeler, February 14, 1903, Wheeler Collection, University of California Archives, Bancroft Library; George Pardee to William Thomas, January 23, February 1, 13, 1903, Pardee Collection; *Water and Forest* 3 (April 1903): 2; the *Country Gentleman* 69 (June 16, 1904): 559–560. In the next four years, Mead's office studied different methods of irrigation in California, assessed the value of various pump systems, and surveyed problems relating to seepage, drainage and reclamation of alkali-choked land. See Elwood Mead, "The Irrigation Investigations in California of the Office of Experiment Stations," *Forestry and Irrigation* 11 (August 1905): 367–369.

The Mead-Maxwell feud can be traced in dozens of letters in Bureau of Agricultural Engineering Records, Irrigation Investigations Division, General Correspondence, 1898–1902, RG 8. The two major national journals concerned with irrigation also joined the battle with *Irrigation Age* supporting Mead and *Forestry and Irrigation* backing Maxwell.

23. The *Sacramento Daily Union*, January 26, 1903. On the rift within the Water and Forest Association, see John D. Works to T. C. Friedlander, March 14, 1904; Works to Elwood Mead, March 17, 1904, "Field Office Correspondence, 1904–1906," Records of the Office of Irrigation Investigations, RG 8. In the first letter Works complained that Smythe and Maxwell were circulating letters urging opponents of the Works Bill to join the Water and Forest Association as one way to *undermine* reform. Also see "California Water and Forest Association," *Forestry and Irrigation* 10 (May 1904): 195; *Water and Forest* 4 (May 1904): 1.

in May 1903 and formed the Commonwealth Club of California. The group included many champions of water law reform including Thomas, Benjamin Ide Wheeler, James D. Phelan, and Governor George Pardee. It also included future governor Hiram Johnson. The club's organizer, Edward F. Adams, edited the *San Francisco Chronicle* and had fought hard for the Works bill in his columns. Adams hoped that the Commonwealth Club would represent the whole state; the Water and Forest Association's members had been drawn largely from northern California. Like most Progressives, he had deep faith in the rationality of man and assumed that people usually disagreed over issues out of ignorance rather than economic self-interest, personal rivalries, or other "sordid" motives. So the club's basic job was to gather "the facts." Individual sections collected information, compiled reports, and drafted bills for consideration by the entire club. Adams believed that the club's "impartial" data allowed it to propose much sounder legislation than members or committees of the railroad-dominated legislature could produce. During the first few years of its life, the Commonwealth Club tackled such hot Progressive issues as civil service reform, the California penal system, the referendum, tax reform, and government regulation of railroads. It first considered water law reform at a meeting held on November 9, 1904.[24]

William Thomas presented the keynote address at that November gathering. By this time, he realized that the legislature would reject any attempt to turn the determination and adjudication of water rights over to an administrative commission. Yet the chief justice of the California Supreme Court had complained specifically about "the large and increasing class of cases arising from disputed water rights." So Thomas suggested that the state, as owner of unappropriated water, appoint a deputy attorney-general to gather information on claims and file suit in the state's name to quiet titles stream by stream. Within watersheds already covered by court decrees, the state's action would be a formality. The Section on Public Laws formally approved the plan, noting:

24. See Edward F. Adam's keynote speech to the Commonwealth Club in the club's *Transactions* 1 (1903–1905): 3–6. The best history of the group's early years is in Frank Adams's biography of his father, *Edward F. Adams: 1839–1929* (Berkeley, 1966), especially chap. 10, pp. 115–122.

"North of Fresno, in spite of innumerable lawsuits, litigation has hardly begun. If nothing is done there will be far more litigation in the North than there ever was in the South."[25]

Thomas's proposal failed to win the approval of the whole club because many southern California members feared reopening litigation under any circumstances and also because of the anticipated cost. Late in 1905, the club's members asked the Section on Public Laws to draft a comprehensive water bill. But soon after the committee began its work, the earthquake and fire of 1906 destroyed most of the club's records and diverted attention to the job of rebuilding San Francisco. For the next few years, the Commonwealth Club looked at municipal issues, including San Francisco's need for a larger water supply. Although the club added a conservation section in 1909, and actively discussed the future of the state's irrigation districts in 1911, it did not return to water law reform until 1912.

Hiram Johnson's election to the governorship in 1910 broke the logjam of sectionalism and special interests that had blocked water law reform since 1900. But the new governor first turned his attention to reviving the irrigation districts and regulating the state's water companies. In the 1890s and after, many schemes were proposed to rescue the districts. For example, in 1900 William Ellsworth Smythe proposed that a state engineering office or certification board review district construction plans and then pledge the state's credit in support of district bonds.

> [The state] could sell its own bonds readily at 3 per cent interest, depositing in its treasury the 5 per cent bonds of the district and making the difference in interest pay all the expenses of administration. It would then be no longer necessary for the district financial agents to hawk their securities in the money markets of the world, selling them at all sorts

25. William Thomas's speech was reprinted in the Commonwealth Club's *Transactions* 1 (1903–1905): 1–10. For the chief justice's statement see p. 12. The comment of the Public Laws Section is on p. 19. Irrigation had developed much more rapidly south of the Tehachapis and the water supply there was much more limited. Consequently, water conflicts in that section erupted earlier than in the north and most of its streams had been adjudicated by 1900. This, of course, helps explain the intensity of opposition to the Works Bill in southern California.

of prices or exchanging them with contractors for doubtful consideration. . . . The State would risk nothing in the operation; the districts would gain everything. The burden of taxation would rest where it belongs—on those who are to receive the benefits. There would be no weary waiting of years for State or Federal schemes to materialize and to reach those remote neighborhoods which have fewer citizens and fewer outside friends. There would be no more heart-breaking private enterprises dealing with undertakings beyond their grasp.

Smythe's idea won little public support because district residents complained that they would be heavily taxed to pay off badly depreciated bonds at par and that such a policy would produce a new crop of speculative irrigation schemes. Moreover, Article IV, section 31 of the state constitution specifically prohibited any state financial aid to "local government," and most state officials considered irrigation districts an institution of government.[26] Smythe proposed a second way to revitalize the irrigation district idea in November 1904, at the irrigation congress meeting in El Paso. He suggested that the Reclamation Service investigate and certify proposed districts and build the irrigation works. Bond issues would pay for construction, but the secretary of the interior would supervise their issuance and sale. When the irrigation system had been completed, control would pass to local landowners. This plan won considerable support from the Reclamation Service, but A. P. Davis warned that the Service could not enter cooperative agreements with the residents of proposed districts without specific authorization from Congress. That authorization never came.[27]

In 1909, the South San Joaquin and Oakdale irrigation districts

26. William E. Smythe, *Report of Irrigation Investigations in California*, U.S.D.A., OES Bulletin no. 100, 110–111. The state legislature did not entertain Smythe's proposal, but it did draft a constitutional amendment, ratified by the voters in 1902, which exempted irrigation district bonds from state taxation.

27. W. E. Smythe, "A Success of Two Centuries," *Out West* 22 (January 1905): 75; "The Reclamation Service," *Forestry and Irrigation* 11 (June 1905): 280. Francis G. Newlands described and endorsed Smythe's plan in his letter to the secretary of the interior, January 26, 1905, "(1420–1904) Miscellaneous Projects: State Irrigation Under National Control, Francis G. Newlands," Records of the Secretary of the Interior, RG 48, National Archives. In the same file, see Charles Walcott's letter to the secretary of the interior, June 3, 1905, drafted by A. P. Davis.

were formed in San Joaquin county, the first districts organized since 1895. Increasing crop prices, the decline of the wheat industry, and a flood of new migrants into the state, all account for the new districts. However, boosters recognized that something had to be done to win the support and confidence of investors, and they hired L. L. Dennett, a representative of the Modesto Irrigation District and a prominent Progressive attorney, to draft a law that would make their bonds more marketable. That law designated the attorney general, the state engineer, and the superintendent of banks as an irrigation district bond commission. It required the commission, when requested by any proposed district, to judge the feasibility of construction plans, assay soil quality and drainage conditions, determine the amount of water available to the district, and estimate the project's cost. Most important, the commission would assess the "reasonable market value" of all district land as well as the value of irrigation works. The district itself would pay for the survey. If the commission deemed the project "feasible" (it could not rule on whether a project was the most efficient scheme or even a good one), the state controller could certify district bonds as a safe investment up to 60 percent of the combined value of district lands, water, water rights, and irrigation works. In addition, the act permitted banks, insurance companies, trust companies, and even school districts to invest in the bonds.[28]

Although the legislature and governor approved the Dennett bill early in 1911, both the state controller and state superintendent of banks considered the new law unconstitutional and blocked its implementation. They opposed permitting savings banks to invest in district bonds until *after* the bond commission had certified that

28. *California Statutes,* 1911, 322; *Sixth Biennial Report of the Department of Engineering of the State of California, December 1, 1916 to November 30, 1918* (Sacramento, 1919), 74; *Call,* July 30, 1910; Ray P. Teele, *Irrigation in the United States* (New York, 1915), 120–121; Frank Adams oral history transcript, Bancroft Library, 232; J. Rupert Mason oral history transcript, Bancroft Library, 106. In 1911, Dennett also drafted a law by which taxes in new irrigation districts, and old ones if a majority of residents approved, were levied solely on the value of farmland, excluding improvements. Large district landowners had paid proportionately lower taxes than smaller farmers, who were taxed for houses, trees, barns, and other property. This law forced many wheat barons to sell their estates, opening up land for smaller farmers. For a contemporary assessment of the law's effects, see the *Modesto Morning Herald,* February 12, 1914.

particular district's securities. By October of 1911 the new commission still had not met, so the law was a dead letter. Meanwhile, the Imperial Irrigation District had been organized in the Imperial Valley; it contained 523,000 acres, the largest district ever formed in the state.[29]

In November 1911, in the midst of a special session of the legislature, Governor Johnson called an irrigation district convention in Stockton. A committee of bankers and financiers representing the Commonwealth Club warned that the "welfare of the entire state" depended "to a great extent" on making irrigation district bonds more attractive. They recommended that the bonds carry higher interest, up to 6 percent; that, with the consent of the commission, districts be able to sell their bonds at less than par; that old as well as new districts be able to have their bonds certified; and that the state commission's responsibilities be spelled out in greater detail.[30] The governor appointed a committee, which included Dennett, the state controller, the state superintendent of banks, and the attorney-general, to repair weaknesses in the law passed earlier in the year. The new statute restricted the value of bond issues to 60 percent of district property; provided for the investigation of extant districts and the certification of their bonds; specified the form to be used by the state controller in reviewing the bonds; and reaffirmed that cities, counties, school districts and municipalities could invest money in district bonds. However, the bond commission still could do little more than vote yes or no on plans submitted to it by district officials; it exercised little initiative or leadership.[31]

The 1911 laws did not give the state commission power to

29. *Call,* September 30, October 1, 18, 21, 1911; *San Francisco Chronicle,* November 16, 1911.

30. See "Marketing Irrigation District Bonds," *Transactions of the Commonwealth Club* (hereafter TCC) 6 (December 1911): 515–583. Many of the Commonwealth Club's suggestions were later written into law by Frank Adams, a prominent member of the club and California's foremost expert on irrigation districts.

31. *Cal. Stats.,* 1911 (Special Sess.), 3. For the proceedings of the Stockton conference held on November 15, 1911, see the *Call, Sacramento Bee,* and *San Francisco Chronicle,* all of November 16, 1911. Also see U. S. Webb, California attorney-general, "State Supervision of Irrigation Districts," in "General Correspondence, Department of Engineering and Irrigation," microfilm reel M–708, DWR Archives.

block the issuance of bonds, but a negative report guaranteed a negative vote at any bond election because district voters recognized that investors would not purchase securities branded as a bad risk by the state.[32] The 1913 legislature went even further by requiring the state engineer to investigate *all* districts prior to organization; no bond election could be held before the state completed its survey. If for any reason the state engineer filed an adverse report, then the district could be formed only with the approval of 75 percent of the district's voters. Only one district, in the Mojave Desert, was organized against the state engineer's advice.[33]

Increasing state supervision, high crop prices during World War I, and the irrigation district's ability to sell electrical power to subsidize the cost of irrigation, all contributed to an explosive growth in the number of districts. In 1909, only 173,793 of California's 2,664,104 irrigated acres were in districts, about 7 percent. By 1928, the state's seventy-three operating districts contained 2,582,316 acres, and twenty more districts were in the process of organization or construction. Only 1,467,500 acres of this land were actually being irrigated, but this constituted nearly one-third of all the state's irrigated acreage. Put another way, the irrigation district was responsible for nearly two-thirds of the expansion of irrigated land from 1910 to 1930. In the latter year, California contained 37 percent of the district land in the entire arid West, more than twice the acreage of its nearest rival, Idaho.[34]

The state's new power over irrigation districts was matched by an expansion of authority over private water companies. In 1911,

32. *Cal Stats.*, 1913, 778. J. Rupert Mason oral history transcript, Bancroft Library, 109–110; Frank Adams oral history transcript, Bancroft Library, 235; Adams, *Irrigation Districts in California* (Sacramento, 1930), 40–41. The bond commission act was revised frequently in succeeding years. For example, see *Cal. Stats.*, 1915, 592; *Cal. Stats.*, 1917, 582; *Cal. Stats.*, 1919, 1207; *Cal. Stats.*, 1921, 1198.

33. *Cal. Stats.*, 1913, 993. The 1913 law required the state engineer to report any conditions he thought might limit the success of a district. In 1915, the legislature amended the law and required the state engineer to report on the overall feasibility of a project, not just its obvious weaknesses or limitations. For a full discussion of the amendments to the Bridgeford Act adopted in 1911, 1913, and 1915, see Frank Adams, *Irrigation Districts in California, 1887–1915* (Sacramento, 1917), 48–57.

34. Wells A. Hutchins, *Irrigation Districts: Their Organization, Operation, and Financing*, U.S.D.A. Technical Bulletin no. 254 (Washington, D.C., 1931), 6, 75, 80–81; A. E. Chandler, *Elements of Western Water Law* (San Francisco, 1913), 144; *Irrigation District Movement in California: A Summary*, State of California, Assembly Interim Committee Reports, vol. 13, no. 5 (Sacramento, 1955), 9.

the legislature created the State Railroad Commission, and two years later gave it the power to set water rates, though mutual water companies and irrigation districts were exempted. The new law also required the Railroad Commission to prevent companies from promising more water than they could deliver and allowed it to require service to additional users if it decided the water supply was adequate.[35] The commission noted in its 1912–1913 report: "Many complaints have been received of discrimination by such utilities in the delivery of water to their consumers. . . . The adjustment of these differences has necessitated the study of the requirements and possibilities of such systems and has led to orders by the Commission correcting such troubles."[36]

The new law required every private water company to submit its rate schedule and regulations for distributing water to the commission for review. The commission, in turn, issued what amounted to cease-and-desist orders when it discovered actions "unreasonable to the consumer." For example, it prohibited companies from requiring liens on the land they served. Such liens had frequently been demanded as a precondition to service. They allowed ditch companies to stifle user complaints and force consumers to buy a certain quantity of water in wet years and dry, whether needed or not. The commission imposed considerable uniformity on the companies by providing utilities and consumers alike with a "model set of rules for the supply of water for domestic and irrigation use." The commission promoted the Progressive goal of efficient resource use by requiring companies to measure water in cubic feet per second (rather than anachronistic, unreliable measures such as the miner's inch), and to use up-to-date water measurement devices. By the early 1920s, the Railroad Commission supervised the operation of over 550 water systems. From July 1922 to July 1923, it conducted 205 formal proceedings against water companies and settled 479 informal user complaints.[37]

35. *Cal. Stats.*, 1913, 84.

36. *Report of the Railroad Commission of California from June 30, 1912 to June 30, 1913* (Sacramento, 1913), 121–122.

37. *Report of the Railroad Commission of California from July 1, 1915 to June 30, 1916*, vol. 1 (Sacramento, 1916), 73–76; *Report . . . July 1, 1916 to June 30, 1917*, vol. 1 (Sacramento, 1917), 104; *Report . . . July 1, 1922 to June 30, 1923* (Sacramento, 1924), 7.

The expansion of state control over irrigation districts and water companies dovetailed with water law reform.[38] Franklin Hichborn, a close student of the California legislature during the Progressive years, claimed that the California Water and Forest Association's model code of 1902–1903 (the Works bill) was reintroduced in 1905 and 1909.[39] If it was, the bill failed to reach the floor of the legislature and attracted no public attention. Nevertheless, few Californians versed in the law denied the need for reform, even though they differed as to the shape it should take.[40] By 1911, several circumstances contributed to a renewed interest in water law reform. A new water code was expected to spur the revival of irrigation districts and prevent power companies from claiming the state's remaining water supply to the exclusion of future irrigators. Perhaps most important, an increasing number of Progressive conservationists deeply feared the monopolization of natural resources by "the interests" and hoped to promote the efficient use of those resources under the supervision of disinterested experts and special commissions. They looked to the state as guardian of the people's patrimony and found a willing champion in Governor Hiram Johnson, who assumed office in January 1911. In his inaugural address, Johnson noted:

> The great natural wealth of water in this State has been permitted, under our existing laws and lack of system to be misappropriated and to be held to the great disadvantage of its economical development. The present laws in this respect

38. For example, one element in the expansion of state control over irrigation districts was giving the state the power to protect the water supply of potential districts from speculators and developers. In 1917, the legislature passed a law that permitted the state engineer to withdraw from appropriation any water that he considered necessary to the success of that district as part of his preliminary investigation into the district's feasibility. See Frank Adams oral history transcript, Bancroft Library, 241. This was, of course, part of the overall expansion of state regulatory power during the Progressive era.

39. Franklin Hichborn, *Story of the California Legislature of 1913* (San Francisco, 1913), 158.

40. One exception was Frank Short, a leading California attorney who represented many large power and ditch companies. In 1906, in an address to the irrigation congress meeting in Sacramento, he declared: "We [in California] simply dispense with the Engineer and Governing Board and let the people settle their

should be amended. If it can be demonstrated that claims are wrongfully or illegally held, those claims should revert to the State. A rational and equitable code and method of procedure for water conservation and development should be adopted.

The threat to bring monopolists to bay squared with the Progressive desire to chastise the wicked and turn politics over to the "right men."

When the 1911 legislature convened, most of its Progressive members had had no more time to study water resource problems than the new governor. Hence, a comprehensive water bill was not proposed. However, the lawmakers did enact three important statutes drafted by a panel appointed by the Republican State Committee. The group included such notable Progressives as Francis J. Heney, William Kent, Chester Rowell, and George Pardee, who served as chairman. One law declared *all* water public property and limited appropriations to generate electricity to twenty-five years. Only publicly owned utility companies, or irrigation districts that generated electricity for use wholly within their boundaries, were exempt. This law built on the questionable assumption that the state could exercise control over water already appropriated, as well as over future claims.[41]

The second California law required hydroelectric companies to file extensive formal applications to appropriate water with a five-member board of control, which included both the governor and state engineer. Progressives throughout the nation believed that monopolies exercised by electric companies posed a great danger to the economic health and future prosperity of the United States. Electricity offered a cheap, clean, abundant energy source, *if* private power companies could be regulated effectively. Under

---

rights in the good old-fashioned Anglo-Saxon way in the courts, and they are settling them, I think, well and satisfactorily and along just lines." *Proceedings of the Fifteenth National Irrigation Congress, 1906* (Sacramento, 1907), 249. In a letter to Milton T. U'Ren, a member of the California Conservation Commission, dated December 8, 1910, in the George Pardee Collection, Bancroft Library, Short commented that "our irrigation laws are almost ideal."

41. *Cal. Stats.*, 1911, 821; George Pardee to Franklin Hichborn, April 12, 1911, Box 125, Franklin Hichborn Papers, Special Collections, University of California, Los Angeles, Research Library.

the 1911 law, the companies were required to file annual financial statements and descriptions of construction work completed during the previous year. The statute allowed the state to charge for water used to generate electricity, and the board could reject or scale down applications deemed monopolistic or extravagant. It also required power companies to obtain permits to build dams. Power "combinations" that restrained trade, as determined either by the board of control or state attorney-general, could be divested of their water rights in the courts. Finally, the act attempted to assert public control over all water previously claimed, but not put to beneficial use, by declaring that water "unappropriated." This represented the first statute to give the state explicit control over the acquisition of new water rights. However, the law did not provide for a full-time board of control, which limited the measure's effectiveness.[42]

The most publicized of the three laws created a three-member state conservation commission and charged it with "investigating and gathering data and information concerning the subjects of forestry, water, the use of water, water power, electricity, electrical

42. *Cal. Stats.*, 1911, 813. On the water laws passed at the 1911 session, see George Pardee to Thomas R. Shipp, April 11, 1911; Pardee to Franklin Hichborn, April 12, 1911; Pardee to Hiram Johnson, April 16, 1911, George Pardee Collection, Bancroft Library. In its first eighteen months of life, the board of control received sixty-six applications to use water and granted fourteen, though many applications were pending.

George Chaffey built the first hydroelectric power plant in California at Ontario in the early 1880s. It generated electricity as a by-product of providing water for irrigation and illuminated street lights in nearby Riverside and Colton. In the 1880s and 1890s, many independent power companies served small communities within a few miles of their power sources. Not until 1895, when a twenty-two mile line was strung from Folsom to Sacramento, was hydroelectric power transmitted any distance. As the technology to send electricity greater and greater distances developed, a consolidation movement began. In 1905, the Pacific Gas and Electric Company was formed, and four years later the Southern California Edison Company was born. By 1930, the P.G. & E. had absorbed dozens of companies and monopolized the generation and sale of hydroelectric power from the Oregon border to Fresno, except in Oakland and San Francisco, which established municipal utility districts. Southern California Edison established the same kind of monopoly south of the Tehachapis, using power plants on the San Joaquin and Kern rivers, though the city of Los Angeles provided its own power after World War I. The consolidation movement and scramble for reservoir sites and water rights alarmed Progressives everywhere. The "power trust" threatened to dominate politics in the early twentieth century as the railroad had in the last decades of the nineteenth century.

or other power, mines and mining, mineral and other lands, dredging, reclamation and irrigation, and for the purpose of revising, systematizing and reforming the laws of this state . . . concerning . . . these said subjects." Johnson appointed ex-governor Pardee to head the commission. The other two members were Francis P. Cuttle, president of the Riverside Water Company and the California Orange Company, and Ralph Bull of Humboldt County. Bull resigned several weeks after his appointment, and J. P. Baumgartner, editor and manager of the *Santa Ana Daily Register,* Orange County's largest paper, took his place. Johnson wisely gave southern California heavy representation on the commission. Doubtless, he remembered that section's overwhelming opposition to the Works bill.[43]

Much of the conservation commission's fieldwork was done by Elwood Mead's old agency in the Department of Agriculture.[44] But the commission's staff also gathered data, including a comprehensive record of all claims to water used for power generation, stream by stream.[45] The commission devoted most of its attention to water law reform. And despite repeated warnings from Cuttle, Pardee decided to try to push a water bill through the legislature during the special session that met from September through December, 1911. Louis Glavis, the commission's secretary, drafted the bill in close consultation with Pardee, and it was introduced without a prior hearing before irrigators, power companies, conservation groups, or other interested parties.[46]

43. *Cal. Stats.,* 1911, 822; *Pacific Rural Press* 79 (March 18, 1911): 210–211; *Call,* April 25, 1911.

44. On the Office of Irrigation Investigations work for the Conservation Commission, see Frank Adams to George Pardee, May 2, 1911; Pardee to Samuel Fortier, May 19, 1911; Fortier to Pardee, May 26, 1911; Adams to Louis Glavis, Secretary of the California Conservation Commission, October 18, 1911, Bureau of Agricultural Engineering Records, Irrigation and Drainage Investigations Division, General Correspondence, 1909–1912, RG 8.

45. Louis R. Glavis to George Pardee, September 11, October 9, 1911; Glavis to the California Conservation Commission, October 9, 1911; Glavis to the Secretary of the Interior, October 9, 1911, Pardee Collection, Bancroft Library.

46. Apparently, Louis Glavis had been hired by several large California timber companies and helped them acquire valuable forested land while working for the commission. This conflict of interest led to his firing, or resignation. Several years earlier, Glavis had precipitated the Ballinger-Pinchot controversy by charging

The legislation was introduced in the assembly on December 11 by William C. Clark of Alameda County. In a letter of the same date to Johnson, Pardee explained that the legislation had been designed to "recover" one-half to two-thirds of the state's water supply and

> clear up and remove all fictitious and speculative filings on and appropriations of water and the use of water—of which there are, we know by actual investigation, something like 20,000 in this State—and give them back to the State, so that they may be reappropriated, under State control and supervision, only by those who will put them to a beneficial use, and not make and hold them in coldstorage for monopolistic and speculative purposes.

The bill declared that all water appropriated but not put to use belonged to the state and created a commission to investigate and quiet water rights on the state's streams. The commission's decisions could be appealed in the superior courts. In many ways, the bill reinforced and expanded the laws enacted during the regular session at the beginning of the year. For example, the Clark bill repeated the twenty-five-year limitation on power permits, though an amended version expanded the permit life to fifty years. The bill also authorized specific filing fees for water rights applicants, ranging from $10 for irrigators to $250 for power companies, as well as graduated fees for the volume of water used. Section 7 of the bill provided for adjudication of water rights by the commission, but did not spell out details of the administrative process. The Clark measure faced opposition even from reformers. Critics complained that such a bill should be introduced only *after* the conservation commission filed its formal report with the legislature in 1913. They also charged that no bill should be enacted that had not been exposed to public criticism; for this reason even the Com-

---

that Ballinger, while commissioner of the General Land Office, had helped private corporations acquire larger tracts of Alaskan coal land than they were entitled to under law. On Glavis's problems in California, see Hiram Johnson to editors of the *Outlook*, January 20, 1913, and Johnson to Meyer Lissner, January 25, 1913, Hiram Johnson Collection, Bancroft Library.

monwealth Club rejected the measure. Pardee and Glavis had made a bad decision. They wanted to move fast to restore water to public control and determine the state's supply of surplus water. In so doing, they made the conservation commission and state board of control—Glavis held seats on both panels—seem arbitrary, capricious, and above public opinion.[47]

Pardee felt betrayed by the legislature, and perhaps by Hiram Johnson, who had refused to support the Clark bill. Ironically, the governor had not been told about the bill until it appeared in the legislature. Still, Pardee blamed "the interests" for its defeat:

> Say what you will, this seems to be the situation to me: The power people are very willing to have any sort of legislation on the books that will not give anybody the machinery to take away from them the water appropriations they are not using. But when it comes to any legislation that will really enable the State to do any real conservation work, then the power people (not unaided and unabetted by some of the irrigation people this time) are not in favor of having anything done.

The ex-governor correctly perceived that the electric companies, a new interest group, had entered the political arena since the defeat of the Works bill, but he failed to own up to his own responsibility for the Clark bill's failure.[48]

In the early months of 1912, the conservation commission revised the Clark bill and held public hearings on the legislation.

47. A.B. 69 (Clark), December 11, 1911, California State Archives. On the Clark bill, see Louis Glavis to George Pardee, December 1, 1911; Hiram Johnson to Curtis Lindley, December 9, 1911; Lindley to Johnson, December 11, 1911; Lindley to Board of Control and Conservation Commission, December 11, 1911; Pardee to Johnson, December 11, 1911, Hiram Johnson Collection, Bancroft Library; Pardee to L. L. Dennett, December 19, 1911; Pardee to Milton T. U'Ren, December 19, 1911; Pardee to Gifford Pinchot, December 27, 1911; Pardee to J. P. Baumgartner, January 4, 1912; Pardee to S. C. Graham, January 4, 1912; and Pardee to R. L. Hargrove, January 16, 1912, Pardee Collection, Bancroft Library. Also see the *San Francisco Chronicle*, November 30, December 5, 14, 15, 16, 1911; *Los Angeles Times*, December 16, 1911; *Call*, December 9, 1911; *Sacramento Bee*, November 30, 1911; *Sacramento Daily Union*, November 30, 1911, December 15, 16, 1911.

48. George Pardee to C. D. Marx, December 19, 1911, Pardee Collection, Bancroft Library.

By the time the 1913 legislature convened, the commission had issued a 502-page survey of California's natural resources. Over half the report concerned water problems. The commission reiterated its call for a wholesale determination of water rights and recommended that the state condemn and purchase all riparian rights. The proliferation of claims filed by power companies made a full determination of rights doubly important. The commission estimated that California's streams could generate over 5,000,000 horsepower of electricity, more than fourteen times the amount produced in 1912. The companies usually included the value of water rights in their capitalization, and water rates reflected these bloated assets even when the rights were not being fully used. The report paid close attention to corporate attempts to "cold storage" water to deprive other companies of its use, reserve a supply for future expansion, or simply reap speculative profits. The commission found that 90 percent of the water that had been claimed for power purposes in Plumas, Butte, Tehama, Stockton, Yuba, Sacramento, Yolo, Tuolumne, and Inyo counties had not been put to use within a reasonable time, as the law required. Instead, companies kept their "rights" alive by doing a small amount of work, or simply by refiling from time to time. The 1911 law, establishing a board of control, allowed the new commission to pass judgment on all *new* applications to use water for power. Unfortunately, it did not permit the board to investigate and weed out claims filed *prior* to 1911.[49]

In revising the Clark bill, the conservation commission was forced to abandon some of its original objectives. For example, the group initially favored regulating underground water when it was diverted onto land not immediately adjoining the well. The commission quickly abandoned this provision because many southern Californians, whose section strongly depended on subterranean water, feared that specific grants of water by the state would promote litigation. And while the conservation commission originally favored the condemnation and purchase of riparian rights, by the

49. *Report of the Conservation Commission of the State of California, January 1, 1913* (Sacramento, 1913), 22–23, 30–31, 38–39. The report neglected many resources. California's forests merited only twenty-nine pages, and fish and game, petroleum, coal and other natural resources received only passing mention, or no attention at all. The fear of water monopolies clearly made water the most controversial resource.

summer of 1912 it considered such a policy infeasible. The cost would be staggering, and contested cases might take years to settle. Even though the commission doubted the legislature's power to limit such rights, it had no alternative but to seek such a restriction.[50]

Hiram Johnson worked hard to pass the conservation commission's water bill; the governor made it the first of his "Ten Commandments" sent to the 1913 legislature. Many of the state's leading newspapers, including the *Sacramento Daily Union*, the *San Francisco Chronicle*, and the *Los Angeles Times*, opposed the legislation and on May 12, 1913, the *Union* described the battle over the proposed law "one of the bitterest fights of the session." The bill was amended four times in the lower house, defeated, then passed by one vote on reconsideration. The arm twisting continued in the senate as Johnson lined up support vote by vote among recalcitrant senators. In most sessions, such technical bills attracted little public attention. But on May 8, the Senate Judiciary Committee's chambers were jammed with spectators and reporters as that committee opened hearings on the legislation. Later, the governor confided to Gifford Pinchot: "In the last Legislature the biggest fight I had was to pass this water bill, and I did it by the narrowest possible margin." Without Johnson's active assistance, the water bill would have suffered the same fate as the Works and Clark bills.[51]

The 1913 law was much more modest than those earlier pro-

50. C. E. Tait to Samuel Fortier, May 21, 1912, Records of Irrigation Investigations, Staff Correspondence 1906, 1912–1915, RG 8; Francis Cuttle to Hiram Johnson, August 18, 1915, Hiram Johnson Collection, Bancroft Library; *San Francisco Chronicle*, May 16, 1912. Tait noted in his letter to Fortier: "Hundreds of new wells are bored every year and these furnish new evidence [in court proceedings]. It is a question if provision can be made under a new system for the frequent reopening of cases after the first adjudication as will be demanded to give justice to all."

51. The quote is from Hiram Johnson to Gifford Pinchot, July 7, 1913, Hiram Johnson Collection, Bancroft Library. Also see Johnson to Francis J. Heney, May 18, 1913, in the same collection; George Pardee to Gifford Pinchot, May 25, 1913, Gifford Pinchot Collection, Library of Congress, Washington, D.C.; Hichborn, *Story of the California Legislature of 1913*, 137. John M. Eshleman, president of the California Railroad Commission and one of the water commission bill's sponsors, reported strong opposition from both power companies and riparian owners. See Eshleman to George Pardee, January 15, February 15, March 7, 1913; Pardee to Eshleman, January 17, February 11, 1913, General Correspondence, Department of Engineering and Irrigation, microfilm reel M–708, DWR Archives; *Sacramento Daily Union*, May 5, 9, 10, 11, 12, 1913. The *Union's* most important editorial against the bill appeared on May 7, 1913. The *Los Angeles Times* published editorials opposing the bill on May 6, 12, 1913.

posals. The most ambitious western water laws, such as Wyoming's, gave a state commission or state engineer power to adjudicate water rights and distribute water, as well as to regulate the acquisition of new rights. The conservation commission's water law created a permanent water commission and a centralized record of claims; provided clear administrative procedures for filing new claims; and sharply reduced the number of "paper rights." The legislature granted the commission power to set time limits within which water claimed had to be put to use, on penalty of forfeiture, and the law's forceful declaration of state sovereignty over water laid the foundation for a state water plan in the future. For the first time, municipal water use was given priority over agriculture and mining, even against prior claims. After ten years, all riparian rights would be limited to beneficial use, just as appropriative rights were. And after twenty years, any rights filed under the new law could be revoked or confiscated by the state water commission, if the state paid for property damages and the cost of distribution works. This step could be taken as an administrative action, without resort to the courts. The expansion of state authority could also be seen in the filing fees and annual charges demanded from new water users. Symbolically, these fees reflected state sovereignty over water, though they also provided a measure of financial independence to the new commission; it did not have to depend exclusively on legislative appropriations. The fees were also expected to conserve water by reducing waste. Irrigators and power companies that paid for their water were more likely to use it efficiently.[52]

Immediately following the governor's approval of the new water law, power and irrigation companies, with the assistance of Miller and Lux and other riparian owners, launched a referendum campaign to nullify the statute. Consequently, in the months preceding the November 1914 election, George Pardee and other members of the Conservation Commission took to the hustings to explain and defend the law. The former governor hammered on two major points: that the law was needed to prevent water monopoly, and

52. *Cal. Stats.*, 1913, 586. For discussions of the bill's provisions, see the Frank Adams oral history transcript, Bancroft Library, 192–222; Alfred E. Chandler's "The Water Bill Proposed by the Conservation Commission of California," *California Law Review* 1 (January 1913): 148–168.

that it would all but eliminate litigation over water rights. He maintained that 90 percent of all the claims ever filed in California had been *exclusively* speculative. Under the new law, no future claim was absolute. In effect, water users would lease water from the state and could only continue to use it if they put it to good use. The fees reflected the fact that ultimate ownership remained with the people. In 1912 only 1,516,000 of the 7,576,000 acres in the San Joaquin Valley were irrigated, and only 123,000 of 3,450,000 acres in the Sacramento Valley. Pardee assumed that virtually all this land was irrigable. "On these eleven million acres," the ex-governor promised, "when they are irrigated, and only when they are irrigated, there will live, in cities, towns, villages and on ten and twenty acre farms, in California comfort, ten millions of happy and prosperous American people. Without the proper, cheap and certain irrigation of this great valley, the full prosperity, development and advancement of California can never be obtained." Similarly, California could generate over ten times the amount of electrical power produced in 1912, but only if the scramble "by three or four large and constantly growing power companies" for water and power sites ceased.

Pardee explained that the adjudication provisions of the 1913 law had been copied from an Oregon statute adopted in 1909. By 1912, Oregon's water commission had adjudicated all 965 rights on fifteen different streams at a cost of $10.50 to each claimant. Nine of these administrative proceedings had already been confirmed by the courts without a single appeal from a disgruntled water user. Pardee boldly predicted: "The proposed water commission law will finally settle in one proceeding in the state courts all the existing water rights on each California stream; will give to every claimant the rights that belong to him, and will assure him those rights so long as he exercises them."[53]

---

53. All statistics and quotes are from George C. Pardee, "Address on the Water Commission Law," November 7, 1913, delivered at the Hanford, California, meeting of the California Development Board, a copy of which is in the California Room of the California State Library, Sacramento, and his "Address on the 'Water Commission Law' delivered before the City Club of Los Angeles, January 31, 1914," in a volume of speeches and pamphlets entitled *California Water and Power Act* at the Bancroft Library. Privately, Pardee expressed less confidence in the 1913 law. See his May 25 and December 7, 1913, letters to Gifford Pinchot, Gifford Pinchot Collection, Library of Congress, Washington, D.C.

Not all reformers shared Pardee's unbridled public enthusiasm. John M. Eshleman, chairman of the State Railroad Commission, helped draft the law and push it through the legislature. Nevertheless, he worried that the water fees were too high and might induce power companies to burn coal to generate electricity. He also warned that the imposition of more than token fees might backfire and reinforce the traditional view of water as private property held by the courts. Along with most friends of the measure, he doubted the constitutionality of the section limiting riparian rights.[54] Frank Adams, the federal Office of Irrigation Investigations chief agent in California and a prominent member of the Commonwealth Club, argued that the Conservation Commission had paid too much attention to fighting monopoly and not enough to settling water conflicts. Adams began his career as one of Elwood Mead's assistants and never lost his sympathy for the "Wyoming System" Mead had created. On June 6, 1913, Adams wrote his boss in Washington: "those behind the measure have not cared to give very serious consideration to the means of accomplishing the things they seek. The detailed procedure which makes up the bulk of the water laws of other States is almost entirely omitted and the desire has been merely to establish the authority of the State and trust any Commission that might be appointed to act wisely." The water commission lacked police powers. It could investigate water conflicts, hold hearings, and provide superior courts with expert information and testimony. But the courts were not bound to consult the commission or honor its findings; the water commission lacked the power to enforce its own decisions, let alone court decrees. In 1913, Adams had won a small victory when he persuaded Governor Pardee to add section 37 to the law, which gave the state power to supervise the distribution of water "when such supervision does not contravene the authority vested in the judiciary of the state." But the state could distribute water only when local water users requested its help.[55] Both Adams and L. L. Dennett also regretted

54. John M. Eshleman to Hiram Johnson, January 13, 1913, Hiram Johnson Collection, Bancroft Library. For the most thorough analysis of the bill by its critics, see *TCC* 8 (February 1913).

55. For Frank Adams's criticisms of the water bill, see his oral history transcript, Bancroft Library, 192–222; idem, letters to Samuel Fortier, February 13, March

that the commission had been given no power over water claimed by cities and that the law did not require cities to follow the same standards of diligence and beneficial use as other water users. No other state gave such free rein to its municipalities. The 1913 law could not have been passed without the support of Los Angeles and San Francisco. Ironically, at the same time Progressive politicians from those two cities were attacking corporate monopoly over natural resources, most supported their cities' efforts to establish monopolies over the Owens and Tuolumne rivers in the Sierra Nevada. Neither city put more than a small fraction of the water claimed to immediate use; Dennett predicted San Francisco would not need its full supply from the Tuolumne for 50 to 100 years. In the meantime, he demanded the surplus water for the Modesto and Turlock irrigation districts (under the 1913 law, San Francisco could dispose of the water in any way it wished).[56]

From the beginning, the 1913 law encountered severe obstacles. The measure took effect in December 1914, following the referendum election, and Hiram Johnson appointed A. E. Chandler, one of the arid West's leading water lawyers, chairman of the new state water commission.[57] Unfortunately, the legislature had approved the law only after slashing the commission's initial appropriation from $75,000 to $25,000, a sum inadequate to pay for a systematic survey of water rights watershed by watershed. Moreover, the commission had a staggering range of responsibilities: preparing a comprehensive record of rights; weeding out bogus or unperfected claims; reviewing applications for new rights; revoking old rights users had allowed to lapse; adjudicating disputed rights; supervising water distribution; determining the volume of unap-

---

22, March 31, April 26, June 6, 1913; idem to W. F. McClure, California State Engineer, February 11, 19, 1913, idem to James F. Farraher, May 5, 1913, Office of Irrigation Investigations, Staff Correspondence 1906, 1912–1915, RG 8. As an engineer, Adams resented the tendency of Progressive reformers to bypass existing institutions and turn water problems over to special commissions and entirely new agencies. He thought that the California State Engineer's office should have been given the responsibility to enforce the new water law. He also strenuously opposed the charges for water levied by the statute.

56. L. L. Dennett to U.S. Senator John D. Works, July 5, 1913, John D. Works Collection, Bancroft Library.

57. On the referendum campaign, see TCC 9 (October 1914): 581–595.

propriated water in California streams; investigating all irrigation and power projects that required large quantities of water; and, finally, performing miscellaneous duties including monitoring salinity in the lower Sacramento River and inspecting the water supply systems of state institutions.

The water commission accomplished little during its first few years of life but won new support during and after World War I. In 1917, the legislature sharply reduced filing fees for water claims and virtually eliminated annual user fees. This was done, in part, because the Commission's members thought that the Railroad Commission should impose and collect user fees.[58] But in the same year, the lawmakers ruled that the board could *reject* applications "detrimental to the public welfare." (The 1913 law simply stated that the commission would "permit the appropriation of unappropriated water.") Four years later, spurred by sharp increases in water rights applications during World War I and a drought that lasted from 1919 to 1921, the legislature went even further. It granted the quasi-judicial commission power to set its own "terms and conditions" for the use of unappropriated water; allocated $50,000 to set up a revolving fund (replenished from filing fees) to pay for "determinations of water rights on the various streams and stream systems of the state"; and permitted the commission to divide the state into water districts and appoint one or more watermasters to supervise distribution when 15 percent or more of the owners of irrigation works in a district appealed for the service.[59]

In part the legislature was responding to court challenges of the Water Commission's authority. The most important case was filed by the Tulare Water Company, a subsidiary of the Kern County Land Company, in December 1919, after the commission rejected the company's application for 2,000 cubic feet per second (c.f.s.) of water from the Kern River. The Water Commission decided that the river carried insufficient unappropriated water to warrant the

58. *Cal. Stats.*, 1917, 195; W. A. Johnstone to J. C. Nagle, June 15, 1917, microfilm reel 1186, DWR Archives.

59. *Cal. Stats.*, 1917, 194, 231; *Cal. Stats.*, 1921, 443, 543, 1638. For other amendments to the 1913 law, see *Cal. Stats.*, 1919, 511, 1193; *Cal. Stats.*, 1921, 482; *Cal. Stats.*, 1923, 51, 124, 161, 162; *Cal. Stats.*, 1925, 586; and *Cal. Stats.*, 1927, 1024, 1668.

new diversion. However, the company's lawyers, led by E. F. Treadwell, argued that the 1913 law had not granted the administrative commission power to reject properly filed applications. The courts, not the commission, should decide how much unappropriated water remained based on the complaints of users. However, the court left a loophole in its decision. Only Judge C. J. Shaw categorically denied that the legislature could confer quasi-judicial powers on an administrative commission. The other judges simply denied that the 1913 law went that far, especially because it did not offer rejected applicants an appeal process. Consequently, the 1923 legislature amended the law so that any applicant turned down by the commission could, within thirty days of the judgment, ask for a review by a superior court. The court had to consider evidence presented by the commission before it confirmed, reversed, or modified the ruling. But since the commission rarely rejected reasonable requests to use water, the courts received few appeals.[60]

The Water Commission, which became the Water Rights Division after 1921, had plenty to do. In 1919, and again in 1920,

60. *Cal. Stats.* 1923, 161; *Report of the Division of Water Rights, November 1, 1922, JCSA*, 45th sess. (Sacramento, 1923), Appendix, 4: 73–75; *Biennial Report of the Division of Water Rights, November 1, 1924*, 46th sess. (Sacramento, 1925), 4: 62–63; Tulare Water Company v. State Water Commission, 187 Cal. 533 (1921); Marion J. Mulkey, "The Division of Water Rights of the Board of Public Works: Forfeiture of Riparian Rights by Non-Users" (J.D. diss., University of California, Berkeley, School of Law, 1923), 29–36.

In 1921 or 1922, the director of public works created a special board of review to hear appeals to the Water Commission's decisions. See R. M. Morton (director of public works) to W. F. McClure (state engineer), July 25, 1923, microfilm reel 1186, DWR Archives. The *San Francisco Chronicle*, December 22, 1927, reported that on December 21 the supreme court, in a case involving the Mojave Irrigation District, had ruled the 1923 legislation unconstitutional. This had little effect because judicial review of new water rights was not the commission's prime function anyway. Of the more than 4,200 claims filed by 1925, most of the 1,500 rejected were for noncompliance with terms of the grant, such as failure to begin construction within a stipulated period or not putting the water to beneficial use. Most claimants feared the litigation that would result from overappropriation and trusted the commission's advice concerning the remaining supply. See Edward Hyatt, "Control of Appropriations of Water in California," *Journal of the American Water Works Association* 13 (February 1925): 125–144. On the commission's powers, also see Hyatt's "Water Conservation and Control in California, Dec. 8, 1925," Folder no. 1, Edward Hyatt Collection, California Water Resources Archives, and "Administration of Stream Flow," July 14, 1927 in the same file.

applicants filed for three times the volume of water claimed in any single year from 1915 through 1918. And after the war, California's half-dozen anticipated storage projects further swelled the number of applications. In 1922, one of eight applications required a field survey. Because the division had only four hydraulic engineers available to perform this task, however, the surveys lagged two to four years behind filings. The need to cooperate with federal officials also slowed down the review process. About 40 percent of the water rights applications laid claim to water within national forests. Before Congress approved the Water Power Act in 1920, the Water Rights Division routinely referred power applications to the supervisor of the appropriate national forest before granting a permit. After 1920, however, the new Federal Power Commission received all filings. State officials claimed sovereignty over all the water within California's borders except the Sacramento River. But they deferred to the F.P.C., if only because it could block any power project by refusing to grant rights of way across government land.[61]

From 1915 to 1928, the Water Rights Division received 6,023 applications to use water. Sixty-one percent pertained to irrigation, 16 percent to power, 11 percent to domestic uses, 9 percent to mining, and the remainder to various municipal uses. In all, the division approved 54 percent of the applications to use water for irrigation, but only 27 percent of the power requests. In 1928, the volume of water claimed totaled more than eight times the amount filed on in 1918, and the division had become increasingly discriminating in reviewing applications. Of each 100 requests, 43 were rejected. Of the 57 approved, 30 were subsequently revoked for noncompliance with terms of the grant, usually simply failing to use the water. Thus only 27 applicants out of every 100 received clear "title." The division measured its grants in both *second-feet*, that is, water measured directly as diverted from a stream, and *acre-feet*, which is gauged as stored water. Of each 100 second-feet of water requested, the division granted only 15 second-feet, and only 7 second-feet were put to beneficial use. Of each 100 acre-feet

61. *Third Biennial Report of the State Water Commission of California, 1919–1920* JCSA, 44th sess. (Sacramento, 1921), Appendix, 1: 25; *Statistical Report of the California State Board of Agriculture for the Year 1921*, 45th sess. (Sacramento, 1923), 3: 26: *Report of the Division of Water Rights, November 1, 1922*, 23, 25, 53, 54.

claimed, only 7 acre-feet were allowed and 6.3 acre-feet of that were revoked for noncompliance. Hence, only 0.7 acre-feet finally received legal sanction. The Division's 1928 report noted: "One of the most fruitful fields of effort . . . is the elimination of proposed projects which for one reason or another have been abandoned. Hopeless and abandoned projects once formed no little obstacle to proposed new development." Equally important, the division showed a clear preference for public water projects. For example, from 1924 to 1926, 78 percent of the direct diversion rights and 88 percent of the storage rights were granted to irrigation districts.[62]

The Division of Water Rights also adjudicated established water rights. It could initiate proceedings on its own, on the request of a superior court, or on appeal from one or more local water users. In each case, the administrative procedure was the same. After the division collected basic information regarding stream flow, diversions, soils, and crops, it asked water users on a stream to submit evidence to support their claims. Subsequently, the division compiled a preliminary abstract of rights and provided each claimant with a copy. Those who objected to the list could demand a formal hearing. Once the division reached a decision, it again notified water users and forwarded its evidence and conclusions to the appropriate superior court. The court entertained appeals before it ratified or amended the judgment in its final decree.[63] An adjudication by the California Water Commission or Division of Water Rights offered many advantages over court determinations. Proceedings could begin *before* conflict erupted, and *all* rights to a stream could be decided, not just those of parties to a suit. Moreover, "disinterested experts" collected the data, and the process cost less, took less time, and better protected the "public interest." Nevertheless, because of the indeterminate nature of riparian rights and the Water Rights Division's limited budget, the 1913 law did

62. *Biennial Report of the Division of Water Rights and State Water Commission, November 1, 1928. JCSA*, 48th sess. (Sacramento, 1929), Appendix, 5: 8–15. The quote is from p. 15. Also see the *Biennial Report of the Division of Water Rights, November 1, 1926*, 47th sess. (Sacramento, 1927), 3: 25–26, 32.

63. The Water Commission explained the adjudication procedure in a pamphlet entitled "Rules and Regulations Governing the Determination of Rights to the Use of Water in Accordance with the Water Commission Act," 1921, Bancroft Library.

not result in a wholesale settlement of water rights. State officials refused to cast the division in the role of a protagonist and, as of 1922, the division had not prosecuted a single adjudication on its own initiative.

Instead, the division usually acted on appeal from the courts. In 1923, the legislature considered a constitutional amendment to abolish the Division of Water Rights and replace it with a special water court. The amendment, prompted by the vast range of interests represented in the Antioch suit discussed in the next chapter, tacitly acknowledged that while an administrative commission could rank appropriative rights, it had no power to define riparian claims. The amendment also implied that the legislature had little confidence in the division. By 1926, the division had launched eighteen adjudication proceedings. Only seven had resulted in court decrees, and these covered only 161 water rights and 15,470 acres of land. The seven streams, remote and probably unaffected by riparian claims, were Willow Creek in Lassen County; San Pedro Creek in San Mateo County; Hat and Burney creeks, and the north fork of Cottonwood Creek, in Shasta County; the west fork of the Carson River in Alpine County and Oak Creek in Inyo County.[64]

In addition to granting new rights and adjudicating old ones, the Water Rights Division provided watermasters to help divide up streams covered by court decrees or informal agreements among water users. Every drought prompted violations of decreed rights,

64. For detailed discussions of adjudication proceedings, see the reports of the State Water Commission and Water Rights Division. For a summary of state actions see *Biennial Report of the Division of Water Rights and State Water Commission, November 1, 1928, JCSA,* 48th sess. (Sacramento, 1929), Appendix, 5: 16. On the Constitutional Amendment no. 28, rejected by the legislature in 1923, see Frank Adams, "Pending Irrigation and Water Legislation," *Pacific Rural Press* 105 (March 31, 1923): 388. One reason the Water Rights Division focused on small streams in the foothills was that miners had claimed the entire volume of such streams before the adjoining land left the public domain, preventing the creation of riparian rights. Most of those appropriative rights were subsequently acquired by farmers. In the Central Valley, the existence of riparian rights prevented the state from making definitive settlements. The only stream of any size adjudicated in the valley was the Stanislaus, and water users there challenged the determination. To avoid agitating riparian users, the Water Rights Division never pressed suit to define that class of rights under the 1913 Water Act, even after the ten-year grace period had elapsed. The supreme court finally overturned this "restriction" on riparian rights in 1935. See Tulare Irrigation District v. Lindsay-Strathmore Irrigation District, 45 Pac. (2d) 972 (1935).

and the courts usually proved ineffective—and always proved slow and costly—in attempting to enforce their decisions. This presented the Water Rights Division with a great opportunity to expand its influence. The sheer mass of new claims, spawned by irrigation districts as well as storage and power projects, demanded an expansion of state authority. Watermasters also played important roles in adjudication proceedings on streams where final decrees had not yet been issued. They could testify to actual conditions in the field, and their testimony often made the court's job easier.

The water code of 1913 said little about state supervision over distribution. Nevertheless, in July 1919, Kings River water users asked the state to appoint a watermaster. The drought contributed to that decision, but they also wanted to clear up water conflicts before beginning the Pine Flat Reservoir Project. The river was ninety miles long and served 625,000 acres of land through forty-five diversion ditches. Court tests began in the 1870s, and by the second decade of the twentieth century 137 different suits had been filed. The obvious futility of litigation persuaded the Kings River claimants to prepare a schedule listing priorities and amounts due to each diverter according to the total volume of water available. The agreement specified that the schedule should be enforced by an agent of the Water Rights Division paid by the water users themselves. The Water Commission had measured diversions from the Kings River since December 1917 and had supervised distribution throughout the 1920s.[65]

Before 1921, the State Water Commission supervised diversions only when it could muster virtually unanimous support from water users on a particular stream. But in that year the legislature enacted a law that permitted the commission to appoint watermasters to administer decreed streams when 15 percent or more of the owners of diversion ditches requested the service. The law required regulated water users to provide lockable headgates and suitable measuring instruments; made interference with the watermaster's work a misdemeanor; and gave him the power to arrest violators. However, the statute said nothing about how the watermaster would be paid. The Division of Water Rights suggested

65. *Third Biennial Report of the State Water Commission, 1919–1920,* 43, 49, 75–76, 133–140; *Report of the Division of Water Rights, November 1, 1922,* 9, 57–64.

that California follow Nevada's lead by allowing county assessors to apportion the cost of supervision according to the amount of water used by individual claimants in relation to the total volume of diversions. The legislature refused, perhaps because Nevada's assessors added the charge to property taxes.[66]

In May 1922, upon petition from 27 percent of the ditch owners on the west fork of the Carson River, the state appointed a watermaster who began his work in August 1922. Voluntary contributions paid his salary in 1922 and 1923. However, in 1924, 1925, and 1926 contributions flagged. In its report for 1926 the Division of Water Rights reported that upstream farmers, "because of their strategic position on the stream, are unsympathetic regarding supervision of diversions." Subsequently, the watermaster abandoned his work and the division refused to appoint watermasters anywhere in the state unless requests for the service were accompanied by an ironclad promise to pay. The division asked California's superior courts to insert a provision in all pending decrees permitting the state to appoint a watermaster on the appeal of one or more water users and requiring all parties to the suit to share the cost. Under such decrees, the state regulated water use in Shasta County on Hat Creek beginning in 1924 and Burney Creek beginning in 1926. The division also took charge of many small streams in Shasta and Modoc counties following the ratification of formal contracts with water users. And after 1924, as a result of the Antioch suit, the state supervised diversions from the Sacramento River at its own expense. By 1929, the division administered fourteen streams which served 1,538,000 acres of land. However, all but 28,500 acres were contiguous to the Sacramento or Kings rivers.[67]

In 1924, the head of the Division of Water Rights, Edward Hyatt, looked back over the previous decade and assessed the significance of the Water Act of 1913:

> The State Water Commission in 1915, when first organized, faced a most difficult situation. Its jurisdiction was limited by

66. Cal Stats., 1921, 543; *Biennial Report of the Division of Water Rights, November 1, 1926,* 48–49.

67. *Biennial Report of the Division of Water Rights, November 1, 1926,* 49–54; *Biennial Report of the Division of Water Rights and State Water Commission, November 1, 1928,* 21–22, and chart on p. 19.

the hundreds of Supreme Court decrees and there were many constitutional questions in the act itself. . . . It must be admitted at the outset that the operations of the State Water Commission and the Division of Water Rights have not fundamentally changed the legal situation, nor has litigation over water matters been done away with. It has, however, been greatly reduced, considering the tremendous increase in the rate of development of water projects in the last few years. . . . New rights to the use of water are under state supervision and are recorded and classified, many old rights have been adjudicated, and distribution of water is being carried on in important areas. The value of the public records maintained will increase with the length of their continuity and as they are gradually extended and made more complete.[68]

California clearly lagged behind Wyoming's and Nebraska's cheap and efficient administrative determination of water rights, as well as behind Colorado's state supervision over distribution. These states, not California, served as the arid West's models of reform.[69] Nevertheless, the 1913 law was more ambitious and successful than many of its critics realized. Wyoming and Colorado did not recognize riparian rights, contained fewer water users, and put their water supply to fewer uses. California's water problems were bigger, more complicated, and more diverse. Hyatt concluded: "The Water Commission Act probably went as far as possible by legislation to establish a complete and efficient code; however the simple fact was that a direct and efficacious solution was not possible in California, as it had been in other Western States."[70]

In short, the 1913 law must be measured by the hopes of reformers as much as by their accomplishments, and by the possible, not the ideal. California historians have all but ignored water law reform. For example, George Mowry's classic, *The California*

---

68. *Biennial Report of the Division of Water Rights, November 1, 1924,* 9–10.

69. Robert G. Dunbar, *Forging New Rights in Western Waters,* University of Nebraska Press, forthcoming.

70. Edward Hyatt, "Review of Work of the Division of Water Rights," February, 1924, Folder no. 1, Edward Hyatt Collection, California Water Resources Archives. Also see the transcript of his speech before the Pomona Kiwanis Club, March 24, 1926, in the same file.

Progressives, devotes only a paragraph to the subject, and Spencer Olin's study of progressivism, *California's Prodigal Sons*, neglects the reform entirely, arguing that "the high tide of reform in the areas of economics, politics and social welfare was not matched by similar advances in conservation and agriculture."[71] Both the irrigation district and water laws enacted in 1911 and 1913 belie this conclusion. The expansion of state power over natural resources, the Progressive challenge to "the interests," the Progressive faith in experts and special commissions, and the Progressive belief in order and planning, all found expression in the quest to manage the state's water supply more effectively. On June 9, 1913, with the new water code still fresh in his mind, Hiram Johnson wrote to fellow Progressive Meyer Lissner:

> Communities will stand just so much reform legislation at one time, and wise is the man who intuitively has some conception of just how far he can go. . . . Indeed, our legislation has brought us to the very verge of disaster, and all over the state the attacks on us have had their effect.[72]

Johnson, Pardee, and other supporters of the new water law were convinced that whatever the statute's weaknesses, they had done as much as they could. Other legislatures, and members of the water commission itself, could be trusted to repair its defects in the light of future experience.

Ironically, as in earlier decades, the fears of reformers were usually exaggerated; battles in the legislature did not fairly represent the state of California agriculture. Despite its outmoded water laws, California grew and prospered during the first decade

71. George Mowry, *The California Progressives* (Berkeley, 1951), 152; Spencer Olin, *California's Prodigal Sons* (Berkeley, 1968), 176.

72. Hiram Johnson to Meyer Lissner, June 9, 1913, Hiram Johnson Collection, Bancroft Library. Also see Johnson to Gifford Pinchot, July 7, 1913, in the same collection. The Progressives' greatest accomplishment may have been in forging a coalition of reformers rather than in their ideals (which often grossly oversimplified complicated resource issues). The Progressive coalition included civic reformers who considered the fight against monopoly one way to purify the political process; engineers who hoped to promote efficiency and expand their role in state government; and bankers and investors who hoped that new water laws would drive up land values at the same time they stimulated irrigation. Holding such disparate groups together was no mean feat.

of the twentieth century. The number of acres cropped in California increased little from the 1880s to the 1920s. But north of the Tehachapis, the number of farms dramatically increased with the breakup of wheat ranches; and from 1900 to 1910, the greatest population gains occurred in heavily irrigated counties. Stanislaus County had the biggest gain, 136.7 percent, largely because of the expansion of irrigation agriculture around Turlock and Modesto. Fresno County's population increased by 99.5 percent, Orange County's by 74.8 percent, Riverside County's by 93.9 percent, San Bernardino's by 103 percent, and Tulare's by 93.4 percent. By contrast, the population of the Sacramento Valley increased by only 34 percent.[73] Between 1900 and 1910, virtually every agricultural statistic spelled prosperity. The average per acre value of the state's farmland increased from $21.87 to $47.16. The number of irrigated farms increased from 25,675 to 39,352. Irrigated acreage rose from 1,446,114 to 2,664,104 acres, an 84 percent increase. The percentage of irrigated farms grew from 35.4 to 44.6 percent. The length of main ditches expanded from 5,106 to 12,599 miles, a 146.7 percent increase. And the total value of irrigation works increased from $19,181,610 to $72,445,669, an increase of 277.7 percent. Of the irrigated land, over 75 percent was in the San Joaquin Valley and southern California; the rich Sacramento Valley contained only 8 percent of the state's irrigated acreage.

In 1900, twenty states ranked ahead of California in population, but by 1910, the state ranked twelfth. Southern California's growth continued to eclipse that of the northern part of the state. From 1900 to 1910, the San Francisco Bay counties increased by 40.6 percent, while the counties south of the Tehachapis grew by 146.9 percent. Not only had the total population of the "cow counties" surpassed that of the Bay Area, but southern California's population exceeded that of the entire northern section, provided the population of the San Joaquin Valley was added to that of the six southernmost counties.[74]

73. Frank Adams, "Irrigation the Basis and Measure of the Present Agricultural Growth of California," *Report of the California State Agricultural Society for the Year 1909, JCSA* (Sacramento, 1912), Appendix, 2: 46.

74. "Statistical Summary of the Production and Resources of California [1911]," *JCSA*, 40th sess. (Sacramento, 1913), Appendix, 2: 222–223. This report was the first complete statistical survey of California agriculture compiled by state officials. After 1911, such surveys were published frequently. The above report should be

In the second decade of the twentieth century, irrigation continued to expand at a rapid rate, particularly during World War I. Irrigated land increased from 2,664,103 to 4,219,040 acres. The second and third decades of the century belonged to the refurbished irrigation districts. From 1910 to 1920, the amount of land in irrigation districts increased by more than 400 percent, from 642,510 to 2,575,198 acres. At the same time, the capital invested in irrigation works swelled from $72,445,669 to $194,886,388. Nevertheless, the precipitous fall of crop prices after the war posed an ominous warning. The total value of California's farm products nearly tripled from 1909 to 1919, increasing from $224,981,000 to $729,661,000. But by 1921, this figure had fallen back to $471,748,000.[75]

During the 1920s, the irrigation lobby continued to push for expanding California's irrigation network even in the face of growing crop surpluses and plummeting prices. But for the first time, providing supplemental water for land already under ditch took precedence. California agriculture was at a crossroads. Virtually all the state's smaller water projects had long since been constructed by private enterprise, mutual water companies, and irrigation districts. By the same token, large storage projects were beyond the means of most farmers. Who would pay for new irrigation works remained an open question, but during the 1920s Californians finally recognized the need for a comprehensive state water plan to guide future development. That plan, adopted by the legislature in 1931, was shaped by the combination of drought, depression, and the state's remarkable economic and population growth since 1900.

---

supplemented with the section on irrigation in the "Statistical Summary of the Production and Resources of California" prepared by the state statistician and included in the *Report of the State Board of Agriculture, 1913, JCSA,* 41st sess. (Sacramento, 1915), Appendix, 3: 166–183. Also see Samuel Fortier, "Irrigated Agriculture Dominant in California," *Pacific Rural Press* 76 (July 25, 1908), 49, 53; *San Francisco Chronicle,* December 14, 1911; *Annual Report, California Development Board, 1913* (San Francisco, 1913), 6; *Annual Reports of the California Development Board, 1918 and 1919* (San Francisco, 1920), 7.

75. California Development Association, *Report on Problems of Agricultural Development in California* (San Francisco, 1924), 5, 7; *Statistical Report of the State Board of Agriculture for the Year 1921, JCSA,* 45th sess. (Sacramento, 1923), Appendix, vol. 3.

12

Toward a State Water Plan: The Genesis of the Central Valley Project

California changed greatly from 1900 to 1925. Heavy migration into the state and an increasingly diversified economy expanded state tax revenue and the state bureaucracy. Until recently, California historians have characterized the 1920s as a period of fiscal retrenchment and shrinking governmental responsibilities.[1] But during the twenties, California spent far more money and attention on water resource surveys and planning than in any previous decade. Even the depression failed to discourage the friends of a state water plan; it simply provided a new opportunity to achieve their objective. In the early 1920s, they promised that sales of hydro-

1. Although he does not consider water planning, for an important revisionist study see Jackson K. Putnam, "The Persistence of Progressivism in the 1920s: The Case of California," *Pacific Historical Review* 35 (November 1966): 395–411. The California state government's increasing interest in a state water project built on the accomplishments of the prewar Progressives, many of whom—including Franklin Hichborn, William Kent, Hiram Johnson, Rudolph Spreckels, and James D. Phelan—became ardent supporters of a state plan and public power. As before the war, those who favored state action included industrialists, bankers, land speculators, and engineers, as well as the lawyers, newspapermen, and small businessmen commonly recognized as the leaders of progressivism. For a fine analysis of resource policies that treats the persistence of progressivism at the federal level, see Donald C. Swain, *Federal Conservation Policy, 1921–1933* (Berkeley and Los Angeles, 1963).

electric power and municipal water would subsidize the soaring cost of irrigation; by the early thirties, they touted the Central Valley Project as a jobs measure as well as an agricultural program. Ironically, the first state plan did not reflect the triumph of science, efficiency, or coordinated planning, though it could never have been drafted without the burgeoning corps of resource specialists in Sacramento. That scheme surmounted the traditional sectional differences and clashing interest groups that had always characterized California water politics by promising something to everyone. As such, it was more the product of political logrolling than of farsighted, disinterested specialists trying to meet a rapidly growing state's future water needs.

The water politics of the 1920s cannot be understood without considering the effects of the droughts that visited California in 1917–1920, 1924, and 1929–1935.[2] From 1900 to 1917, most parts of the state enjoyed above average, if not abundant, rainfall. But in its reports for 1919–1920, the Department of Engineering warned:

> The succession of dry years experienced by the State, beginning with 1917, attended by unprecedentedly small flow of water in the streams, has demonstrated most forcibly that irrigation development has reached its limit unless conservation of the water resources can be established. This can be brought about partly by learning and applying more economical use[s] of water. . . . But the greater conservation is in controlling the wild destructive flood waters by storing them for summer use.[3]

Though California led the West in miles of ditches and canals, reservoir construction, vital to both irrigation and power generation, had lagged behind Colorado during the first two decades of the twentieth century. By August 1920, the water supply of many communities around San Francisco Bay was, according to San Francisco's city engineer, M. M. O'Shaughnessey, "in a very precarious

2. See the chart entitled "Precipitation Variability" in William Kahrl et al., *California Water Atlas* (Sacramento, 1978), 7.

3. *Seventh Biennial Report of the Department of Engineering of the State of California, 1919–1920, JCSA*, 44th sess. (Sacramento, 1921), Appendix, 5: 137.

condition." For example, the C & H sugar refinery at Crockett, on the north arm of the bay, was forced to import fresh water by ferry from Marin County. On August 19, 1920, the Power Administration of the State Railroad Commission ordered a 20 percent cut in the use of electricity. The power shortage, which created profound difficulties for farmers who pumped their irrigation water from streams or subterranean sources, became so severe that Governor William D. Stephens warned that advertising signs and lights in store display windows might have to be shut off.[4]

Everywhere in California dry years culminated in painful water shortages, touching off new conflicts and reviving old ones. In the Sacramento River basin, subnormal rainfall, the cultivation of rice, and reclamation of swampland intensified rivalries among up-stream and downstream water users. In 1910, California farmers planted only 100 acres to rice. This figure increased to 15,000 acres in 1914, soared to 83,000 acres by 1917, and peaked at 164,700 acres in 1920. Acreage irrigated from the Sacramento River increased from 206,000 in 1902 to 640,000 in 1919. In the latter year, rice sold for a record $5.93 per hundredweight, and while the price fell off sharply by 1921, to $2.56, it remained a popular crop. The Sacramento Valley produced over 95 percent of California's annual rice harvest, in part because the Sacramento River provided plenty of water. Since rice required "flood irrigation," it used far more water than wheat or alfalfa; in the Sacramento Valley, the average seven acre-feet of water per acre per year needed to cultivate rice was over twice the amount required for most other crops. Unfortunately, as diversions increased, the flow of water into San Francisco Bay decreased. Moreover, the "return flow" was often heavily contaminated with alkalis and salts, which prompted the state engineer to observe in 1925: "The quality of any return irrigation water is poor, and the return water from rice irrigation on the heavy lands of the upper valley is especially so, since the water has stood on the rice fields for many weeks practically stagnant and when re-

---

4. M. M. O'Shaughnessey to Robert B. Marshall, August 30, 1920; "Sacramento Valley Irrigation and Water Conference: Synopsis of Proceedings, February 26 and 27, 1920," Marshall Plan file (615.021), DWR Archives, Sacramento. At the same repository, see the unpublished report, "Emergency Water Conference," December 29, 1920. For accounts of the February water conference, see the *Sacramento Bee* and *Sacramento Daily Union*, February 26, 27, 1920.

leased contains not only rotted vegetation but the accumulation of salts leached out from the lands, making it an undesirable domestic supply even when filtered." Pollution became particularly acute in dry years when most of the river consisted of return flow.[5]

In the early months of 1920, the drought and increasing demands of irrigators contributed to saltwater "intrusion" from San Francisco Bay. Teredos, or marine borers, attacked piers and moorings ever farther upstream, and the city of Antioch stopped using the salty water as a domestic supply. In February, the state water commission warned rice farmers to restrict their planting because the winter's rainfall had been unusually light; stream flow at Red Bluff measured only one-third of normal. Subsequently, water users along the river held several meetings, and some farmers did limit their sowing. But since no court decree covered the stream, no easy way could be found to decide who should plant less.[6]

On July 1, 1920, the city of Antioch and ninety-seven delta landowners pressed suit against a multitude of reclamation districts, irrigation districts, irrigation and land companies, and individuals, demanding a flow of 3,500 cubic feet per second past the city; the flow had dropped to less than 500 cubic feet per second at Sacramento during the fall of 1920. On January 7, 1921, the Alameda Superior Court granted a temporary injunction prohibiting upstream diversions. The defendants represented 452,584 acres of land, including 136,581 acres planted to rice and 72,618 acres of other irrigated crops. Even though the Sacramento-San Joaquin delta included 400,000 acres of land, owners of only 40,000 acres joined the suit. The city of Antioch's lawyers claimed both riparian and appropriative rights. They argued that an appropriation of

5. The quote is from Edward Hyatt, "Control of Appropriations of Water in California," *Journal of the American Water Works Association* 13 (February 1925), 142. On rice cultivation see the *Third Biennial Report of the State Water Commission of California, 1919–1920, JCSA,* 44th sess. (Sacramento, 1921), Appendix, 1: 58; *Report of the Division of Water Rights, November 1, 1922, JCSA,* 45th sess. (Sacramento, 1923), Appendix, 4: 95; Frank Adams, "The Water Supply of Sacramento Valley," *California Cultivator* 66 (January 30, 1926): 125; and *Proceedings of the Sacramento River Problems Conference, January 25, 26, 1924* (Sacramento, 1924), 73.

6. *Third Biennial Report of the State Water Commission of California, 1919–1920,* 152–155; "Engineer's Reports on Salinity Tests in the San Joaquin and Sacramento Rivers," unpublished report of the Department of Water Resources (hereafter DWR), February 21, 1921, DWR Archives.

water for domestic purposes included sufficient "extra" water to deliver a *pure* supply, not just the actual quantity taken from the river. The defendants countered that Antioch owned only a sixty-foot riparian lot and diverted no water there. They also claimed that appropriative rights could not exceed the amount of water actually diverted. As for the charge of polluting the stream, the upstream interests responded that legally they could not be charged with contaminating the river unless they *consciously* added harmful ingredients. Nature herself contributed many unpalatable and destructive chemicals to the water, some of which leached into the stream from riverbanks. The suit was not a simple contest between upstream and downstream farmers, or even between domestic water users and irrigators. The city of Sacramento had been named as a defendant, and what court would limit that city's water supply to protect the rights of delta farmers and residents of Antioch? In 1922, the California Supreme Court rejected Antioch's argument. However, as a result of the suit the state Department of Engineering began to discuss the feasibility of a saltwater barrier to block the intrusion of salt water and store fresh water for farms and towns surrounding Suisun Bay.[7]

The next dry year, 1924, spawned additional lawsuits, including conflicts between upstream irrigators; so the California Water Rights Division called a "River Problems Conference" at the end of January which was well attended by different water users along the stream. It resulted in the formation of a "Permanent River Problems Conference" sponsored by the Water Rights Division and Sacramento Chamber of Commerce. In April, the conference appointed a watermaster to measure and monitor diversions from the river. By reducing waste, the watermaster won a temporary truce.[8]

7. Antioch v. Williams Irrigation District, 188 Cal. 451 (1922); *Pacific Rural Press* 120 (December 18, 1920): 781, 794; *Seventh Biennial Report of the Department of Engineering of the State of California, 1919–1920*, 51–53, 77. The California Supreme Court decided that if it upheld Antioch's arguments, any claimant along the lower river could "appropriate or control 3,080 second-feet of water . . . to supply his pipe with [an] infinitesimal quantity, and in that way . . . keep more than 300,000 acres of fertile land in the valley above dry and unproductive" (p. 461). The justices suggested that the city simply move its intake pipe farther upstream.

8. *Biennial Report of the Division of Water Rights, November 1, 1924, JCSA*, 46th sess. (Sacramento, 1925), Appendix, 4: 100–118; *Sacramento Bee*, January 25, 26,

The Army Corps of Engineers' attempt to keep the Sacramento River open to shipping added another element to the controversy. The Army had no formal policy, but it had tried to maintain a depth of seven feet from the river's mouth to Sacramento; a four-foot depth from Sacramento to Colusa; and a three-foot depth from Colusa to Chico Landing—in all, 200 miles of waterway. Before 1915, the Corps confined its work to clearing snags and drifts and building jetties. In 1915, the cost of the channel dredging was less than $2,000. However, this sum increased to an average $70,000 in 1918 and 1919, when even Sacramento would have been inaccessible to river traffic without dredging. Army officials noted that from 1912 to 1922 river tonnage increased by 170 percent, and the value of freight transported more than tripled during World War I.[9]

For thirteen weeks during the summer of 1920, the Sacramento River was closed to shipping above the capital city. The War Department's desire to maintain a minimum flow in the river, as it had before 1917, was more than a commitment to cheap river transportation at the expense of agriculture. The Corps pointed out that the irrigation of rice fields in the Upper Sacramento Valley often interfered with irrigation in the delta. The interests of delta farmers were perfectly consistent with maintaining the river's navigability, and in 1920 and 1921 the four delta counties produced harvests worth ten times the value of the rice sold in those years. And while a diversion of one second-foot of water irrigated only 43 acres of rice land in the upper valley, the same flow served 120 acres in the delta. The Corps of Engineers favored new state laws to prevent water waste, reservoirs to augment the supply, and "canalization" with locks to reduce the water needed for efficient shipping. It got none of these, so in each dry year its representatives complained about the high cost of dredging and raised the specter of federal sovereignty. Nevertheless, after 1924, when the state

---

December 11, 1924; *Sacramento Daily Union,* July 27, November 2, December 12, 1924. In June 1924, the watermaster warned all irrigators that he would restrict diversions in proportion to the amount of water they wasted. In the following month, he prohibited the irrigation of pastureland and asked farmers to follow an irrigation schedule to reduce demands on the river. The DWR Archives contain several reports filed under the "Sacramento-San Joaquin Water Supervisor," from 1924–1928.

9. *Seventh Biennial Report of the Department of Engineering of the State of California, 1919–1920,* 53, 70–74.

watermaster appointed by the River Problems Conference began to monitor diversions, the Corps promised water users along the upper Sacramento and its tributaries that it would follow "an administrative policy by which the recognized requirements of navigation would be limited so as to conform to the general good of the community."[10]

On California's second most important stream, the San Joaquin, conflict over water rights occurred less frequently during the 1920s. Though early state laws declared the river navigable to Tulare Lake, they were never enforced, and by the 1870s irrigation had eclipsed navigation. Because irrigation developed much earlier along the San Joaquin than in the Sacramento Valley, conflicts between farmers and power companies were less common; the hydroelectric companies often bought their water rights from farmers and ditch companies. And, since litigation over water began relatively early in the San Joaquin Valley, by the 1920s the region's water users had settled many differences out of court. The popularity of irrigation districts in the San Joaquin, as opposed to the Sacramento Valley, testified to the spirit of cooperation that had developed among litigation-weary residents of the southern valley. Since ships rarely carried freight past Stockton, roughly forty-five miles above the mouth of the river, in 1917 the Corps of Engineers' ranking officer in California concluded: "The paramount interest in this valley is irrigation rather than navigation. There is insufficient water even if economically used to supply the area that may ultimately become available for irrigation." On the San Joaquin as well as the Sacramento, canalization, which the Corps had already started on the Ohio River, offered the prospect of expanding the Corps' work and influence without antagonizing local interests.[11]

10. The quote is from the *Biennial Report of the Division of Water Rights, November 1, 1924,* 106. The War Department's position on the navigability of the Sacramento River can be seen in three documents at the DWR Archives. See U. S. Grant III to the Consulting Board, Survey of Water Resources, July 27, 1922, in "Water Resources Investigation: Copies of Letters Addressed to J. C. Forkner, Chairman, Consulting Board," microfilm reel 1033; the letter from Grant to Forkner, March 28, 1922, reprinted in the unpublished report by J. J. Jessup entitled "California Water Resources Investigation: Sacramento-San Joaquin River Delta Salt Water Investigation"; and Grant's address to the water conference of 1924 in *Proceedings of the Sacramento River Problems Conference, January 25 and 26, 1924,* 119–127.

11. *Seventh Biennial Report of the Department of Engineering of the State of California, 1919–1920,* 58, 68–70, 76–77.

In the San Joaquin Valley, where one-third to one-half of irrigation was from underground water sources, the declining water table posed a much greater danger than the conflict between irrigation and navigation. From 1909 to 1919, land irrigated from underground water in California increased from 32,539 to 299,841 acres, and most of that increase came in the San Joaquin Valley. As the use of electricity and gasoline engines spread, the number of pump wells increased from 597 in 1906, to 5,000 in 1910, to 11,000 in 1920, and to 23,500 in 1930. By the end of the 1920s, drought forced farmers to abandon 20,000 acres in the southern San Joaquin Valley and to curtail irrigation on many more. In Tulare County, where farmers pumped an average 800,000 acre-feet each year while nature returned only 300,000 to the aquifer, the assessed value of property fell by $1,000,000 in one year alone. In the middle and late 1920s, as the cost of pumping water soared, 400 wells were abandoned in the county. In 1927, one Tulare farmer complained that his water table had declined from eight feet below the surface in 1917 to forty-four feet in 1927. The 200 oak trees on his 960-acre farm were dying, and the sunbaked soil was so hard and dry that much more water had to be turned on the land to accomplish the same results as a decade earlier.[12]

From 1917 to 1924, water shortages in the Sacramento and San Joaquin Valley spawned a half-dozen major reservoir projects, and California entered a new epoch of irrigation development. At Pine Flat on the Kings River, twenty different canal companies, irrigation districts, and associations of water users agreed to build a 600,000 acre-foot capacity structure capable of serving 1,000,000 acres. On the Tuolumne River, the Modesto and Turlock Irrigation districts voted $5,000,000 in bonds to pay for a 270,000 acre-foot reservoir designed to serve 200,000 acres. The Madera Irrigation District acquired San Joaquin River water rights from the Miller and Lux estate and planned a 575,000 acre-foot reservoir at Friant, expected to water 300,000 acres. In Tulare County, a 292,000-acre water storage district was created to build a 300,000 acre-foot res-

12. "Irrigation and Agriculture," *Transactions of the Commonwealth Club* (hereafter TCC) 20 (November 24, 1925): 359, 374; S. T. Harding, *Water in California* (Palo Alto, Calif., 1960), 93; K. E. Small, *History of Tulare County, California* (Chicago, 1926), 302–303; William Kahrl et al., *The California Water Atlas*, 47; *Chico Record*, July 28, 1929.

ervoir on the Kern River at Isabella. The Merced Irrigation District voted to construct the Exchequer Dam with a 280,000 acre-foot capacity. Outside the San Joaquin Valley, storage projects were planned for the Sacramento River at Iron Canyon, for the Mojave River to serve land near Victorville, and for the Colorado River to serve 900,000 acres in the Imperial and Coachella valleys through an "All-American" canal. (The virgin soil of the Coachella Valley was particularly attractive because half of it was still part of the public domain.) Frank Adams, one of the closest students of irrigation development in California, noted in the fall of 1919: "Thus far storage has had only a relatively small part in California irrigation development. Except where irrigation use is coupled with domestic use and hydro-electric development, we have not over a half dozen important irrigation storage reservoirs in the State; yet each year enough water susceptible of economic storage goes to waste in [the] Sacramento and San Joaquin valleys to water at least one-half of the Sacramento Valley, and to add at least one million acres to the irrigated portion of the San Joaquin Valley."

Not surprisingly, the cost of irrigation had soared since 1900, when the cost of ditches and canals averaged between $5 and $10 an acre, excluding distribution works. By 1909, this figure reached about $20 an acre, and by the 1920s, largely due to the cost of storage works, the cost increased to $50 to $100 an acre or more. Without the revenue from the incidental sale of hydroelectric power, most of these projects would have been infeasible. For example, proponents of the Pine Flat Dam expected that power sales would pay half the structure's cost (as well as provide cheap power to those who pumped underground water).[13]

Such large storage projects also would have been inconceivable without greater cooperation among water users. As mentioned in the last chapter, the irrigation district was enormously successful

13. The quote is from Frank Adams, "Pending and Proposed New Irrigation Development in California in 1919," a memorandum prepared in October 1919, Document no. 55, Frank Adams Collection, California Water Resources Archives, University of California. Also see his "California Irrigation Development," *TCC* 15 (December 1920), 333, 335; J. B. Lippincott, "Recent Irrigation Development in California," *California Cultivator* 62 (January 26, 1924): 91, 107, 111; "Water Conservation for Irrigation in California," ibid., 64 (May 30, 1925): 615, 631; and "California Has Water Enough for Millions of Acres," *Pacific Rural Press* 101 (January 1, 1921): 17.

during World War I and the 1920s. Four districts were formed in 1915, three in 1916, five in 1917, seven in 1918, seven more in 1919, thirteen in 1920, seven in 1921, six in 1922, six in 1923, seven in 1924, and eight in 1925. By November 1926, California's districts included 3,583,284 acres and had issued $136,053,841 in bonds. The districts contained over twice the acreage of *all* the West's federal irrigation projects and smaller-sized farms as well. The Reclamation Bureau acknowledged the district's success as a cooperative institution during the middle and late 1920s by urging farmers on its projects to form districts to replace less formal water users' associations.[14]

Despite the popularity of irrigation districts during the 1920s, large landowners continued to chafe, as they had during the late 1880s and 1890s, under the restrictions imposed by this institutional form. They found an alternative way to organize and raise money in the water storage district, which built on principles embodied in the state's swampland reclamation district statutes. The legislature adopted the first water storage act in June 1915, at the urging of the powerful Iron Canyon Association.[15] However, critical parts of that law, as well as provisions in a 1917 act sponsored by proponents of the Pine Flat Reservoir on the Kings River, were subsequently invalidated by the state supreme court. The lawmakers were more careful when they drafted a new law in 1921. It permitted the owners of a majority of land within any proposed storage district, or 500 landowners representing title to at least 10 percent of the land, to petition the state engineer to organize a district. The new law gave the state substantially greater power over storage districts than the state engineer exercised in evaluating irrigation districts. Petitions to organize irrigation districts went directly to county boards of supervisors, not Sacramento, but the act of 1921

14. Frank Adams, "Irrigation Development through Irrigation Districts," *Transactions of the American Society of Civil Engineers* 90 (June, 1927): 773–790; *Report of the Division of Engineering and Irrigation, November 1, 1926, JCSA,* 47th sess. (Sacramento, 1927), Appendix, 5: 11, 17–18, 19; Harding, *Water in California,* 86.

15. *California Statutes,* 1915, 1173. The law created a "state irrigation board" to confer with Interior Department officials and designate the boundaries of water districts served by new dams constructed by the Reclamation Bureau. The State Department of Engineering would supervise the distribution of water from these federally constructed dams "under such contracts and regulations as may be made by the board, or agreed upon between the board and the United States."

required the state engineer to approve all petitions to include or exclude land from the district as well as pass judgment on the feasibility of dams. Large landowners would clearly dominate the new districts because voting in all storage districts was proportional to land ownership; landowners received one vote for each $100 in assessed value of their property. This contrasted sharply with the one-man, one-vote philosophy of the irrigation district. So did the requirement that tax assessments to pay off bonds vary according to benefits received rather than apply uniformly to all land within a district. In addition, the landowner within a storage district did not have to live within the district to vote in district elections; the land or ditch company, or any corporation for that matter, could exercise control from afar. While the irrigation district had revolutionary implications for the size of landholdings, the pattern of farming, and the political leadership within a particular region, the water storage district built on the status quo. It permitted the construction of reservoirs without tampering with existing water distribution systems.[16]

By November 1922, the state engineer had received three petitions to form storage districts, all in the San Joaquin Valley, and all engineered by large landowners, ditch companies, or other corporate interests. For example, the famous Kern County Land Company, which dated back to the 1870s, still owned much land adjoining the Kern River. It petitioned for the formation of a 250,000-acre storage district in May 1922. The election to create a district was held on November 10, 1923, and carried by the overwhelming vote of 68,465 to 21,929. However, since the land company held nearly half the votes, the election did not represent a fair test of public opinion in the Kern Valley. Without the company's votes, the election would have lost by 10,222 to 21,929.[17]

In 1923, the legislature enacted another important law, though

16. For the law itself see *Cal. Stats.,* 1921, 1727. Also see Harding, *Water in California,* 84; idem, oral history transcript, 73–75; Frank Adams transcript, 255–258. Copies of both transcripts are at the California Water Resources Archives as well as at the Bancroft Library.

17. S. T. Harding discussed the history of these water storage districts in his oral history transcript, 78–116. Also see *Report of the Division of Engineering and Irrigation, November 1, 1922, JCSA,* 45th sess. (Sacramento, 1923), Appendix, 5: 25–26.

it received much less use than the water storage act of 1921. This act permitted the formation of "water conservation districts" to unite storage, reclamation, irrigation, or drainage districts into broader governmental units. Water users along the Kings River sponsored the legislation in the hope it would facilitate the construction of the reservoir at Pine Flat. They could have used the 1921 law, but then district assessors would have imposed taxes according to benefits; the 1923 law left that job to the assessors in each constituent district. Like water storage districts, conservation districts were formed on appeal to the state. The state engineer and two lieutenants assumed the title of "state irrigation board," and enjoyed even greater power than the state engineer did in supervising water storage districts. The board was charged to survey proposed irrigation works, estimate the cost of construction, schedule bond elections, and apportion costs among the different districts. It also decided how those districts would share the stored water and revenue from hydroelectric power. In effect, this law appointed the state as an arbitrator to settle anticipated future water conflicts among the multitude of water users on the Kings River.[18]

The water storage and water conservation district legislation continued the expansion of state authority and responsibilities that began during Hiram Johnson's administration. By the beginning of the 1920s, the state had a fledgling "water bureaucracy." The legislature had abolished the first state engineer's office in 1889, but created a new Department of Engineering in 1907, mainly to assist in rebuilding San Francisco following the earthquake and fire of 1906.[19] In 1921, the lawmakers consolidated the Department of Engineering, Highway Commission, Water Commission, State Land Settlement Board, and Carey Act Commission into a Department

18. *Cal. Stats.*, 1923, 978. Also see Frank Adams, *Irrigation Districts in California* (Sacramento, 1930), 25–28; S. T. Harding oral history transcript, 117; Maass and Anderson, . . . *And the Desert Shall Rejoice* (Cambridge, Mass., 1978), 254. The Pine Flat Reservoir was finally completed in 1954 by the Army Corps of Engineers under authority of the 1944 Flood Control Act.

19. *Cal. Stats.*, 1907, 215. Among other responsibilities, the law gave the Engineering Office authority to carry out flood control investigations and joint hydrographic work with federal agencies. The first state engineer, Nathaniel Ellery, served from 1907 to 1912. He was followed by W. F. McClure (1912–1926); Paul Bailey (1926–1927); and Edward Hyatt (1927–1950).

of Public Works.[20] The new department included five divisions: Architecture, Land Settlement, Highways, Water Rights, and Engineering and Irrigation. The last division investigated the feasibility of plans for irrigation and storage districts, vouched for the security of their bonds, and supervised the construction of all district works. It also reviewed plans for bridges over navigable streams and for all reservoirs not built by municipalities or public utility districts. Finally, it assisted federal agencies in gauging streams, preparing topographic maps, and determining the amount of water needed to grow different crops in different soils. In 1927, Frank Adams noted that "90 percent of the time of the State Engineer, all the time of one irrigation engineer, and occasional part time of several others is consumed in investigation and supervision of irrigation and water storage districts."[21]

The growth of state bureaucracies coincided with a revival of interest in a state water plan. The need for a coordinated state policy had been recognized as early as 1856 by California Surveyor-General John A. Brewster, and reiterated by the federal Alexander Commission in 1874 and by William Hammond Hall in the late 1870s and 1880s.[22] But the legislature did not act until 1915 when it authorized a water conference to prepare "a unified state policy with reference to irrigation, reclamation, water storage, flood control, municipalities, and drainage, with due regard to the needs of water power, mining and navigation."[23] However, when the State Water Problems Conference, which met in 1916, published its for-

20. *Cal. Stats.*, 1921, 1022–1027, 1039–1042. In 1929, the Department of Public Works was divided into the Division of Water Resources, Division of Highways, and Divison of Architecture. The head of Water Resources was designated as the state engineer, and the functions of the former Divisions of Engineering and Irrigation and Water Rights were merged. See *Cal. Stats.*, 1929, 695.

21. Frank Adams, "Irrigation Development Through Irrigation Districts," 774; *First Report of the Division of Engineering and Irrigation, November 1, 1922*, 11, 57–58; *Report of the Department of Public Works of the State of California, November 1, 1922, JCSA*, 45th sess. (Sacramento, 1923), Appendix, 2: 5. The *California Water Atlas* contains a chart that lists the creation of districts decade by decade (p. 63). By the end of the 1960s, 895 different entities had been formed.

22. *Annual Report of the State Surveyor-General, 1856, JCSA*, 8th sess. (Sacramento, 1957), Appendix, 24–26; 43d Cong., 1st sess., 1974, H. Ex. Doc. 290 (serial 1615).

23. *Cal. Stats.*, 1915, 514.

mal report, it failed to provide a plan, largely because its members considered riparian rights an insurmountable barrier to planning.[24]

The legislature's failure to act provided a welcome opportunity to Robert Bradford Marshall, who prepared California's first comprehensive blueprint for water development during World War I. Marshall was a geographer who began his career in 1889 as a surveyor for the U.S. Geological Survey's irrigation investigation in Colorado and Montana. In 1890 or early 1891, he was assigned to California and by 1903 supervised all topographic work in the state. In January 1908, he became chief geographer in charge of all topographic work in the United States. But he was much more than a geographer, and he formed close friendships with many leaders in the Progressive conservation movement ranging from Gifford Pinchot to John Muir. His deep interest in the national parks won him the post of superintendent of National Parks in 1916, but in the same year Congress created the Park Service and Stephen Mather was selected to head the new agency. Disappointed, Marshall returned to the U.S.G.S. and supervised military mapping from June 1917 to April 1919. In 1919, he left the survey to give full attention to publicizing his "Marshall Plan."[25]

In 1937, Marshall remembered that inspiration for his project came in November 1891, as he gazed at the Central Valley from a vantage point at Folsom, not far from Sacramento:

> I saw a lot of water in the American River as I crossed it at Folsom. Was any one using it for irrigation? The next morning early, the road leading to Galt, the next stop for the night, was along the bluff south of Folsom. The morning was bright, sunshine, blending the brown grasses and the few scattered oak trees under a blue canopied sky—a dream landscape— and west, south and north I saw the [Central] valley of California, a natural bluff canvas of endless beauty as far as the eye could see. There were strips of green shades and here

24. *Report of the State Water Problems Conference, JCSA,* 42d sess. (Sacramento, 1917), Appendix, vol. 6.

25. See Robert Marshall's typescript "Autobiography," Marshall Collection, Bancroft Library; A. D. Edmonston (California state engineer) to Ralph M. Brown, February 21, 1951, Marshall Plan File (615.021), DWR Archives.

and there green splotches, mixtures of yellowish grass-stubble fields, scattering oaks and strings of green along streams and stream-beds under the sky blue canopy—all aglow with the untempered light of that November sun—what a country! Then and there I paused, overpowered by the picture—an endless plain with not a house in sight. In my mind came the thought—irrigation, Maj. Powell's talks, alfalfa along Fountain Creek [in Colorado], of farms, colonial houses, fruit trees and vines, happy laughing children, health, happiness, wealth, contentment—a new world lay before me. I pledged my effort, that something must be done to reclaim those brown fields—endless. Thus, in November, 1891, was born in my soul the reclamation of the Valley of California, embodied in my Marshall Plan given to the people of California, without cost, in 1919.[26]

Others told a different story. In 1957, Louis Bartlett, a prominent figure in California's water history in his own right, recounted a trip to the Sierra in 1890, one year before Marshall's "revelation" at Folsom. At Tuolumne Meadows he shared a campfire with Marshall and a survey party:

> [Marshall] had been working with the government for some time, had gone through the Sacramento and San Joaquin Valleys, and also the mountains that fed them with water, and he gave us a picture of what could ultimately be done in the Central Valley if the mountain streams were harnessed and the water retained until summer and then released to the valleys for irrigation. He gave us so interesting and graphic a picture that I have never forgotten it.[27]

Frank Adams recalled hearing Marshall speak several times in Berkeley in the 1890s or opening years of the twentieth century. Marshall had an office on the top floor of the civil engineering building on the University of California campus. The panoramic

---

26. Robert B. Marshall, "California As I Have Seen It," typescript manuscript, · March, 1937, Box 7, Marshall Collection, Bancroft Library.

27. Louis Bartlett oral history transcript, Bancroft Library, 138–139.

view, according to Adams, prompted the geographer to speculate on the vast quantity of water "flowing out of the Golden Gate from the Sacramento and the San Joaquin and this idea came to him, it was an inspiration. Over the years he kept it in mind and finally outlined the plan and proposed it to the governor."[28]

While the precise date of Marshall's vision is uncertain, he unveiled his plan in September 1919. It included a huge dam on the Sacramento River upstream from Redding at Kennett as the capstone of his project; a "west side" canal along the Coast Range from the Kennett dam to Dos Palos in the San Joaquin Valley (along with a siphon at Benecia to carry the water under the Carquinez Straits); a second canal along the east side of the Central Valley from Kennett to about Madera or Fresno; a third aqueduct along the flank of the Sierra from the Stanislaus River to Buena Vista Slough on the Kern River, twenty-five miles south of Tulare Lake; and a fourth, horseshoe-shaped channel from the San Joaquin River south along the Sierra, then north along the Coast Range to Dos Palos. Gravity dictated against the construction of two north-south canals spanning the entire Central Valley. The Marshall Plan also included storage reservoirs near major canals, but even though Marshall claimed to have surveyed the sites for sixty-five large structures and thirty or forty smaller ones, he did not identify specific locations. He promised that his scheme would reclaim 12,000,000 acres of land in the Sacramento, San Joaquin, Santa Clara, Livermore, and Concord valleys and increase the value of that land by $500 an acre. While farmers in some parts of California paid $15 an acre per year to irrigate their land, Marshall promised that, under his comprehensive project, economies of scale would reduce the price to $1 an acre.

Marshall also promised more water to the counties surrounding San Francisco Bay and in the Los Angeles basin. A branch aqueduct would carry water from the west side canal to farms and communities near the bay, and Los Angeles would be served by a tunnel and aqueduct from the Kern River through the Tehachapi Mountains. This watercourse would connect with the Owens Val-

28. Frank Adams oral history transcript, Bancroft Library, 279–298.

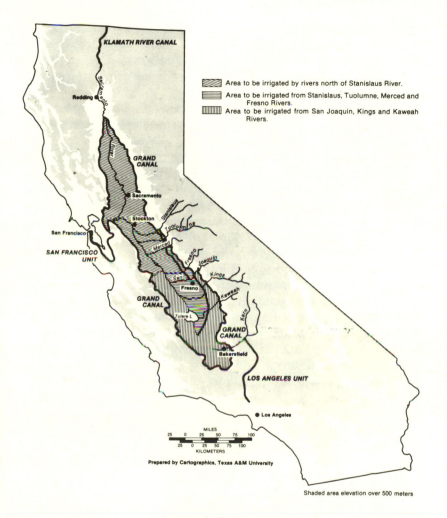

The Marshall Plan, 1919.

ley aqueduct and provide four times more water than the eastern Sierra source. To replace water lost to Kern River users, the Klamath River in northwestern California would be turned into the channel of the Sacramento near Shasta Springs to augment the water supply moved into the San Joaquin Valley.

The entire project hinged on damming the Sacramento River at Kennett. The new reservoir, according to Marshall, would store enough water to cover the Sacramento Valley's irrigable land to a depth of three feet; maintain shipping from Red Bluff to San Francisco Bay; improve the navigability of the bay itself; protect delta farms and communities from saltwater intrusion; and create enormous amounts of hydroelectric power to lure factories and canneries into the valley and revive the mining industry. In particular, Marshall noted that the development of iron and copper mines near Redding depended on cheap power to run smelters.

Marshall promised something to everyone and to every section of the state; he recognized the political value of the multiple-use concept. But the popularity of his scheme derived from more than its grand scale and anticipated benefits to water users. The construction work would provide thousands of jobs, and the reclaimed land thousands of homes, for returning veterans; so the Marshall Plan could help California escape the expected postwar economic slump. Moreover, the project did not require tax revenue. Marshall pledged that a bond issue could be retired by direct state sales of water and power to farms and municipalities. The Marshall Plan tacitly assumed that the cost of reclamation had reached or surpassed the immediate agricultural return from reclaimed land. Consequently, urbanites would bear part of the cost of rural improvement. Every gallon of water or kilowatt of electricity they bought would aid California agriculture, even though the cost of power would decline. Marshall estimated the entire project's cost at $700,000,000 to $800,000,000. He expected the San Francisco Bay Area to pay $100,000,000 toward the project and Los Angeles another $50,000,000. Finally, Marshall sold his plan as a practical flood control project. He challenged the "foolish levee policy" of the Corps of Engineers, arguing that a network of storage reservoirs in the foothills, and canals that could double as overflow channels, would furnish better protection. "It is a great scheme," Marshall crowed in February 1920. "It is a monumental scheme. It can all be done in ten years and it will pay a fine investment on the money."[29]

The Marshall Plan received a mixed reception. Its public debut came at an auspicious time. The drought's effects were felt everywhere. Food and land prices were soaring, there were no fears of overproduction, and the project's enormous cost seemed less of a burden than it would by the middle 1920s. The Antioch suit raised the specter of a new era of water litigation, litigation far more complicated, expensive, and protracted than earlier rural suits restricted to irrigators. Even though Marshall introduced his plan too late for consideration by the 1919 legislature, he hoped to enlist public support before the lawmakers reconvened in 1921.

A Shasta County newspaper, the *Fall River Tidings*, described the Marshall Plan as "a scheme so huge as to stagger conception, and yet so comparatively simple as to command itself to every sensible man and woman." The *Sacramento Daily Union* commented:

> His plan is gigantic in its scope and tremendous in its results; for he would solve at once the problems of flood control, irrigation and navigation, and provide a domestic supply of water for all the large cities. He would bring 12,000,000 acres under cultivation by placing water upon lands now useless. He would develop hydroelectric power on a scale heretofore undreamed of. He would provide farm homes for 3,000,000 people, and increase production to an incalculable degree. He would add six billion dollars to the assessed valuation of the state.[30]

29. "Irrigation of Twelve Million Acres," pamphlet dated November 1920, in folder entitled "Manuscript and published text of 'Marshall Plan,'" Box 6, Marshall Collection, Bancroft Library. *Sacramento Daily Union*, September 29, 1919, carried a full description of the Marshall Plan along with a map. For the Marshall quote, see "Sacramento Valley Irrigation and Water Conference: Synopsis of Proceedings, February 26 and 27, 1920," Marshall Plan File (615.021), DWR Archives.

30. *Fall River Tidings*, October 3, 1919; *Sacramento Union*, September 29, 1919. For editorials reflecting varying degrees of support for the Marshall Plan, see *San Francisco Chronicle*, September 1, 27, 1920; San Francisco *Call*, February 10, 1921; *Santa Barbara Press*, February 23, 1921; *Modesto Morning Herald*, February 17, 1921; Oakland *Enquirer*, September 10, 1920, January 6, 1921; *Fresno Daily Republican*, December 4, 1920; and Stockton *Daily Record*, December 4, 1920. Also see the pamphlet "What They Say of the Marshall Plan," December 1920, in the bound volume *Marshall Plan Pamphlets* at the Bancroft Library.

By January 1921, the plan had won endorsements from the San Joaquin Valley Water Conservation and Development Association, the Fresno Realty Board, the Visalia Board of Trade, the Lodi Business Men's Association, the Arbuckle Chamber of Commerce, the California League of Municipalities, the American Legion, and a multitude of San Joaquin Valley farm organizations. Support for the scheme centered in the San Joaquin Valley, the section hit hardest by the drought.[31]

The strongest opposition to the Marshall Plan came from established irrigation districts, private power companies, and professional engineering societies. Residents of the Modesto and Turlock districts worried that the scheme might drive down the value of land already under irrigation and slow the rate of settlement. Such districts already enjoyed an ample water supply.[32] The San Francisco section of the American Society of Civil Engineers, which included some of California's best-known irrigation engineers, described the plan as "physically, legally and financially impossible of accomplishment and the move to promote it, or any consideration of it by the legislature either directly or indirectly may prove inimical to public interests." The engineers maintained that there were no storage sites on the Sacramento River capable of capturing all the river's flood water. A complete storage system would still permit one-third to one-half of the water to escape into the bay, and *all* the "surplus" water would be needed in the future to develop several million acres of irrigable land in the Sacramento Valley. In effect, there was no water to transfer south into the San Joaquin Valley, and the engineers predicted that any interbasin diversion would produce endless litigation. They also chided Marshall for not providing a detailed statement of the project's cost, suggesting that the price of irrigation would probably run about double the $50 an acre he predicted. The plan had already delayed the construction of pending irrigation projects because promoters feared—or hoped—that the legislature might adopt the plan at its 1921 session. It had also encouraged land speculation, particularly on the west side of the San Joaquin Valley. The San Francisco

31. *San Francisco Pacific Builder,* January 3, 1921.

32. *Modesto Morning Herald,* February 17, 1921.

Engineering Council, representing local chapters of the American Society of Civil Engineers, the American Institute of Electrical Engineers, the American Society of Mechanical Engineers, the American Institute of Mining Engineers, and other professional engineering organizations, echoed these criticisms. Other critics noted that while Marshall assumed that riparian rights could be restricted to beneficial use, the state would probably have to purchase those rights, which would inflate the project's cost. The Marshall Plan also threatened to undermine the existing flood control program on which millions of dollars had been spent. The landowners themselves paid for most of the levee work, and they were reluctant to spend their money if the state would pay to protect them. Whether storage reservoirs and canals could do the job was open to doubt. Many critics believed that large storage reservoirs could not be constructed at a low enough elevation to provide flood protection. Thus, as the 1921 legislature began its work, the Marshall Plan enjoyed substantial popular support but little favor from the experts.[33]

Nevertheless, Marshall had the support of a powerful lobby. In July 1920, he formed the California State Irrigation Association to publicize his scheme. The association rented a two-story brick building across from the capitol and spent $9,000 a month "educating" the public. The group contained many warmed-over Progressives who favored expanding state control over natural resources. Most were dedicated to efficiency and opposed the increasing power of "the interests," specifically hydroelectric power companies. Some were also devotees of "central planning." The

33. The quote is from a resolution unanimously adopted by the San Francisco chapter of the A.S.C.E., February 15, 1921, reprinted in its undated "Statement of Action on the 'Marshall Plan'," Frank Adams Collection, California Water Resources Archives. For a good summary of the engineers' objections, see the untitled memorandum describing the San Francisco chapter's meeting, February 4, 1921, Adams Collection, File no. 376. In the same file, see "Report of Meeting [of] Civil Engineers with L. C. Davidson [February 11, 1921]." Davidson was a booster and lobbyist for the Marshall Plan. Also see Clyde L. Seavey, California State Board of Control, to Marshall, June 10, 1919, Marshall Collection, Bancroft Library. The *Sacramento Bee* was one of the plan's most vociferous critics. For example, in its March 7, 1921, edition, it called the scheme "the bunko game of a few real estate speculators, who care more for their own profit than for benefit to their customers, or the prestige and permanent welfare of the state."

members included M. M. O'Shaughnessey, Harris Weinstock, Joseph H. LeConte, Elwood Mead, David Starr Jordan, Ray Lyman Wilbur, Chester H. Rowell, and William Kent. An advisory board included State Engineer W. F. McClure and State Highway Engineer A. B. Fletcher.[34]

A bill authorizing a thorough examination of the Marshall Plan passed the state senate in 1921 but failed by a narrow margin in the assembly.[35] Nevertheless, the legislature did approve a $200,000 appropriation for a more general hydrographic survey. Marshall took full credit for the measure, but forty years later, Frank Adams claimed that the Commonwealth Club and Adams himself deserved greater recognition than Marshall:

> I asked the state engineer how much he thought he could use profitably in the biennium [for an investigation] and he said $200,000, so we prepared a bill appropriating that amount to the state engineer's department to make such a study. I took it up to Sacramento and showed it to Mr. Bradford Crittenden—whether he was then senator or assemblyman I don't remember. The club had already authorized us to promote that legislation. Mr. Crittenden said, "That'll be my bill."

The Commonwealth Club's original bill failed, but the legislation adopted included many of its features, including the $200,000 appropriation.[36]

The law required the state engineer to prepare a water plan for consideration by the 1923 legislature. The plan had to provide for the irrigation of *all* irrigable land and maximum flood protection. While Marshall's scheme emphasized maximum *water* utilization, the 1921 statute focused mainly on *land* development.[37] The state

34. See the folder entitled "California State Irrigation Association, Reports 1921–1922," Marshall Collection, Bancroft Library.

35. Mary Montgomery and Marion Clawson, *History of Legislation and Policy Formation of the Central Valley Project* (Berkeley, 1946), 21–22.

36. Frank Adams oral history transcript, Bancroft Library, 302.

37. *Cal. Stats.*, 1921, 1685. Partly as a concession to Marshall and the California Irrigation Association, the legislature authorized the governor to appoint a consulting board "composed of citizens of special and technical qualifications, to serve in an advisory capacity . . . in making the above investigation" (p. 1686). The nine-

engineer launched the investigation in August 1921, following organization of the Department of Public Works. The survey included
gauging stream flows; searching for reservoir sites; classifying reservoirs according to cost and benefits; mapping the land irrigated
in 1920; determining the total amount of irrigable land in California
and classifying it according to quality and yield; determining the
water requirements of that land; investigating the feasibility of
diversions of water from water-rich to water-deficient areas; estimating the future water needs of California cities and possible
sources of supply; determining the effect of reservoir construction
on flood control; estimating the potential power development on
California's streams; recommending ways to prevent saltwater encroachment; and assessing the effects of deforestation on stream
flow.[38]

By the spring of 1921, a rift appeared in the ranks of the
California State Irrigation Association. In 1919, the Marshall Plan
had been introduced to the public largely as an irrigation and flood
control scheme. The sale of electric power by the state was a vital
feature of the project, but Marshall did not crusade for public power
per se. In the wake of the 1917–1920 drought, however, calls for

---

man committee was organized on September 19, 1921, and held fifteen meetings
throughout the state from December 1921 to December 1922. The group included
Marshall; California Attorney-General U. S. Webb; California Bank Superintendent
Jonathan F. Dodge; Professor B. A. Etcheverry of the University of California's
engineering department; H. D. McGlashan, a U.S. Geological Survey engineer;
H. Hawgood, a Los Angeles engineer; O. B. Tout, a newspaper editor; H. A. Kluegel
of the State Water Commission; and two farmers, J. C. Forkner of Fresno and Peter
Cook of Rio Vista. Transcripts of meetings are in Box 9, Marshall Collection, Bancroft
Library; and in file entitled "Consulting Board, Public Meetings, Water Resources
Investigation," microfilm reel M–653, DWR Archives. The state engineer also received advice from a special engineering committee which included such well-
known hydraulic engineers as William Mulholland, W. L. Huber, and F. C. Hermann.

38. *Report of the Division of Engineering and Irrigation, November 1, 1922*, 20–
22. This was not the first investigation undertaken by the state. The state engineer's
office had already completed studies of water use in Kern and Tulare counties as
well as in the Victor and San Jacinto valleys. It had also surveyed reservoirs on the
Calaveras, Kings, Kaweah, Tule, Kern, and Sacramento rivers, the last an investigation of the Iron Canyon Project's feasibility. On the nature and scope of the
hydrographic survey, see Paul Bailey's statement before the Consulting Board of
the Water Resources Investigation, December 1, 1922, file entitled "Water Resources
Investigaton: Transcripts and Minutes of Meetings, 1922," microfilm reel 1029, DWR
Archives. Bailey revealed that the staff of the Department of Engineering and Irrigation grew from 20 to 110 after passage of the 1921 law.

public power echoed throughout the state. Farmers who hoped the state would one day provide them with cheap water for irrigation also remembered bitterly how the cost of electricity used to pump water had soared during the drought. And while utility users complained of high rates, advocates of public power pointed with pride to the government electric system in Ontario, Canada, where consumers paid one-third the rates charged by private companies just across the border in the United States. Many Californians agreed with the State Railroad Commission which, in an October 24, 1920, report to the governor, argued that the time had passed "for spasmodic development of [the state's] water resources by individuals. The problem as a whole must be solved by the state."[39]

The power companies were very active during the 1921 legislative session. They blocked a bill providing for the appointment of a state hydroelectric power commission, under whose direction cities could have joined together to develop public power projects, just as they helped defeat the bill to investigate the Marshall Plan. Nevertheless, under strong public pressure from such civic groups as the League of California Municipalities, the lawmakers approved a measure allowing the formation of municipal utility districts with the power to issue bonds to pay for hydroelectric systems and levy taxes to help pay off the principal and interest.[40]

The five-man executive committee of the California State Irrigation Association, which claimed over 4,000 active members at the beginning of 1921, included two staunch advocates of a statewide public power system, J. F. Mallon of Colusa and State Senator L. L. Dennett of Fresno. It also included two members who opposed public power: W. O. McCormick, a vice-president of the Southern Pacific Corporation, and Alden Anderson, a prominent Sacramento banker who had served as speaker of the California Legislature and state bank commissioner. C. A. Barlow, chairman

39. The State Railroad Commission's report is quoted in the *San Francisco Examiner*, May 17, 1922. The Ontario system is described in many documents in the Franklin Hichborn Collection, University of California, Los Angeles, Research Library; and in the bound volume of pamphlets entitled *California Water and Power Act* at the Bancroft Library.

40. *Cal. Stats.*, 1921, 245. The League of California Municipalities included civic representatives from 240 towns and cities. It launched its crusade for public power in 1919.

of the association's executive committee and a close friend of Marshall's, tried to mediate between the two factions even though he shared Marshall's belief that state power sales could help subsidize the staggering cost of the Marshall Plan. Nevertheless, at a joint meeting of the California Irrigation Association and League of Municipalities held on May 20, 1921, Barlow had a change of heart. The meeting had been called to discuss an initiative campaign to put the power issue before the voters, but on May 24, 1921, in a statement published in the *Sacramento Bee,* Barlow joined McCormick and Anderson in denouncing public power. Two years later, Eustace Cullinan, a spokesman for California's power companies, admitted before a legislative committee that he had struck a bargain with Barlow by which the California State Irrigation Association dropped its support for public power in exchange for a monthly subsidy from the power companies. What part Marshall himself played in courting the new alliance is uncertain.[41]

Barlow, if not Marshall, had decided that the Marshall Plan could never be constructed without at least acquiescence from the power syndicate, but his decision forced many leading members of the California State Irrigation Association—including L. L. Dennett, Rudolph Spreckels, J. R. Haynes, William Kent, James D. Phelan, Louis Bartlett, William Mulholland, and Franklin Hichborn—to abandon the group and form a new organization, the California State Water and Power League. Dennett drafted a water and power act and patiently defended it at a series of public meetings. Kent and Spreckels bankrolled the campaign to get the initiative measure on the ballot and sell it to the public. The final version was ready in August 1921. It authorized a $500,000,000, fifty-year bond issue bearing 6 percent interest. It also provided for the creation of a California Water and Power Board with full power to do "any and all things necessary or convenient for the

41. Franklin K. Hichborn, "The Strange Story of the California State Irrigation Association," a pamphlet written for the California State Water and Power League at the Bancroft Library. Also see Hichborn's undated, unpublished statement, "Water and Power: California State Irrigation Association," Box 75, Franklin Hichborn Collection, University of California, Los Angeles, Research Library. On the joint meeting of the California State Irrigation Association and League for Municipalities, see the *Sacramento Daily Union* and the *Sacramento Bee,* May 20, 21, 1921, and the *San Francisco Chronicle,* May 21, 1921.

conservation, development, storage, and distribution of water, and the generation, transmission and distribution of electric energy." The board would attempt to set water rates just high enough to pay the interest and principal on the bonds, but state funds could be used, if necessary, to supplement this uncertain source of revenue. No more than 20 percent of the power could be sold to private companies, and contracts to furnish power to such companies would be limited to five years. Predictably, the power companies opposed the measure. So did many irrigation districts and local officials. The opposition of irrigation districts has already been explained. The opposition of local officials grew out of their business ties or support for municipal utility districts. In either case, the Water and Power Act promised to give the state vast new powers and lead to a "centralization" which had become increasingly unpopular following the Russian Revolution. Proponents of the act promised that the state would distribute water through local districts and political subdivisions. Nevertheless, the act did not preclude direct state power sales to individual consumers. Given the frequent public charges of corruption and incompetence leveled against the Railroad Commission, which was responsible for regulating utility rates, this possibility seemed all the more dangerous. The power companies spent their way to victory as they would so many times in the future. They raised over $500,000 to defeat the measure, by far the largest sum expended in an initiative campaign before 1922. In the fall election, the Water and Power Act won approval from only about 30 percent of those who voted.[42]

Robert Marshall's position on the Water and Power Act took a bewildering series of turns from 1922 to 1926. In 1922, he opposed the measure; two years later he supported a virtually identical bill; then in 1926 he returned to the opposition camp. He justified his opposition in 1922 on grounds that the legislation was premature

---

42. The Water and Power Act—actually a constitutional amendment—was reprinted in *TCC* 17 (July, 1922), 183–187. Provisions of the act are discussed on pp. 181–183 and 188–294. For an able summary of the arguments for and against the act, see Frank Adams, "The California Water and Power Act," October, 1922, Document no. 82, Frank Adams Collection, California Water Resources Archives. For newspaper coverage see the *San Francisco Examiner*, April 21, June 23, September 6, 27, 1922; *San Francisco Chronicle*, April 25, May 11, 13, 27, 31, June 4, August 16, 18, September 18, October 27, November 6, 1922.

in light of the legislature's $200,000 appropriation to prepare a state water plan. "If I had three and one-half million votes to cast next November," Marshall remarked in a speech at Porterville on June 10, 1922, "I would cast them all against the Water and Power Act."[43]

The Water and Power Act met the same fate in 1924 and 1926. In both years the issue of public power overshadowed the Marshall Plan, and in both years that issue turned on the pervasive fear, well cultivated by the power monopoly, of "sovietization," which later came to be called "creeping socialism."[44] Of course, by the middle 1920s fears of agricultural overproduction and declining land values in parts of the state made a comprehensive irrigation plan less attractive than in 1919. C. A. Barlow, who managed the California State Irrigation Association through its stormy early years, conceded in 1924 that the plan "cannot be made sufficiently clear to the average business man to get him to tackle it as a real proposition. They all like its idealization and it interests them but to really feel, as I have felt, that on its success depends the future of the State and the happiness of its people, that point you cannot get them to appreciate."[45]

Marshall's abortive run for the Republican nomination to a seat in the California Senate in 1926 reflected the waning interest in his plan. By that time he had broken formally with Barlow and the California State Irrigation Association to form the California

43. See the undated broadside of the California State Engineering Association, on which Marshall is listed as a "consulting engineer," in the Marshall Plan File (615.021), DWR Archives. Also see Louis Bartlett's oral history transcript, 157–176; Frank Adams's transcript, 311–312; and William Durbrow's transcript, 117, all at the Bancroft Library. In 1924, in a speech explaining his new support for the Water and Power Act, Marshall declared: "There is a concerted effort on the part of the corporate interests of the state, in fact of the nation, to kill the development of water power in California. These corporations have used the natural, God-given advantages of the state to their own benefit and for their own profit. It has been said many times, and correctly, that the corporation that controls the water and power resources of the state controls the state." Stockton *Daily Record*, October 3, 1924. Why Marshall returned to the opposition camp in 1926 is unclear.

44. On the 1924 and 1926 water and power acts, see the *San Francisco Chronicle*, April 8, May 13, August 22, 28, October 22, 31, November 6, 7, 1924; ibid., April 30, May 14, July 12, September 24, October 8, 13, November 12, 1926. For the Water and Power Act of 1924, see "Water and Power" in *TCC* 19 (October, 1924), 355–452.

45. C. A. Barlow to Robert B. Marshall, May 3, 1924, Marshall Collection, Bancroft Library.

Water Resources Association.[46] His district included Stanislaus, Merced, Madera, Calaveras, Tuolumne, and Mariposa counties, and one of his two opponents, a director of the Turlock Irrigation District, pointed to the thousands of idle acres within established irrigation projects as the best argument against a comprehensive state irrigation plan. Marshall countered that overproduction resulted mainly from planting the wrong crops, but the voters remained skeptical and he finished a distant third. Then, in 1927, tragedy compounded misfortune. Following a siege of hoarseness, perhaps exacerbated by the tough campaign, doctors removed Robert Marshall's larynx. This incapacitated him for administrative work and prevented him from defending his plan from the rostrum. By the late 1920s Marshall's dream had been all but forgotten.[47]

Meanwhile, in 1923 the division of Engineering and Irrigation presented the results of its first statewide hydrographic survey to the legislature. Its report predicted that 18,000,000 acres could be irrigated in California at an average cost of $80 an acre.[48] In all, the division collected evidence concerning 1,270 reservoir sites and directly examined 3,500 miles of streambed and 176 potential sites. The ultimate development of California's irrigation system, it suggested, would require 260 reservoirs.[49]

The report considered California "as a virgin territory with its waters and soils unsegregated in private ownership." In other words, the survey ignored the cost of acquiring water rights and litigation (though it tried to integrate existing irrigation works into the proposed system). The recommended plan reflected several assumptions: that gravity-fed canals were impractical and too ex-

---

46. Robert B. Marshall to Paul Bailey, April 19, 1926, file entitled "Water Resources Investigation: General Correspondence, January 1, 1926–December 31, 1927," microfilm reel M–644, DWR Archives.

47. For a typical campaign debate over the Marshall Plan and agricultural overproduction, see the *Turlock Journal*, August 2, 1926.

48. The 18,000,000 acres was three times the land irrigated in California in 1919 and more than all the irrigated land in the entire arid West. Nevertheless, some of the state's irrigation boosters predicted that as much as 25,000,000 acres could be irrigated.

49. *Water Resources of California: A Report to the Legislature of 1923*, Department of Public Works, Division of Engineering and Irrigation Bulletin no. 4 (Sacramento, 1923) 18, 43, 44.

pensive; that some way had to be found to reconcile the conflicting needs of irrigators and power companies; that the project's cost had to be kept to a bare minimum to make it financially feasible; and that any state water project should be built in sections, not all at once as Marshall had urged.

The first state plan's major features included a dam across the Carquinez Straits of San Francisco Bay to prevent saltwater incursions into the delta, help reclaim the tidal flats along the margin of Suisan Bay, provide unlimited fresh water for communities and farms in the Bay Area, and serve as a bridge for automobiles and trains. More important, much of the water captured behind this structure would be diverted into a 200-mile-long aqueduct linking the Sacramento-San Joaquin delta with Tulare Lake. Essentially, this was the west side canal promoted by the San Joaquin and Kings River Canal and Irrigation Company and the Grange in the 1870s, except that the flow of water moved south instead of north. Marshall's gravity-fed canals followed the contours of the Coast Range's foothills, but this canal ran along the valley floor. The report concluded that gravity-fed canals had to be very large and

> tortuously [follow] a grade contour on steep mountain hill-sides and [wind] in and out around every rocky spur and into each receding ravine. The total length attained in its devious route would double or treble the air line distance of five hundred miles between the source of supply in the Sacramento River and the extreme southerly lands to be watered. The cost of constructing crossings for a gravity canal at the innumerable drainage channels that it would intercept, alone would probably exceed the total cost of all the works of the comprehensive plan.

The division proposed nine pumping plants to lift the water up the valley's gradual grade. Tulare Lake's vast capacity as a storage reservoir would reduce the size of the canal and pumping plants because they could operate eleven months a year, not just during the irrigation season. The canal would serve over 2,000,000 acres.[50]

50. *Water Resources of California: A Report to the Legislature of 1923,* 47, 48. The quote is from p. 47.

The report paid scant attention to power revenue, probably because most of the energy would be used to run the pump stations. The state engineer decided that if all storage reservoirs were built below 2,500 feet, the needs of agriculture could be reconciled with those of hydroelectric power companies upstream. In particular, water used to generate power in the winter could be captured for reuse by farmers during the growing season. In addition, by building dams in the foothills, no pump station would be located farther than 100 miles from a power station, and the branch canals that connected dams and valley farmland would be shorter and cheaper.

The water plan also mentioned other possible diversions, including a tunnel to carry water from the Eel and Trinity rivers into the Sacramento Valley; an All-American canal from the Colorado River into the Imperial and Coachella valleys; an aqueduct from Mono Lake to Antelope Valley north of Los Angeles; and a canal from the Carquinez dam through Contra Costa County into the rich Santa Clara Valley. The largest proposed watercourse, aside from the west side ditch, stretched from the Merced and Tuolumne rivers to the south end of the San Joaquin Valley. The plan also included many shorter canals, as well as "spreading grounds" to replenish southern California's underground water.[51]

The fifty-five page report was almost as sketchy as the Marshall Plan, which it never mentioned.[52] Most of the work conducted by the Department of Engineering and Irrigation since 1921 had been devoted to compiling rainfall records, determining the flow of California's largest 260 streams, and assessing the water needs of the state's different soils. Much information on potential reservoir sites had been gathered, but no formal surveys had been made, no construction plans had been drafted, and no test borings

51. *Water Resources of California: A Report to the Legislature of 1923,* 48–49. Southern California offered few good reservoir sites, and those reservoirs could capture little of the heavy rainfall which came intermittently. "Spreading grounds" offered one solution. They were ponds designed to increase the supply of ground water through "percolation."

52. More significant than the sketchy state water plan of 1923 was the evidence upon which it was based. In 1923, the Division of Engineering and Irrigation also published *Flow in California Streams* as Bulletin no. 5 (Sacramento, 1923). This volume contained 300 pages of data, the most extensive compilation of stream flow statistics ever compiled in California.

had been made. The state now knew how much water was available for storage on each stream, but not the cost of storing it. Moreover, the report failed to consider the vast array of legal problems involved in implementing a state water project or the project's possible effects on flood control and navigation. For those like Marshall, committed to the *total* development of California's water supply, the document was a profound disappointment. Nevertheless, many of its features, including the saltwater dam and pump aqueduct, helped shape future state proposals and public discussions.

The economy-minded 1923 legislature refused to fund further studies of California's water supply. However, in September 1924, the San Francisco and Los Angeles chambers of commerce gave the Division of Engineering and Irrigation $90,000 to study an immediate problem: Tulare County's rapidly declining water table. Kern County had experienced the same problem, but Tulare County suffered most because it contained only two small streams, the Kaweah and the Tule. The Tulare County survey rejected the west side canal as too expensive. Like the 1923 report, it assumed that individual landowners would have to pay the entire cost of dams and canals, and the Tulare farmers could not afford to transport water 200 miles and lift it 250 to 350 feet. So the state engineer proposed an elaborate water exchange. First, a gravity canal would divert water from the upper San Joaquin River at Friant south to the Kings River. This water would serve the two largest users of the Kings River, the Fresno Irrigation District and the Consolidated Irrigation District. In turn, the saved Kings River water would be stored in a reservoir at Pine Flat for exclusive use in Tulare County. So much for Tulare County, but what about those farmers who used the San Joaquin? Their water would be replaced by water from the Sacramento River delivered through the channel of the San Joaquin by an elaborate dam and pump system that would literally reverse the stream's flow. This defiance of nature promised to aid in the reclamation of swampland adjoining the river and render the stream, through an elaborate system of locks, navigable for 160 miles upstream from the delta. The state engineer estimated the price of these works at $12,876,800, or $107.30 an acre, excluding the cost of reservoirs. He also extolled the value of reservoirs for flood control, suggesting that they would cut maximum floods in half without reducing the water supply of irrigators. However, little evidence was offered in support of this contention. The division had just begun to study ways to coordinate the use of reservoirs for different purposes.[53]

In 1925, following the severe drought of 1924, the legislature appropriated $150,000 and ordered the Division of Irrigation and Engineering to prepare a broader water plan that considered *all* uses of water, not just irrigation and power generation. Drought would continue to ravage California agriculture unless the cost of a state water project could be spread out among the greatest number of different water users.[54]

Just before the Department of Engineering and Irrigation presented its formal report to the 1927 legislature, the California Supreme Court undermined the foundation of the state water plan that was slowly taking shape. On December 24, 1926, the court issued its ruling in *Herminghaus* v. *Southern California Edison Company*.[55] From 1902 to 1906, the forerunner of the Southern California Edison Company, headed by Henry Huntington, formulated a comprehensive plan to develop hydroelectric power on the San Joaquin River and deliver it to the burgeoning city of Los Angeles. In all, the company hoped to build fourteen reservoirs, capable of storing 700,000 acre-feet of water, on tributaries of the San Joaquin. However, Miller and Lux and others held extensive riparian rights along the stream, so in 1906 the company bought consent to its plans from the owners of 200 of the river's 225 miles of frontage. Amelia Herminghaus did not sign the 1906 contract. She owned an 18,000-acre estate on the south side of the river above Fresno Slough with about 775 acres contiguous to the stream. Most of the land had been leased to heirs of the Miller and Lux estate as pasture for $1 to $2 an acre per year. In August 1924, Herminghaus filed suit to prevent the electric company from storing water at Shaver

53. Paul Bailey, California State Engineer, *Supplemental Report on Water Resources of California*, Division of Engineering and Irrigation Bulletin no. 9 (Sacramento, 1925), 15, 17–18, 28, 44, 45.

54. Testimony of Paul Bailey in "Transcript of the Hearings of the Joint Committee [of the Legislature] Concerning Proposed Constitutional Amendments Affecting Water Resources of the State of California," January 26, 1927, Robert Marshall Collection, Box 11, Bancroft Library. For both 1925 laws see *Cal. Stats.*, 1925, 1013.

55. Herminghaus v. Southern California Edison Company, 200 Cal. 81 (1926).

Lake on Stevenson Creek. She claimed that during the drought the dam prevented the San Joaquin from overflowing its banks and producing lush spring pastures. The power company refused to pay property damages.[56]

The electric company claimed that the production of power was a higher water use than raising natural grasses for the consumption of cattle. It also demanded the right to store water because it owned riparian land upstream; the company's lawyers pointed out that the riparian doctrine had been modified many times to suit climatic conditions in California, including the right of riparian owners to irrigate. Furthermore, it noted that the Water Law of 1913 restricted riparian owners to "reasonable use" and barred flood irrigation because Section 42 of that law prohibited the use of more than 2.5 acre-feet of water per acre per year for irrigation. The Herminghaus attorneys countered that storage violated the central precept of riparian rights and that the river should be allowed to flow as under natural conditions. The case boiled down to two issues: should any standard of reasonable use apply to riparian rights, and did those rights include the privilege of storing water, especially to generate electricity (which purpose was usually considered inferior to irrigation).[57]

California officials recognized the many dangers posed by this contest between riparian owners. In 1919 riparian rights served less than 6 percent of California's irrigated land. The state's brief pointed out that the courts traditionally limited riparian owners to diversions "by the ordinary and usual methods"; flooding, it argued,

56. Herminghaus v. Southern California Edison, opening brief for appellant filed August 22, 1925, p. 14 and passim, California State Archives, Sacramento. The Herminghaus file contains many other briefs filed by both sides.

57. For legal discussions of the Herminghaus case, see S. T. Harding's oral history transcript, 122–135; *Water in California*, 41–42; and his "Report on the Effect Which Would Result from the Establishment of the Right to Store as a Part of the Rights of Riparian Owners on the Feasibility of Proposed Irrigation Projects, Particularly Projects on San Joaquin and Kings River," unpublished document, October 1, 1925, DWR Archives and also in File no. 40, S. T. Harding Collection, Water Resources Archives. Harding prepared the State of California's brief. Also see Samuel C. Wiel, "The Pending Water Amendment to the California Constitution, and Possible Legislation," *California Law Review* 16 (March, 1928), 171–173; Wells Hutchins, *The California Law of Water Rights* (Sacramento, 1956), 13; Maass and Anderson, . . . *and the Desert Shall Rejoice*, 231.

was not a "usual" or "ordinary" method. A decision in favor of the Herminghaus interests would be wasteful for several reasons. Clearly, raising forage was not a particularly valuable use of water. In addition, flood irrigation was inherently wasteful because the diversion was not controlled or measured. Finally, demanding the full flow of a stream to transport the small amount of water that overflowed riverbanks obviously hurt upstream users.

The power company's claims were even more revolutionary. Power companies, miners, and municipalities had traditionally obtained their water through appropriation; every one of the seventy-six reservoirs in California designed to generate power relied on such rights. The Southern California Edison Company's claims threatened a race among riparian owners to acquire storage rights. It also challenged the legal premise that riparian rights were limited or defined by geography. If such rights could "shift" upstream, then no rights along that stream were safe. Moreover, the efforts of the Division of Water Rights to regulate storage through permits would be undermined and the permits already issued rendered worthless. The already difficult task of determining the state's surplus water supply would become impossible.[58]

Two federal agencies took nearly as much interest in the suit as the state. The Reclamation Bureau opposed the power company's claim because it indirectly threatened the Orland Project's water supply as well as potential federal storage projects on the Kings and San Joaquin rivers. In addition, the Bureau had filed an adjudication suit on the Carson River, a stream that originated in California but served the Truckee-Carson Irrigation Project in Nevada. Riparian owners along the Carson eagerly awaited the Herminghaus decision and refused to cooperate with the Bureau. Meanwhile, project officials in Nevada worried about what would happen to their water supply if the California Supreme Court supported the power company. Ironically, the Reclamation Bureau did not have the Justice Department's assistance. The Federal Power Commission had granted the electric company the right to build

---

58. *Biennial Report of the Division of Water Rights, November 1, 1926,* 20–22, 63; and "Brief on Behalf of the State of California in the Supreme Court of the State of California in the case of Amelia Herminghaus et al. v. Southern California Edison Company," Bancroft Library and California State Archives.

the dam and power plant, and to flood government land in the process. Even though Southern California Edison's claim threatened power projects previously approved by the F.P.C., that agency demanded legal protection for its permit. Apparently, the F.P.C. wanted to assert or maintain federal sovereignty over water used to generate power.[59]

The supreme court upheld the Herminghaus interests.[60] It ruled that flood irrigation was a useful and reasonable and beneficial use of water and that the river's full natural flow was needed for this purpose. The court strongly implied that no legal distinction existed between normal stream flow and the floodwaters of spring, at least as they related to riparian rights. The judges did not rule on the section of the Water Act of 1913 that limited riparian rights to beneficial use, but invalidated Section 42 on grounds that that provision gave an administrative agency the judicial power to destroy vested property rights. In effect, the court ruled that 98 percent of the volume of a stream could be preserved from use so that the remaining 2 percent could be used. This controversial decision, the most significant water case since *Lux* v. *Haggin* nearly fifty years earlier, had vast implications. State Engineer Paul Bailey predicted that the state would have to buy up all riparian rights before reservoir construction could begin.[61] On November 6, 1928, California voters approved a constitutional amendment limiting riparian rights to "reasonable use," but defining "reasonable" proved difficult and

59. For the U.S. Reclamation Bureau's position, see the series of letters in The Reclamation Bureau Records, "(032) General Water Rights: Settlement of Herminghaus, etc.—Southern California Edison Case," RG 115, National Archives.

60. The Supreme Court had several precedents for its ruling. For example, in Miller and Lux v. Madera Land Company, 155 Cal. 59 (1909), Justice Sloss said of water rights: "Neither a court nor the legislature has the right to say that because . . . water may be more beneficially used by others [that] it may be freely taken by them. Public policy is at best a vague and uncertain guide, and no consideration of policy can justify the taking of private property without compensation." For press comment on the Herminghaus verdict, see the *Fresno Bee*, December 27, 29, 1926; January 14, 1927; *San Francisco Chronicle*, December 30, 1926, January 28, 1927; *Sacramento Daily Union*, January 14, 1927; *Sacramento Bee*, January 28, 1927; *Oakland Tribune*, January 5, 29, 1927; *San Francisco Examiner*, December 25, 1926; January 2, 1927.

61. Testimony of Paul Bailey in "Transcript of the Hearings of the Joint Committee [of the Legislature] Concerning Proposed Constitutional Amendments Affecting Water Resources of the State of California," 84–86.

ultimately the amendment provided less of a limitation than its draftsmen had hoped.[62]

Meanwhile, the 1927 legislature considered revisions to the state water plan introduced in 1923. By 1927, the scheme included dams on the Sacramento, Feather, Yuba, Bear, American, Trinity, upper San Joaquin, and Kings rivers, the cost of which constituted 80 percent of the project's anticipated price tag of $358,000,000. The 1923 plan included fewer dams and longer canals, and the canals cost as much as the reservoirs. But reservoirs served more purposes than canals. The structures planned for the Yuba, Feather, Bear, and American rivers, for example, would trap mining debris, helping to revive that moribund industry, as well as provide flood control and water for irrigation. By 1927, the Department of Engineering and Irrigation had completed its first studies of the effect of reservoirs on flood control. It found that floods followed fairly predictable patterns: severe floods usually occurred during years of heavy stream flow generally; the heaviest flooding came in late January or early February as a result of torrential, warm, winter rains; and the most frequent floods occurred in May or June due to melting snow. Storage for irrigation and power generation could be regulated to reduce the volume of a "twenty-five-year flood" by half, but how much water to impound and when to release it would be critical questions because the requirements of irrigation, flood control, and power generation often clashed. More reservoirs also meant more opportunity to generate electrical power. State Engineer Paul Bailey noted that "the plan proposes to operate those reservoirs for the first period of years in a manner that will produce the greatest revenue from power" and estimated that the proposed dams would yield as much electricity as all of northern California's

62. The 1928 constitutional amendment became Article 14, Section 3 of the California Constitution. See *Cal. Stats.*, 1927, 2373. Both the California and United States Supreme Court later ruled that since the amendment destroyed private property, riparian owners deserved compensation. This limited its effectiveness even more. For discussions of the amendment's implications, see Hutchins, *The California Law of Water Rights*, 230–234; S. C. Wiel, "The Pending Water Amendment to the California Constitution, and Possible Legislation," *California Law Review* 16 (March 1928); Harding, *Water in California*, 42–43; Edward F. Treadwell, "Developing a New Philosophy of Water Rights," *California Law Review* 38 (October 1950): 527–587; and Frank Adams, "The Water Situation in California," *California Cultivator* 70 (January 28, 1928): 93, 107.

power plants generated in 1927. The new emphasis on upstream reservoirs also convinced Bailey to drop the $45,000,000 to $90,000,000 saltwater barrier ·on the upper arm of San Francisco Bay from the state plan. Many engineers questioned the feasibility of such a novel structure, and its cost nearly matched that of the proposed Kennett Reservoir on the upper Sacramento.[63]

The 1927 water plan rested on several assumptions. The central assumption was that the benefits of a state water project would outweigh the costs, at least to the economy as a whole. However, this was far from a foregone conclusion. No study had been made of how much the cost of litigation would add to the project's price, and since the future growth rate of California's cities, farms, and industries was uncertain, so was the future demand for power. The vast new power supply produced by state dams was likely to create at least a temporary market glut and depress the price of electricity, reducing state revenue from this source. Of course, the amount of revenue also depended on who distributed the power and who sold it to consumers; this was an explosive issue the state engineer had ducked in his engineering reports. A second major assumption was that the state water project would coordinate and supplement discrete existing water systems rather than replace them with a modern, comprehensive, integrated system. In short, the state plan would not make the best possible use of California's limited water supply in the light of future needs. Rather, it would serve the immediate demands of those who needed water most— implicitly those who could bring the greatest political pressure to bear in Sacramento. Finally, Bailey and most members of the legislature assumed that the federal government should pay at least part of the project's cost, if only for flood control and navigation benefits. By 1927, many of California's leaders doubted that a state

63. Paul Bailey, *Summary Report of the Water Resources of California and a Coordinated Plan for Their Development*, Division of Engineering and Irrigation Bulletin no. 12 (Sacramento, 1927), 31, 33, 37, 38, 41, 42. The report included a map of suggested reservoirs on p. 26. Also see Bailey's *The Control of Floods by Reservoirs*, Bulletin no. 14 (Sacramento, 1928), 13, 14. The 1927 report was accompanied by several volumes of engineering data: Bulletin no. 13 entitled *The Development of the Upper Sacramento River*; Bulletin no. 15, *The Coordinated Plan of Water Development in the Sacramento Valley*; Bulletin no. 16, *The Coordinated Plan of Water Development in the San Joaquin Valley*; and Bulletin no. 17, *The Coordinated Plan of Water Development in Southern California*.

water project could be constructed without substantial federal aid.

The 1927 legislature, working under the shadow of the Herminghaus decision, formed a joint legislative committee to study the legal and financial problems involved in constructing a state water project, issues that had been largely ignored by the Department of Engineering and Irrigation. The committee, in turn, quickly recommended that the state create a special administrative tribunal to speed up the process of condemning private land and water rights and determine proper compensation. It also proposed that the state reserve much of California's unappropriated water for future use.[64] The legislature complied with the second request and authorized the Department of Finance, on the recommendation of the Department of Public Works, to withdraw water for up to four years and renew those withdrawals as necessary. On July 30, 1927, the Finance Department filed sixteen permits to use water for irrigation and another nine to reserve water for power generation. The streams involved included the Trinity, Pit, Sacramento, Feather, Yuba, Bear, Cosumnes, American, Mokelumne, Calaveras, Stanislaus, San Joaquin, Kings and Kern rivers and their tributaries. The legislature also appropriated $200,000 for the purchase of reservoir sites.[65]

64. "Central Valley Project Documents," 84th Cong., 2d sess., 1956, H. Doc., 416, 1: 186–187. For the legislation creating the joint legislative committee see *Cal. Stats.*, 1927, 2393.

65. *Cal. Stats.*, 1927, 508; *Biennial Report of the Division of Water Rights and State Water Commission*, November 1, 1928, 28–29; Adams, "The Water Situation in California," 93, 107; *San Francisco Chronicle*, April 23, 1927; *Modesto Morning Herald*, May 18, 1927. S. T. Harding noted in *Water in California*, p. 47, that by January 1959, the state had filed 142 water and power applications using the 1927 statute.

By reserving water and giving the legislature the power to renew the withdrawals every four years, the state circumvented the legal principle of "diligence"—putting claimed water to use, or at least beginning the construction of diversion works, within a reasonable period. It did the same thing when it passed the "Watershed Protection Act" (1927) and "County of Origin Law" (1931). These laws were inspired both by the efforts of Owens Valley residents to prevent diversions by Los Angeles and by the fear of water users in the Sacramento Valley that the water transfers contemplated in the state water plan would retard their section's economic growth. The 1927 law promised "all of the water reasonably required to adequately supply the beneficial needs of the watershed area or any of the inhabitants or property owners therein." Erwin Cooper, *Aqueduct Empire* (Glendale, Calif., 1968), 415. Apparently these laws, which have obvious implications for California water policies today, have never been tested in the courts.

The joint legislative committee submitted a formal report to the legislature on January 18, 1929. Since the state engineer's office had paid scant attention to the legal ramifications of a state water plan, the committee had asked a group of prominent California water lawyers, including S. C. Wiel and E. F. Treadwell, to survey potential legal obstacles.[66]

The issues considered by the legal subcommittee were extraordinarily complicated. The conflict between state and nation over California's water involved far more than navigation on the Sacramento River. For example, could the federal government extinguish riparian rights on navigable streams by administrative decree (even if the state could not)? What about riparian rights on the tributaries of navigable streams? The Reclamation Bureau had consistently opposed the riparian doctrine in the West, but the Corps of Engineers' position was not so clear. Moreover, state and federal laws differed over the terms and conditions of power permits. The Federal Power Commission allowed fifty-year grants while the state allowed only forty. Under state law, the construction of power plants had to begin within six months of the approval of a permit, while federal law permitted the grantee two years. What rules should the state follow in implementing its water plan? What if the state approved the construction of a dam and the federal government rejected it?[67]

Nor did the state's police powers offer an attractive tool to curb riparian rights. By their very nature, storage reservoirs interfered with the normal flow of streams and challenged riparian rights, even if the state sought merely to impound the stream's unappropriated water. Most lawyers conceded the state's right to control the acquisition of water rights, regulate the distribution of water, and even prevent waste. But did its authority to prevent waste extend to limiting riparian rights to beneficial use without

66. S. C. Wiel, "The Recent Attorneys' Conference on Water Legislation," *California Law Review* 17 (March, 1929), 197–213.

67. For the legal committee's report to the joint legislative committee, dated October 27, 1928, see *Report of the Joint Committee of the Senate and Assembly Dealing with the Water Problems of the State, January 18, 1929, JCSA*, 48th sess. (Sacramento, 1929), Appendix, vol. 4. The discussion of navigable streams is on pp. 29–31. The legal committee's report is also reprinted in "Central Valley Project Documents," H. Doc. 416, 1: 202–219.

condemnation or compensation? The 1913 water code suggested as much, but the legal committed noted:

> There can not be much serious question that the Legislature has full power to require such economy in the use of water as is necessary, at all events, in order that all those entitled to use the water may enjoy it. But, whether it can make like regulations for the purpose of making a more extended use of the water by a general plan for the conservation and use of the waters of the state in the interest of the public presents a different question.

Clearly, the police power often extended beyond mere regulation. For example, city, county, or state limitations on the height of buildings also imposed a limitation on the value of pieces of property; land zoned for commercial development usually sold for more than residential property. But even assuming that the state could restrict riparian rights to beneficial use without compensation, only the courts could define *beneficial* (or *reasonable*).[68]

Condemnation offered a surer alternative, but it, too, raised vexing questions. Aside from the obvious question of who would conduct the proceedings—the superior courts, the Division of Water Rights, or a special court or administrative tribunal—the legal commission doubted the feasibility of condemning part of a riparian right. If riparian owners were limited to the amount of water they actually used, what responsibility would the state have in dry years when there was not enough water to go around? By defining riparian rights as specific quantities of water, the state might render itself liable to *guarantee* those amounts. Yet the cost of condemning those rights in conflict with a state water plan would be enormous, perhaps prohibitive. These problems barely scratched the surface. How to restrict riparian rights used to generate power or, as in the delta, to prevent saltwater incursion, posed even thornier issues.[69]

The report of the joint legislative committee contained several surprises. First, the committee supported the saltwater barrier even

68. *Report of the Joint Committee . . . January 18, 1929*, 28–29. The quote is from p. 29.

69. *Report of the Joint Committee . . . January 18, 1929*, 31–34.

though the state engineer had rejected such a structure in his 1927 report.[70] A Reclamation Bureau engineer had argued that the barrier would be vital to any state water project. With or without the Kennett Reservoir, he argued, diversions from the Sacramento River into the San Joaquin Valley would reduce the flow into San Francisco Bay.[71] The joint committee also promised that a barrier at Point San Pablo would capture enough water to irrigate 51,000 acres of marshland and 48,000 acres of high land around San Pablo Bay. If constructed at the Carquinez Straits, it would irrigate 70,000 acres of marshland and 93,000 acres of high land surrounding Suisun Bay. The appreciation in land values would exceed $8,000,000. Moreover, 169,000 acres of delta land threatened by salt water would be spared, litigation reduced, and industrial development stimulated. The joint committee warned that if northern California failed to provide a sufficient industrial water supply to attract new industries, Seattle or Portland could and would.[72]

A second major recommendation of the joint legislative committee involved the sale of electrical power. The committee relied heavily on consulting engineer Lester Ready's survey of Kennett Dam's potential and the anticipated power needs of northern California. Ready estimated that Kennett would yield an average 1,217,000,000 kilowatt hours of power each year, about 38 percent of the electricity used within a fifty-mile radius of San Francisco in 1927. He predicted that the future needs of the Bay Area, Kennett's major market, would increase from 3.2 billion kilowatt hours in

70. The joint committee correctly recognized that river improvements and storage reservoirs in the San Joaquin Basin had contributed to salinity as well as drought and irrigation in the Sacramento Valley. The construction of levees and bypass channels in the first decades of the 20th century produced better drainage but reduced the river's volume in the summer. Dredging also contributed to salt-water incursion. During the 1920s, the Merced Irrigation District completed a 278,000 acre-foot reservoir; the Modesto and Turlock districts a 290,000 acre-foot reservoir; and San Francisco's dam at Lake Eleanor, and the Hetch-Hetchy Aqueduct, virtually eliminated all but return flow in the Tuolumne River during the late summer. The 1929 *Report of the Joint Committee* stated that over two-thirds of the 4,000,000 acre-feet of storage space on streams that flowed into San Francisco Bay had been created since 1920, virtually all on the San Joaquin River and its tributaries (pp. 159, 160).

71. Walker R. Young, *Report on Salt Water Barrier*, Division of Engineering and Irrigation Bulletin no. 22, 2 vols. (Sacramento, 1929), 55.

72. *Report of the Joint Committee . . . January 18, 1929*, 161–167.

1927 to 5.3 billion kilowatt hours in 1936 (the earliest year Kennett could be ready). In short, California's increasing appetite for power insured that the price of electricity would remain stable or increase. The state could expect an annual revenue of $4,250,000 if it sold the power at the dam, or $5,300,000 if it delivered the electricity to the Bay Area. However, the cost of stringing transmission lines and constructing relay stations would add $110,000,000 to the dam and power plant's $70,000,000 to $80,000,000 cost.[73]

The eventual decision to sell Kennett's power to private utility companies at the dam, or "switchboard," was not a sellout. State officials recognized that such companies might provide potent opposition to a state water plan, as they had helped block the Water and Power Act in 1922, 1924, and 1926. The support, or at least acquiescence, of these corporate interests was a prerequisite to winning the approval of California voters at the polls. The joint committee also concluded that the cost of building a "power grid" would exceed the revenue returned to the state, and that duplicating existing transmission works would be wasteful. What if the state could not sell electricity at a price low enough to compete with the private companies? By selling the power under contract, the state reduced or eliminated many risks. The power sold at Kennett would pay more than 90 percent of the principal and interest on forty-year bonds—and secure valuable political support.[74]

The legislative committee issued a supplemental report in April 1929, which summarized its conclusions and recommendations. It proposed that the legislature accept the state engineer's Bulletin number 12 issued in 1927 as the state water plan, but without the saltwater barrier the lawmakers had recommended in January. The committee recommended submission of a $100,000,000

73. Lester S. Ready, *Report on Kennett Reservoir Development: An Analysis of Methods and Extent of Financing by Electric Power Revenue,* Division of Engineering and Irrigation Bulletin no. 20 (Sacramento, 1929). Ready summarized his conclusions in a cover letter dated October 23, 1928, which accompanied the report. The letter was reprinted on pp. 7–10. Ready's conclusions were also reprinted on pp. 49–51 of the joint committee's report.

74. *Report of the Joint Committee . . . January 18, 1929,* 18, 101. Also see the elaborate charts on 58–68, 69, 74, 79, 89, 94, and on the insert facing 96. The report implied that power generated at other state dams would also be sold to private companies if it was not needed to move water.

bond issue to the voters to pay for the immediate construction of Kennett Dam, the San Joaquin Valley diversion system, and a flood control project on the Santa Ana River. The decision to drop the barrier was largely political. Southern California's leaders recognized that years might elapse before construction began on the state project, and they decided that local water users could afford to build the Colorado River Aqueduct from Parker Dam without state aid. But with only the Santa Ana flood control project left to benefit Southern California, what incentive did its voters have to support the plan? The barrier was sacrificed not just because it was expensive and might interfere with navigation, but also because it promised to aid the industrial development of northern California at the expense of the southern counties. Without the saltwater barrier, more political support could be expected south of the Tehachapis.

The supplementary report concluded that the cost of the Santa Ana flood control project should be shared equally by the state and the districts directly benefited. It also recommended that the state assure Sacramento Valley residents that that section's "ultimate" water needs would be satisfied before any water was transferred into the San Joaquin Valley. If the joint committee had its way, all future water projects would conform to the state water plan; the state would distribute all water; costs would be apportioned according to "benefits received"; and federal aid would be solicited "upon the basis of navigation, flood control, irrigation, and other benefits that would accrue to the nation at large."[75]

75. *The Supplemental Report of the Joint Legislative Committee,* April 9, 1929, is reprinted in "Central Valley Project Documents," H. Doc. 416, 1: 228–247.

Perhaps in reaction to the joint committee's decision, some proponents of a saltwater barrier suggested linking it to an aqueduct through the San Joaquin Valley to Los Angeles. Vincent Wright, an Alameda engineer, argued that southern California would receive only 4.4 million acre-feet of water from the Colorado River Aqueduct, most of which would go to the Imperial Valley. On the other hand, 37 million acre-feet ran to waste in San Francisco Bay. C. W. Schedler, president of the Contra Costa Industrial Water Users Association, and Thomas Means, an engineer who had studied the barrier in detail, concluded that water could be delivered to Los Angeles from the bay barrier for less than $125,000,000, about $25,000,000 less than the cost of the Colorado River Aqueduct. See Vincent Wright to C. C. Young, February 15, April 8, 1929; and Thomas H. Means to Wright, March 12, 1929, in "Salt Water Barrier, General Correspondence, Sept. 1, 1923 to July 8, 1929," microfilm reel M–707, DWR Archives. Also see the *San Francisco Chronicle,* March 25, 1929.

The 1929 legislature refused to authorize construction of the 1927 water plan without more extensive data on the cost of construction, without a firm promise of federal aid, and without an investigation of potential water projects outside the Central Valley. State Engineer Edward Hyatt worried particularly about this last requirement.[76] It was a sound political decision because members of the joint legislative committee could pacify constituents by holding hearings to discuss local water problems and needs, even if there was little chance that the state could or would offer a solution. These meetings served the purpose of convincing Californians that their representatives in Sacramento looked out for their interests. Public meetings might also win votes for a statewide project from those who lived outside the Bay Area or Central Valley but hoped that one day their own county would be blessed with a state dam or aqueduct. From an engineering standpoint, however, this course made the state water project even more of a political football. Isolated units, however attractive to local residents, might not fit well into a state plan and might not be cost effective. Nevertheless, the lawmakers appropriated $450,000—cut to $390,000 by Governor C. C. Young—to pay for the joint committee's investigation, as well as surveys of the saltwater barrier, Santa Ana flood control project, Mojave River project, and snow surveys. State engineers hoped that by surveying the snowpack each winter, they could predict spring runoff, allowing farmers to plant fewer acres or less thirsty crops in the spring and summer of dry years.[77]

The $390,000 included $25,000 to pay for the work of a special committee to coordinate state and national water planning and win federal financial help. The group, headed by former Governor George Pardee, found a sympathetic audience in Washington, including President Herbert Hoover, Secretary of the Interior Ray

76. Edward Hyatt, Memorandum: Water Resources Investigation, 1929–1930, July 15, 1929, in "Water Resources Investigation: Minutes of Hearings," microfilm reel M–669, DWR Archives.

77. Edward Hyatt, California State Engineer, "Resumé and Present Status California Water Plan Investigation," November 5, 1930, Edward Hyatt Collection, File no. 1, Water Resources Archives; *Report to the Legislature of 1931 on State Water Plan, 1930*, Division of Water Resources Bulletin no. 30 (Sacramento, 1931), 18; *Modesto Tribune*, June 21, 1929.

Lyman Wilbur, and Commissioner of Reclamation Elwood Mead. Hoover had been educated at Stanford, and his engineering background gave him a good understanding of the technical problems of arid land reclamation. Like Hoover, Wilbur was an Iowa native who moved to California as a young man. He graduated from Stanford in 1896, in the class right behind Hoover's, and became president of the institution in 1916. Mead had vast experience with California water problems stretching back to his survey of water rights at the turn of the century, to his years as professor of irrigation at the University of California, and to his directorship of the California Land Settlement Board from 1917 to 1923. As secretary of commerce, Hoover repeatedly supported federal dam building in California, particularly in a speech delivered in Seattle in August 1926. He met with Governor-elect C. C. Young in November of the same year to discuss California's water shortages, and one month later wrote a confidential letter urging the new governor to write President Coolidge "to the effect that you desire to secure coordination of these many activities [flood control, irrigation, power generation, navigation, etc.] and that you propose to appoint a coordinating committee . . . and would like to secure his approval and direction to the various departments of the Federal Government, [and] that they should instruct their representatives to accept membership on such a committee. . . . The purpose of this committee would be to consider not only broad policies of development of the state's water resources but pass upon and advise with respect to each single project which arises." In August 1929, after he became president, Hoover wrote to Joseph M. Dixon, assistant secretary of the interior, suggesting that the Reclamation Service join "with the States and local communities or private individuals" to build irrigation projects which would be administered entirely by the states. "It is only through the power of the states," the president noted, "that reclamation districts can legally be organized which would incorporate the liability of privately owned lands for irrigation expenditures." Hoover's proposal was only partly a result of his sympathy for California's water problems. He favored a "new federalism," with increased cooperation between the nation and states and better coordination among the resource agencies in Washington. The president maintained a lifelong interest in using engineering methods to solve economic problems and to promote efficiency in administration. Thus, ironically, the Hoover-Young Commission was Hoover's child.[78]

The president quickly appointed a committee consisting of Commissioner of Reclamation Elwood Mead, Lieutenant Colonel Thomas M. Robins of the Corps of Engineers, and Frank E. Bonner, executive secretary of the Federal Power Commission, to confer with Pardee's California delegation. Pardee's committee included William Durbrow, manager of the Nevada Irrigation District, and for years president of the California Irrigation Districts Association; B. A. Etcheverry, professor of irrigation engineering at the University of California; Alfred Harrell, director of the State Chamber of Commerce; W. B. Mathews, a member of the Colorado River Commission and chief counsel for the Los Angeles Bureau of Water and Power; Warren Olney, former associate justice of the California Supreme Court; and Frank E. Weymouth, chief engineer of the Metropolitan Water District of Southern California. Several state officials, including Director of Public Works B. B. Meek, and State Engineer Edward Hyatt, also served, along with a handful of members from the Joint Legislative Water Committee. Only Robins participated in all the meetings between the two groups in 1929 and 1930, but the other two federal officials supported his recommendations.

The Hoover-Young Commission largely ratified the water plan recommended by the joint legislative committee in 1929. Kennett north of Redding on the Sacramento River, and Friant, on the San

78. Herbert Hoover to Governor-elect C. C. Young, December 15, 1926, and Hoover to Joseph M. Dixon, August 21, 1929, "Hoover-Young Commission Correspondence, December 15, 1926 to April 30, 1930," microfilm reel 1032, DWR Archives. On the same reel, see Young to Hoover, January 13, 1927; and Young to Ray Lyman Wilbur, April 28, 1929.

In a July 28, 1929, letter to Wilbur, on the same reel, Young wrote: "As I told you a month ago in your office at Stanford University, I feel that it is practically impossible for California to solve her water problem except with the cooperation of the federal government. I have been from the first exceedingly pleased with President Hoover's assurance that he believed such cooperation was both feasible and right. I have understood first, from the President's speech of acceptance, and subsequent utterances on the subject, that he believes in a comprehensive program of federal aid for public works in various portions of the country, and that development of water resources in California is sufficiently important to form part of that program."

Joaquin River 15 miles north of Fresno, were the prime reservoir sites. Kennett, the commission predicted, would nearly pay for itself from power sales and store sufficient water to maintain a minimum flow of 3,300 second-feet of water past Antioch year-round. This volume of water would achieve the purposes of the saltwater barrier at a small fraction of the cost, including supplemental water for the San Joaquin Valley and communities surrounding San Francisco Bay. The commission concluded that a direct water transfer from the Sacramento Valley to drought-stricken Kern and Tulare counties would be needlessly expensive and complicated. In 1923, the Department of Engineering and Irrigation had recommended pumping water through a 200-mile uphill aqueduct from a bay barrier to Tulare Lake. By 1925, it decided on a water exchange by which San Joaquin River water would be moved from Friant into the Kings River basin to replace water diverted through a canal from the Pine Flat Dam into Tulare County. For lower San Joaquin River users, a substitute supply would be obtained by reversing the river. By 1929, the state engineer decided on a canal directly from Friant into Kern and Tulare Counties (today's Friant-Kern Canal). The Hoover-Young Commission approved this transfer but suggested that a pump canal from the delta to Mendota, on the great bend of the San Joaquin, would be cheaper and involve less litigation than reversing the river.

The Hoover-Young Commission concluded that the project could not be built if the interest rate on state bonds ran higher than 3.5 percent, and only the federal government could borrow money at such low rates. If the state issued 3.5 percent bonds guaranteed by the federal government, and amortized them over fifty years, the net annual cost would be less than one-third the cost of 4.5 percent bonds amortized over forty years. "We recommend, therefore," the commission concluded, "that the project be constructed by the Federal government, it meeting the cost thereof, in the first instance, and that the works, when completed, be operated by the State as far as practicable." Since the Reclamation Bureau was already building Boulder Dam, the commission recommended against federal construction of the proposed Colorado River Aqueduct to carry water to the farms and cities of southern California. But it did suggest that the state adopt a constitutional amendment allowing it to guarantee the Metropolitan Water District's bonds. The commission also supported a constitutional amendment guaranteeing repayment for Kennett and Friant dams and a third to create a special tribunal to appraise the value of condemned water rights and other property. The group did not support an interest-free federal loan. The Central Valley Project was being sold as a relief measure, not a scheme to reclaim more arid land; the commission predicted that 200,000 acres in the San Joaquin Valley would return to desert without additional water. The idea of subsidizing agriculture using the interest-free provision of the Reclamation Act of 1902 had not yet achieved the popularity it would after World War II. Besides, if the federal government provided interest-free loans, it might also insist on operating and administering the water project—an expansion of federal authority deeply feared by most Californians.[79]

Meanwhile, in 1929 and 1930 the water plan faced strong opposition throughout the state. Although the saltwater barrier had been dropped from the project in 1929, many opponents of the scheme still feared it would be built later, perhaps after the completion of Kennett and Friant reservoirs. Boosters in Solano and Contra Costa counties formed the Salt Water Barrier Association late in 1929 to publicize the benefits of such a structure, and the War Department, the Coast Geodetic Survey, the Geological Survey, and the state health department all conducted feasibility studies in 1929 and 1930.[80] Moreover, despite the joint legislative

79. *Report of the California Joint Federal-State Water Resources Commission, 1930,* JCSA, 49th sess. (Sacramento, 1932), Appendix, vol. 5; and also on microfilm reel 1033, DWR Archives. The quote is from p. 12. Also see "The State Water Plan," *Transactions of the Commonwealth Club* 26 (June 2, 1931): 73. By October 1930, the War Department had decided that the federal flood control and navigation contribution to the construction of Kennett Reservoir would be $10,000,000 of the estimated cost of $73,000,000; power revenue would make up the difference. However, since power revenue from the Friant Dam would be far less than that from Kennett, San Joaquin Valley irrigators would have to repay most of the cost of that structure, though Elwood Mead had already considered building the dam under the "interest-free" terms of the Reclamation Act of 1902. See Mead to "Dent" (U.S. Reclamation Bureau), October 2, 1930 in "Hoover-Young Commission Correspondence, October 7–30, 1930," microfilm reel 1032, DWR Archives.

80. For a detailed discussion of the controversy over the saltwater barrier before 1931, see Alan M. Paterson and W. Turrentine Jackson, *The Sacramento-San Joaquin Delta: The Evolution and Implementation of Water Policy, An Historical Perspective* (Davis, Calif., 1977), 1–28.

committee's decision to drop the scheme, state officials seemed encouraging. In a May 1930 speech delivered in the Suisun Bay community of Pittsburgh, Bert B. Meek, California's director of public works, described the barrier as "an important part of the whole great scheme of water conservation." "When this program is completed," said Meek, "we are coming down here and ask your people to help us put the program over with a bang, and I know you are going to do it as you will be directly benefitted."[81] Meek assured his audience that Governor Young was not, as commonly reported, opposed to the barrier. State Engineer Hyatt went even further. At a State Realtors Association convention in Brentwood, just south of Suisun Bay, Hyatt called the barrier "a great thing" and an "ultimate necessity" which would be constructed "at the proper time." Finally, in June 1930, gubernatorial candidate James Rolph endorsed the barrier. Whether these pronouncements were sincere or simply an attempt to win votes, they frightened barrier critics.[82]

Bay Area proponents of the dam claimed a natural alliance with delta farmers and communities upstream on the Sacramento and San Joaquin rivers. They promised that the barrier would protect delta farmland from saline intrusion, reduce litigation among upstream and downstream users, and promote trade on the two streams. But critics of the barrier pointed out that the scheme's strongest support came from land developers, real estate companies, and local chambers of commerce, not from those directly affected by water shortages. An engineer who represented the owners of 60,000 acres of delta farmland warned the Hoover-Young Commission that potential flood damage from heavy runoff trapped by the barrier far outweighed the occasional damage from salt water intrusion. The barrier would also maintain a higher year-round level of water in the delta, eroding levees. Moreover, land in the delta was already heavily taxed to pay for reclamation works, and

81. *Martinez Standard* (Martinez, California), May 26, 1930.

82. For Hyatt's comments see the *Pittsburgh Dispatch* (Pittsburgh, Calif.), June 7, 1930; also see the *Dispatch* for April 28, 1930. For Rolph's position see the *Pittsburgh Post*, June 6, 1930. The scrapbooks in the Edward Hyatt Collection at the Water Resources Archives are a rich source of newspaper clippings from all parts of the state during the late 1920s. Many of the newspaper articles cited for 1927 and after can be found there.

farmers there expected that they would have to help pay for the barrier. Kennett was a far better alternative, if only because power revenue and the federal flood control contribution would pay the cost. Stockton also offered substantial opposition to the bay barrier. The chairman of the Stockton Chamber of Commerce's "marine committee" charged that the structure's lock system would discourage river transportation and retard or prevent the city's development as an inland port. Stockton had begun construction of a deep water channel at a cost of $6,000,000, half of which was paid by the War Department. Of course, the city's economic health also depended on the prosperity of the irrigation districts upstream on the San Joaquin River and their water claims had been contested by the towns surrounding Suisun Bay and by the delta farmers. Many other criticisms were raised against the barrier in northern California. The War Department warned that interference with the Sacramento River's natural flow would contribute to siltation and an accumulation of debris dangerous to shipping. Others claimed that the barrier would increase pollution. Chemicals, sewage, and other contaminants dumped into the river upstream from the barrier would be trapped, killing fish and creating the danger of epidemics in communities that drew their drinking water from the stream. Similarly, because sewage and many other effluents were dumped into San Francisco Bay, the lack of a "flushing effect" might destroy the bay's fishing industry. But most important, according to the critics, was the assumption that Kennett would provide the same benefits as a barrier at a lower cost and with fewer possible side effects.[83]

The trump card of supporters of the barrier was that the cheap supply of fresh water would promote industrial development. This explains the support of the San Francisco Chamber of Commerce and the business associations in Oakland, Berkeley, and Napa, as well as in Antioch, Martinez, Pittsburgh, Benecia, and other communities near the proposed barrier sites. At a time when San Fran-

83. Martinez *Daily Gazette*, February 11, 1930; *Lodi Sentinel*, February 18, 1930; *Red Bluff Morning Times*, April 4, 1930; *Fresno Bee*, April 9, 1930. For arguments for and against the saltwater barrier, also see "Salt Water Barrier, General Correspondence, Sept. 1, 1923 to July 8, 1929," microfilm reel M–707; "Saltwater Barrier Engineering Reports, 1922 to 1929," microfilm reel 1185; and "Transcript of Hearing before the Joint Committee of the Legislature of the State of California on Water Resources, February 7, 1930," microfilm reel M–669, all at the DWR Archives.

cisco businessmen fretted about being eclipsed by competitors in Los Angeles, the prospect of new industries flocking to the Bay Area proved very attractive. Nevertheless, skeptics challenged even this assumption. A special committee—consisting of the deans of Stanford's Graduate School of Business and U.C. Berkeley's College of Commerce, along with a consulting engineer—reported that if the dam was paid for by local taxes, those taxes might double and retard industrial growth. The water supply was only one reason companies decided to build in one place and not another; fuel and labor costs were much more important. The committee found no case where a company rejected the upper Bay Area because of an inadequate water supply. It concluded: "The upper San Francisco Bay area affords a most attractive combination of location factors for industries which require relatively large tracts of land, low cost transportation, proximity to large cities and, at the same time, the comparative isolation of a country site." That the south shore of Suisun Bay offered the cheapest land around the Bay counted for far more than the quality or quantity of fresh water.[84]

Within the Central Valley, there were many more critics of the state water plan. Many upper San Joaquin Valley farmers worried that diversions from Friant Dam south would reduce their ample water supply even while they paid higher state taxes to "subsidize" the state project. Other farmers wondered why reservoirs they planned to build to supplement an inadequate water supply, such as the one at Pine Flat on the Kings River, were not included in the plan. Sacramento Valley water users charged that San Joaquin Valley irrigators wasted water. They also denied that there was any surplus water to export and worried that litigation over water rights would reduce the value of their land or make it unmarketable. The parochialism and sectionalism characteristic of California water politics since the 1870s often reappeared. For example, an Oroville newspaper resented any effort to bail out Tulare County:

> The individual frequently guesses wrong and invests improperly, but seldom is the state asked to penalize itself to

84. *Industrial Survey of Upper San Francisco Bay Area with Special Reference to a Salt Water Barrier* (1930), *JCSA*, 49th sess. (Sacramento, 1932), Appendix, 5: 6, 45, 62. The quote is from p. 62. This survey was paid for from the $390,000 appropriation made in 1929.

make good for the incorrect guess of the individual or the group. The Tulare County farmers guessed that their water supply was good, but they guessed wrong and invested improperly. Should the state be asked to make good their investment now that their land is preparing to return to the desert? Should the state be asked to make a great river flow uphill and send water hundreds of miles from a land not yet developed but which may in the near future need this same natural resource for its own development? The farmers of rich Stanislaus County are in great fear that rising water tables will destroy the fertility of their soils. We do not hear them asking relief of the state, and yet if Tulare can do it what is to prevent Stanislaus at some future time asking relief on the same precedent?

The Oroville-Wyandotte Irrigation District's chairman of the board commented: "The entire population of Tulare county could be moved up here and every resident given a farm for the amount of the bonds which it would be necessary to issue for the project."[85]

Not surprisingly, supporters of the Iron Canyon Project also opposed the state water plan. They wanted to use all the Sacramento River's surplus water at home and warned that the project's enormous cost, conflicts between different parts of the state, competition between rival groups of water users, and the threat of litigation would prevent the plan from getting off the drawing board. W. A. Beard, head of the Iron Canyon Association, favored a series of local works instead of a coordinated state scheme. Since the Iron Canyon site had been surveyed and inspected many times, work there could begin at once; the Reclamation Bureau would build the structure just as it was building Boulder Dam. The Iron Canyon Dam, along with the saltwater barrier and several flood control reservoirs on tributaries of the Sacramento, would solve the valley's water problems.[86] Beard found a welcome ally in A. M.

85. *Oroville Register*, August 2, 1929, January 9, 1930.

86. See the testimony of W. A. Beard in "Proceedings Before Joint Committee of the Senate and Assembly, Dealing with the Water Problems of the State of California, Held in Red Bluff, California, January 6, 1930," file entitled "Water Resources Investigation: Minutes of Hearings, Transcript of Hearing, January 6, 1930," microfilm reel M–669, DWR Archives. See especially pp. 12–39, 44–46, 55–58.

Barton, head of the State Reclamation Board, who predicted that the state project would take twenty years to complete. In the meantime, he urged the state to construct reservoirs on the American River at Auburn, Folsom, and Coloma after negotiating repayment contracts with beneficiaries. Barton noted: "Construction of storage works by the state is so uncertain and so subject to delay as to warrant its rejection as a possible solution. The introduction of a political football into the situation will in no measure benefit the parties now injured or subject to injury."[87]

In southern California, attention focused on the Santa Ana Flood Control Project and the Colorado River Aqueduct. Between 1925 and 1929, the state spent $65,000 studying the Santa Ana Basin's water problems. Unfortunately, it found only two feasible dam sites, and they could not capture enough water to provide flood control and eliminate water shortages. As an alternative, state engineers favored canalization of the river and spreading ponds to replenish the underground supply.[88] But this was a pale copy of the works planned for northern California and did not stir up much enthusiasm among southern California voters. Many northern and southern Californians alike urged the state to include the Colorado River Aqueduct in the state water plan. But southern Californians feared that construction of the aqueduct would be delayed for years if it was tied to the state project. And if the state underwrote the project, it might insist on setting water rates and parceling out the new supply. California officials also considered underwriting the construction of the All-American Canal which had been included in the Boulder Canyon Act of 1928. But the state could not compete with the Reclamation Bureau's no-interest loans and, in any event, the handful of voters in the Imperial Valley could provide little help in winning approval of the state water plan. The feasibility or actual need for component projects in the plan often

87. A. M. Barton, "Memorandum to the Joint Legislative Committee Pertaining to the American River," *Central Valley Project Documents*, 225; *Sacramento Bee*, March 8, 1929.

88. For an overview of the state's Santa Ana Basin work, see the undated memorandum entitled "Santa Ana River Investigation: Memorandum for use of Legislative Committee on Water Resources" in the file entitled "Santa Ana Investigation, General Correspondence, June 4, 1925–Sept. 29, 1930," microfilm reel 1185, DWR Archives.

mattered less than where the votes were located and in what numbers.

State Engineer Hyatt forwarded an updated water plan to the legislature on March 4, 1931, culminating ten years of study at a cost of more than $1,000,000. Though loosely based on the 1927 water plan proposed by Paul Bailey, it was far more comprehensive and detailed: twenty-four reservoirs with an aggregate capacity of 17,817,000 acre-feet of water; six major canals; and a price tag of between $500,000,000 and $600,000,000 (depending on whether the dams included power stations). On August 5, 1933, the governor signed a bill providing for a $170,000,000 bond issue to pay for the project's first units. Most of these works were familiar to Californians who had followed the evolution of the state water scheme: Kennett Dam, a cross-delta canal to carry water from the Sacramento to the San Joaquin River, a pump system to reverse the San Joaquin River; and an aqueduct from the Friant Reservoir to Bakersfield. The most notable change was the addition of the Contra Costa Canal from the delta fifty-eight miles to a reservoir at Martinez. This aqueduct, designed to serve the homes, farms, and industries along the south shore of Suisun Bay, was an obvious sop to those who had favored the saltwater barrier.[89]

The enabling legislation signed by the governor in August 1933 provided for state construction of the project's hydroelectric plants and transmission lines. Consequently, the powerful Pacific Gas and Electric Company, which led opposition to the Water and Power Act in 1922, 1924, and 1926, launched a referendum campaign to rescind the legislation. The company's prime objection, of course, was to public power. But its campaign literature emphasized broader arguments: that California already produced too many crops; that southern California had been slighted in favor of northern California; that the state had no firm promise of federal aid; that the project would create a dangerous "political machine"

89. Edward Hyatt outlined the formal state water plan in "Report to the Legislature of 1931 on State Water Plan, 1930," *JCSA*, 49th sess. (Sacramento, 1931) Appendix, vol. 3. The 1930 plan was presented in greater detail in Division of Water Resources Bulletin no. 25. Even though most southern Californians shunned state aid, the complete plan, including the Colorado River Aqueduct, was expected to cost $198,600,000.

Major Units of State Water Plan, 1931.

in Sacramento and a sprawling new bureaucracy; that the bond issue would usher in an era of reckless state spending on a wide variety of unnecessary and expensive public works projects; and that ultimately the state would be forced to dip into general tax revenue to pay for the project. Aside from California's immediate and future water needs, and their relationship to the state's economic growth, proponents of the measure used the depression to win votes. State Senator Bradford Crittenden noted in his ballot argument for the bonds:

> Closely linked with President Roosevelt's National Recovery program, this act provides for the construction of a great water conservation project which will give immediate employment to more than 25,000 men for at least three years, thereby affording a livelihood for approximately 100,000 persons. This tremendous aid to unemployment relief and economic recovery in California will be accomplished without a single dollar of cost or obligation to California taxpayers! All costs of the entire program will be defrayed through Federal aid and a revenue bond system, or self-financing plan, which requires that revenues of the project pay for the project.

Though the bond issue carried, the final vote, 459,712 to 426,109, reflected the persistence of sectionalism in California water politics. Los Angeles County rejected the plan two to one, but the population of southern California was not large enough to overcome the huge positive vote north of the Tehachapis. In northern California, the water project attracted strong urban as well as rural support. San Francisco voters approved the bonds by nearly two to one; Sacramento County by nine to one; Shasta County by eighteen to one; and Tulare County by twenty to one. Unquestionably, adoption of the measure depended as much on drought as depression. As if to underscore the need for a state project, California's most persistent drought descended on the state in 1929 and lingered on until 1935. Many voters assumed that the future vitality of California's economy depended directly on the health of its farms. They did not reflect the New Deal's faith in central planning or conservation, or even the Progressives' devotion to efficiency and the "common good." Most simply voted their pocketbooks.[90]

Partly because of the depressed bond market, and partly because of the expectation of federal aid, the bonds were never put on sale. Congress authorized the Reclamation Bureau to begin construction of the Central Valley Project in 1935; the first $20,000,000 appropriation came from the Federal Emergency Relief Appropriation Act of 1935.[91] Construction began in 1937, but the Shasta Dam (at Kennett) was not completed until 1944 and the Friant-Kern and Delta-Mendota canals did not open until 1951, ten years after the end of the depression and sixteen years after the end of the Great Drought. After nearly three decades, the Reclamation Bureau finally got its chance to build a substantial project in California, though the 1928 Boulder Dam Act paved the way. Just as the Reclamation Service had abandoned the principle that federal reclamation should benefit primarily public lands early in its existence, in the late 1920s and 1930s it discarded the requirement that the land benefited should bear the cost of reclamation. By the 1930s the Reclamation Bureau fully grasped the implications of the doctrine of multiple use which freed it from an economic straitjacket. The answer to the increasing cost of arid land reclamation was to build bigger projects to serve a variety of needs, urban as well as rural. The exact cost to individual groups of water users, such as irrigators, became largely irrelevant to many Bureau planners be-

90. On the bond issue, see the *Sacramento Bee* and *Sacramento Daily Union*, both of December 20, 1933. Bradford Crittenden's statement is in the file entitled "Arguments for and Against Water and Power Referendum Voted on at Special Election Held Dec. 19, 1933," microfilm reel 1033, DWR Archives. This file contains a good sample of campaign literature for and against the bond issue.

91. Several historians have suggested that the state accepted federal aid only because the market for state bonds disappeared during the early years of the depression. For example, Michael G. Robinson, in his *Water for the West: The Bureau of Reclamation, 1902–1977* (Chicago, 1979), notes that "voters approved the $170 million project in December 1933 and authorized the issuance of revenue bonds to be retired by water and power sales. However, because of the Depression the bonds did not attract investors. Desperate for support, the State sought assistance from the Federal Government" (p. 67). As noted in the text, most state officials recognized that the state water project could not be built without federal aid as early as 1926 or 1927. They did not, however, anticipate that the Reclamation Bureau would build and finance the entire project.

cause entire regions benefited from the multiple use schemes launched in the 1930s and after World War II. The Reclamation Bureau also profited from the success of the Tennessee Valley Authority. The TVA demonstrated that increasing the productivity of the soil was an essential part of regional development.

The only history of the Central Valley Project noted that "by 1933, long planning had produced a fine plan."[92] But the scheme was neither complete nor scientific. The 1931 project focused mainly on the needs of agriculture and all but ignored municipalities, industry, and recreation. Moreover, it left plenty of loose ends. No practical method had been devised for extinguishing riparian rights or speeding up the process of condemnation; no formula had been devised to charge water users for benefits received; no way had been found to market the power; and no method had been found to win federal aid without federal conditions, such as the 160-acre limitation. Science and efficiency had played a part in molding the state water plan, but politics entered every phase of decision making.

The story of the evolution of the state water plan cannot end without a postscript concerning Robert Bradford Marshall. In January, 1950, on the eve of the opening of the Central Valley Project, Commissioner of Reclamation Michael W. Straus declared that "California's Central Valley Project will stand, for ages to come as a monument to the far sighted vision and untiring efforts of Col. Marshall. . . . I know of no other individual who could rightfully claim greater personal credit for this development." Frank Adams echoed Straus's praise nine years later in his oral history. "It wasn't

---

92. Robert de Roos, *The Thirsty Land: The Story of the Central Valley Project* (Stanford, California, 1948), xi. There is no comprehensive, balanced history of the Central Valley Project, in part because historians have largely ignored the genesis of the project in the 1920s. However, useful studies for the 1930s and after include Mary Montgomery, "Central Valley Project: Highlights of its History," *Land Policy Review* 9 (Spring 1946): 18–21; Paul S. Taylor, "Central Valley Project: Water and Land," *Western Political Quarterly* 2 (June 1949): 229–254; S. T. Harding "Background of California Water and Power Problems," *California Law Review* 38 (October 1950): 547–571; Arthur D. Angel, "Political and Administrative Aspects of the Central Valley Project of California" (Ph.D. diss., University of California, Los Angeles, 1944); Jack T. Casey, "Legislative History of the Central Valley Project, 1933–1945" (Ph.D. diss., University of California, Berkeley, 1949); Charles E. Coate, "Water, Power, and Politics in the Central Valley Project, 1933–1967," (Ph.D. diss., University of California, Berkeley, 1969); and Marion Clawson and Mary Montgomery, *History of Legislation and Policy Formation of the Central Valley Project* (Berkeley, 1946).

a plan, it was an idea," explained Adams, "but we need inspirations of that kind. As a result of his proposals great sentiment was created for a state study, right in the grass roots up and down the state. So I give credit to Colonel Marshall for that, as well as for finding Kennett Reservoir." Marshall did not live to see the first water from the Sacramento Valley splash into the Delta-Mendota Canal. From 1928 to 1937 he worked as an obscure landscape architect for the Division of Highways, and he died all but forgotten on June 21, 1949. Deprived of speech, Marshall poured out feelings of bitterness and disillusionment on paper. For example, in 1935 he wrote:

> Today, as I sit on the side lines I am amused that the State Engineer calls the project the State Central Valley Water Plan. Never has there been given me the faintest suggestion of credit for originating the idea as briefly outlined in the Marshall Plan data. Thank God I am not small enough to resent the intentional discourtesy. I know I did my duty. I am happy that our people will benefit by my unselfish effort in their behalf.[93]

Marshall exaggerated his contribution. The Central Valley Project dated at least from the 1870s and had many fathers. By the 1920s, master builders like William Mulholland and Robert Marshall were a dying breed giving way to technicians more interested in getting a job done than in the glory of a dream or boldness of a vision. Such men became anachronisms as water projects became more and more complicated. The advent of huge hydraulic schemes, such as the Central Valley Project, represented a new era in the history of the West; agriculture, as well as the economy and society so closely linked to it, would never be the same.

---

93. For Michael Straus's comments, see the Bureau of Reclamation Press Release, January 19, 1950, Marshall Plan File (615.021), DWR Archives. Marshall outlined his objections to the state water plan in a letter to Matt I. Sullivan, October 22, 1931, in the same file. The Marshall quote is from his letter to Lloyd McAulay, October 17, 1935, Robert Marshall Collection, Water Resources Center Archives. Also see Marshall to Earl Lee Kelley, California Director of Public Works, March 5, 1937, in the same collection.

13

Conclusion: The Lost Dream

All of American history reflects conflict between rural and urban values, between an often idealized life close to the earth and the impersonal, complicated social and economic relationships intrinsic to a commercial and by the late nineteenth and twentieth centuries, industrial nation. Californians were both more and less nostalgic than other Americans. Since statehood, their lives had been dominated by "the city," San Francisco, so they never witnessed at close hand the waning influence of the country vis-à-vis the metropolis. Nevertheless, the absence of strong rural traditions made those traditions all the more attractive. From the 1870s to the 1920s, leaders of the irrigation movement in California agreed that the state's cities were overcrowded and growing much too fast; that large farms created a wide range of problems from tenantry to an inadequate rural transportation system; that more had to be done to make country life attractive; and that the family farm was the foundation of a stable society and the wellspring of republican virtues. They saw irrigation as a way to return to a more homogeneous, virtuous, middle-class society, but also as a way to make farming more efficient and profitable. Irrigation promised to promote the cultivation of high value fruits, nuts, and vegetables; reduce the size of farms; stimulate community life; arrest the migration from country to city; and attract new immigrants from outside the state.

The year 1931 represents a convenient dividing line. By that time most Californians acknowledged, though not always directly, that the health of their economy and society did not depend on the existence, perpetuation, or proliferation of the family farm. They might exhibit a sentimental attachment to "small town America," if not farming itself, but they also recognized that the values represented by the small freehold—widespread property ownership, high wages, and economic independence, to name a few— were anachronistic, however attractive. Californians had begun to see their state as "the great exception," and even to revel in its exceptional qualities, rather than try to recapture the institutions and life-style that had prevailed in the Midwest or "back East." By the 1930s, and especially after World War II, irrigation entered a new phase. It was no longer an agent to transform society, but an ally of the agricultural establishment.

During World War I, the crusade for the family farm enjoyed a renaissance in many parts of the United States, but nowhere more than in California. A special California commission on land colonization reported in 1916:

> Within the last five years questions of land tenure and land settlement have assumed a hitherto unthought of importance in the United States. The causes for this are the disappearance of free, fertile public land; the rising prices of privately-owned lands; the increase in tenant farming and a clearer recognition of its dangers; and the increasing attractions of city life which threaten the social impairment of rural communities by causing young people to leave the farms. . . . In [some] countries the state has taken an active part in subdividing large estates and in creating conditions which will enable farm laborers and farmers of small capital to own their homes. They have adopted this policy because experience has shown that nonresident ownership and tenant farming are politically dangerous and socially undesirable; that ignorant and nomadic farm labor is bad; and that the balance between the growth of city and country can be maintained only through creating rural conditions which will make the farm as attractive as the office or factory for men and women of character and intelligence.

The problem of land monopoly and nonresident ownership was particularly acute in California where 310 property owners held over 4,000,000 acres of prime farmland suited to intensive cultivation, land capable of providing 100,000 40-acre farms and sustaining 500,000 additional rural residents. The Southern Pacific Railroad owned over 500,000 acres; four Kern Country land companies owned over 1,000,000 acres, or more than half the county's privately owned land; and in Merced County, Miller and Lux owned 245,000 acres. Most midwestern states had relatively homogeneous populations, few very large or very small farms, few nonresident owners, and few very rich or very poor farmers. However, in many parts of California, rural society was characterized by wealthy, nonresident land barons and migratory farm laborers or tenants who had no allegiance to place or sense of civic responsibility. The commission on land colonization argued that land monopoly undermined democratic values and political stability just as it retarded the state's agricultural development.[1]

The interest in restoring the family farm also grew out of a deep fear of the "yellow peril." By 1920, the Japanese, who had begun to migrate to California in great numbers in the 1890s to replace the excluded Chinese as field hands, had acquired over 500,000 acres of land, most of it reclaimed swamps in the Central Valley. They dominated the production of rice and tomatoes, and their success raised the prospect that they would one day displace the white farmer. Given the prevailing belief in white supremacy, many Californians assumed that democracy was possible only in a homogeneous society. They favored measures to restrict Japanese immigration, segregate Japanese schoolchildren, and prohibit alien landownership. They also hoped to lure more white farmers onto the land.[2]

Aside from the problem of land monopoly and the "yellow peril," Californians shared the assumption of many other Americans that a massive economic slump, if not depression, would

1. "Report of the Commission on Land Colonization and Rural Credits of the State of California, November 29, 1916," *JCSA*, 42d sess. (Sacramento, 1917), Appendix, 2: 5, 8.

2. Roger Daniels, *The Politics of Prejudice: The Anti-Japanese Movement in California and the Struggle for Japanese Exclusion* (Berkeley, 1962); Lloyd Fisher and Ralph L. Nielson, *The Japanese in California Agriculture* (Berkeley, 1942); Masakazu Iwata, "The Japanese Immigrants in California Agriculture," *Agricultural History* 36 (January 1962): 25–37; Adon Poli and W. M. Engstrand, "Japanese Agriculture on the Pacific Coast," *Journal of Land and Public Utility Economics* 21 (November 1945): 355–364.

follow hard upon the end of World War I. Helping returning soldiers acquire a farm could soften the economic impact of demo-
bilization, reward faithful service to the nation, and reverse the
migration from country to city. "Nothing short of ownership of the
land one toils over," Elwood Mead remarked, "will suffice to overcome the lure of the city." Federal reclamation had tried, but failed,
to turn the tide. The United States population increased from
76,000,000 to 106,000,000 from 1900 to 1920, but the Reclamation
Bureau had managed to provide rural homes for only about
1 percent of these new people. To make matters worse, easier living
conditions in the cities and high wages, especially during World
War I, led to dramatic increases in tenant farming and the aban-
donment of many farms in the older agricultural regions. Michigan
alone contained 19,000 idle farms in 1920.[3]

In California, one of the prime obstacles to taking up a farm
was the high price of land and water. The average price of unimproved farmland increased from $27.63 an acre in 1900 to $116.84
in 1920. The value of the average farm tripled in the same period,
and the price of improved land increased even faster. By 1920, most
irrigated land sold for $100 to $500 an acre. And the cost of setting
up a new farm, including the price of livestock, machinery, barns,
and fences, made farming even more expensive.[4]

In 1915, the California legislature created a Commission on
Land Colonization and Rural Credits; Hiram Johnson appointed
Elwood Mead chairman. The irrigation engineer had just returned
from Australia where from 1907 to 1914 he served as chairman of
the Victoria Rivers and Water Supply Commission overseeing arid
land reclamation and colonization in that province. After a survey

3. Elwood Mead, *Helping Men Own Farms* (New York, 1920), 10; "How California is Helping People Own Farms and Rural Homes, August, 1920," University
of California, College of Agriculture, Experiment Stations Circular no. 221, Bureau
of Reclamation Records, "(512) Correspondence re Plans and Methods Employed
by States in Colonizing and Settling Lands—Thru 1924," RG 115, National Archives,
Washington, D.C.; and the comments of J. C. Forkner, chairman of the board of
consultants to the State Department of Engineering and Irrigation, in the transcript
of the board's meeting, November 4, 1922, file entitled "Water Resources Investigation: Transcripts and Minutes of Meetings, 1922," microfilm reel 1029, DWR
Archives.

4. Frank Beach, "The Economic Transformation of California, 1900–1920: The
Effects of the Westward Movement on California's Growth and Development in the
Progressive Period" (Ph.D. diss., University of California, Berkeley, 1963), 106.

of thirty-two California land settlement schemes in 1916, which demonstrated the lack of planning in rural settlement, the commission concluded that the state was ripe for a "demonstration in scientific colonization" based on the Australian experience. Irrigation was a vital part of the "demonstration" because small farms could not be created without it. The Commonwealth Club of California drafted a bill containing most of Mead's recommendations, and the legislature passed it in 1917 with a $250,000 appropriation to launch the program. The lawmakers expected this money to be repaid, along with 5 percent interest, within fifty years.[5]

As the home of the first colony, Mead and the California Land Settlement Board selected a 6,239-acre tract at Durham, in the middle of the Sacramento Valley just south of Chico. The land cost an average $88 an acre. Durham represented a sharp break with the policies of the Reclamation Bureau and private land development companies. The site had been chosen over thirty-nine others for its relatively low price, soil quality, ease of preparation for irrigation, and access to transportation (rather than for the political reasons that too often influenced the choice of sites for federal reclamation projects). The land had been tested and graded according to its productive capacity by soil scientists from the University of California and priced from $48 to $225 an acre. Instead of the uniform-sized farms found on federal projects, the 110 units at Durham varied in size from 8 to 300 acres, depending on whether the land was best suited for raising forage or fruit. There were also 26 two-acre farm laborer plots. When the first settlers arrived in 1918, they found roads, irrigation ditches, barns, houses, and fences ready. In fact much of the land had already been seeded to pasture.

The average cost of reclamation at Durham ran $80 an acre, including $25 to $83 an acre simply to level the land and prepare it for irrigation. This was, as Mead freely acknowledged, more than the price of improved acreage in the East or South. However, the

5. For general discussions of the Mead colonization scheme in California see James R. Kluger, "Elwood Mead: Irrigation Engineer and Social Planner" (Ph.D. diss., University of Arizona, 1970), 105–134; Gerald D. Nash, *State Government and Economic Development: A History of Administrative Policies in California, 1849–1933* (Berkeley, 1964), 344–347; F. L. Tomlinson, "Land Reclamation and Settlement in the United States," *International Review of Agricultural Economics* 4 (1926): 255–272; and Roy J. Smith, "The California Land Settlement Board at Durham and Delhi," *Hilgardia* 4 (1943): 399–492.

farmer had twenty years to repay the cost of the land (5% down and 5% per year at 5% interest). To protect against land speculation, which contributed to the problems experienced on federal reclamation projects, settlers had six months to move onto their farms and were required to reside there for at least eight months a year for the first ten years. For the first five years, they could not mortgage, transfer, or sublet any part of their land without approval from the board. The state also provided a range of ready-made house plans drafted by architects in the state engineer's office. Settlers decided on the size and floor plan when they filed for their farm, left construction to the land board, and found farmhouses ready for occupancy when they moved onto the land. The state loaned up to 60 percent of the cost and up to $3,000 for other improvements, equipment, and supplies.

Mead often commented that federal reclamation had been too concerned with constructing dams and canals, but little interested in the farmers it served. The settlement of government projects had been random and haphazard, with little attention to community life or cooperative institutions. Durham, by contrast, was settled systematically, created as an irrigation community rather than a loose collection of individuals. The land settlement board carefully sifted through 1,000 applications for the 100 farms and 150 applications for the 26 two-acre farm laborer plots. Those accepted were required to have some agricultural experience (except the laborers), an ability to work hard, and a minimum of $1,500 capital. Married men with families and former tenant farmers received preference. The board turned to landscape architects at the University of California for a townsite plan, which contained plenty of room for schools and public buildings on a twenty-two acre "commons" in the middle of the settlement, just as it relied on the Office of Good Roads and Rural Engineering of the U.S.D.A. to furnish plans and supervise construction of the irrigation and drainage systems. The board helped organize cooperative marketing associations that sold poultry and dairy products; provided heavy farm machinery; bought barbed wire, cement, pipe, lumber, and other supplies in wholesale lots to reduce the cost to Durham residents; provided a superintendent to advise farmers on crops to plant and help solve agricultural problems; and even established a mosquito abatement district to reduce the danger of malaria. In short, the land settlement board anticipated and provided for nearly every need.[6]

One of many early writers who visited Durham provided an idyllic picture for *Collier's* readers in 1925:

> Here some 150 families have been making a prosperous living and having a bully time in the very same hard-boiled period when a million people in a single year left our farms, starved off. Nobody wants to leave Durham. . . . Then what's the secret? No pioneering. Cooperation. Expert advice. Available capital. And no isolation. Unless you are naturally a crab you won't be dull in Durham. Farmers don't live miles apart but in a close-knit, pretty township covering 6,000 acres, laid out logically for business and pleasure. No tiring trips after chores to go to a dance, see the school movie, hear a radio, drop in at the club, take a dip in the swimming hole in Butte Creek, join a picnic, or attend the fair.

Durham, the writer gushed, was "the seed of a new land policy for the United States."[7]

The early success at Durham was not matched by a second colony established in Merced County in the San Joaquin Valley in 1920. The 4,800-acre Delhi Colony, built on land acquired from the Turlock Irrigation District, suffered crippling problems from the start. After grading and the installation of irrigation pipe, the farms cost an average $230 an acre, substantially more than at Durham.

---

6. Elwood Mead, "The Relation of Land Settlement to Irrigation Development," January or February, 1923; "(516) General Correspondence re Soldier Settlement Plans and Methods thru 1925," RG 115; Mike Commons to Mead, May 3, 1924, in "(512) Correspondence re Plans and Methods Employed by States in Colonizing and Settling Lands—thru 1924."

7. Gertrude Matthews Shelby, "Nobody Wants to Leave this Town," *Collier's* 75 (May 30, 1925): 26, 34. For other laudatory articles on the Durham settlement, see "The Most Helpful Experiment in the Settlement of Irrigated Lands," *Engineering News-Record* 81 (December 5, 1918): 1013; "Developing Irrigated Land with Selected Settlers," 1014–1018; Elwood Mead, "Farm Settlement on a New Plan," *American Review of Reviews* 59 (March 1919): 270–277; H. A. Crafts, "Back to the Land: California Pioneers the State Land Settlement Plan," *Scientific American* 121 (August 23, 1919): 185; "The State and the Farmer: Successful Development of the California Land Settlement Scheme" 123 (November 13, 1920): 494, 507–508; W. V. Woehke, "Food First: Work of the California Land Settlement Board," *Sunset* 45 (October 1920): 35–38; "California's Farm Colonies," *The American Review of Reviews* 64 (October 1921): 397–404

Moreover, much of the soil was poor and required frequent, heavy doses of fertilizers. It was so sandy that some of the first alfalfa crops literally blew away. The project opened just as the agricultural boom of World War II ended. With inadequate capital and falling crop prices, more than 90 percent of the settlers fell behind in their payments to the state. By 1925, only about 57 percent of the land at Delhi was under cultivation, and 38 percent of the farms had returned no income at all.[8]

California voters turned down a $3,000,000 bond issue to aid settlement in 1922, and Governor Friend Richardson strongly opposed any new colonies. In 1925, he charged that "the sandy wastes at Delhi was [sic] purchased by the projectors of this amazing scheme [the land settlement board] at much more than its worth; settlers without experience and without funds were lured to the colony by glowing advertisements; advice regarding crop planting by alleged experts proved worthless, and money was squandered on railroad switches, town site and other unnecessary dreams. The result has been disastrous to the settlers. For two years we have been trying to salvage this wreck left as a legacy from the Delhi projectors."[9] Elwood Mead, who had resigned from the board at the end of 1923, engaged in a running dispute with the governor. Richardson, he claimed, had encouraged the complaints of Delhi residents and replaced Mead with a real estate agent who opposed state-sponsored colonization on principle. In Mead's words, Delhi "would have succeeded if it had been supported by Governor Richardson as the Durham settlement was supported by his predecessor."[10] Nevertheless, the legislature shared the governor's concerns. The lawmakers investigated conditions at Delhi in 1923, 1925, and 1927, and extended relief to the Delhi settlers by writing off much of their debt. Subsequently, farmers at Durham demanded the

8. Robert Welles Ritchie, "The Rural Democracy at Delhi," *Country Gentleman* 85 (November 27, 1920): 8, 36; Nathan A. Bowers, "California's Land Settlement at Delhi," *Engineering News-Record* 88 (July 23, 1925): 143–145.

9. *San Francisco Chronicle*, April 11, 1925.

10. *Orland Register*, May 29, 1925. In a letter to E. F. Benson, March 13, 1926, "(516) General Correspondence re Soldier Settlement Plans & Methods 1926 & 1927," RG 115, Mead wrote: "One has to know how personally vindictive he [Richardson] is to understand his attitude on settlement. . . . His personal animosities govern his public acts. He is an enemy of land settlement as he has been of all the state's social and economic activities."

same concessions. In 1929, the state worked out a schedule reducing the cost of farms on the two projects from inflated World War I prices to the prices prevailing in the late 1920s, and in the following year turned management of the projects over to the farmers themselves. The *Washington Post* characterized both schemes as failures, even though Mead stalwartly insisted that Durham had been a success. The newspaper blamed their failure not just on the high price of land during the war, but also on "incompetent [soil] surveys," inadequate screening of prospective settlers, and general "mismanagement." The *Post* story ended with a comment that must have stung Mead. Mead, as Commissioner of Reclamation after 1924, had tried to introduce many features of his California program into the federal projects. "The experience of California," the newspaper concluded, "is similar to that of the Federal Government on some of its ill-chosen reclamation projects."[11]

During World War I, Mead predicted that the land settlement ideas he had worked out in Australia would ultimately result in the investment of $300,000,000 in planned colonies and the addition of 250,000 farmers and their families to California's rural population. He had hoped that Durham and Delhi would serve as a model to private land developers. "State settlements should . . . be considered as an educational agency, as examples of correct methods and policies and as places where records can be kept of costs and results," he wrote in 1923. "The greater part of the area to be developed must be developed by private enterprise or by a different type of organization."[12]

Mead believed that the day of the land speculator and unplanned development had passed, that land colonization was the wave of the future. But his attempt to reconstitute rural America using irrigation was anachronistic and impractical. Private colonization had worked well on the Kings River near Fresno in the 1870s, and in southern California in the 1880s. It promoted intensive farming, the subdivision of large estates, a population boom, a dramatic increase in property values, and relatively homogeneous communities of small freeholders. But conditions had changed by the mid-1920s. From 1909 to 1919, an average of 28,000 acres of trees and vines were added to California's horticultural productivity

11. *Washington Post*, March 26, 1929.
12. Mead, "The Relation of Land Settlement to Irrigation Development."

each year. Then, from 1919 to 1925, an average 112,000 acres were planted each year. At the same time, the value of most fruits dropped sharply. From 1919 to 1923, the price of oranges fell by 35 percent, almonds and apples by 50 percent, raisins and peaches by about 70 percent, and apricots by nearly 80 percent. The value of land planted to fruit declined by half during the same years, and these figures were adjusted for postwar inflation. Frank T. Swett, president of California's Pear Growers' Association, commented in 1925: "The fruit industry, ultimately, will get over its present troubles, if the Bullfornia unscrupulous land peddlers, boomers, and poets and painters of rainbows will let it alone for a while." Such observers became prime critics of the expansion of irrigation.[13]

In 1925, Frank Adams noted that more than 400,000 acres under ditch within the state's irrigation districts remained to be settled and perhaps 1,000,000 acres in the state as a whole. Adams agreed with Mead that the $5,000 or more needed to buy and develop a 40-acre irrigated farm exceeded the resources of most prospective settlers. He concluded that "obtaining settlers is the most urgent need and not bringing more land under irrigation projects."[14] Nevertheless, Adams believed that the state's agricultural problems resulted as much from underconsumption or lack of adequate transportation as from over-production. He did not worry about the irrigation projects planned for the San Joaquin Valley; they would take years to complete and the nation's rapid increase in population would provide new markets in the future. California's population increased by 44 percent during the second decade of the twentieth century, and in 1923 the Division of Engineering and Irrigation predicted that by 1940 the demand for California farm products would triple. Moreover, even during the early 1920s, demand for vegetables increased—demonstrating that not all crops suffered uniformly—and irrigated land sold briskly

13. "Irrigation and Agriculture," *Transactions of the Commonwealth Club* (hereafter TCC), 20 (November 24, 1925): 346, chart on 348, 351. The Swett quote is from p. 351.

14. Frank Adams to John Gabbert, October 22, 1925, Frank Adams Collection, California Water Resources Archives. California's irrigated land increased from 1,466,000 acres in 1900, to 2,664,000 acres in 1910, to 4,219,000 acres in 1920. However, during the 1920s the increase was a more modest 528,000 acres, and in the following decade only 323,00 acres were added to the total. Then, from 1940 to 1950, the total mushroomed from 5,070,000 to 6,599,000 acres, as Shasta Dam and other units in the Central Valley Project began to operate.

in many parts of the state. The Commonwealth Club's Section on Irrigation observed that only 75 percent of the state's readily irrigable land was under ditch opposed to 87 percent in Colorado and more than 80 percent in both Utah and Idaho.[15]

Since they faced distinct competitive disadvantages, the question of whether California *needed* more irrigated family farms was largely academic anyway. For example, as California farmers tapped national and international markets, they recognized the value of flexibility, of being able to raise different crops from year to year to anticipate gluts. But different crops usually required new harvesting machines, which dramatically increased the cost of farming. Similarly, as the San Joaquin Valley's water table fell during the 1920s, the powerful pumps needed to draw water to the surface became more expensive to buy and operate. William L. Preston has shown that in Tulare County irrigation contributed to a substantial increase in ten- to forty-acre farms from 1895 to 1925, especially during the boom years of World War I. But, he concludes, "by 1926, fluctuating prices, increasingly expensive irrigation water, migrant-labor unrest, rising land taxes, and the growing profitability of machine harvested crops had tipped the scales in favor of large farmers once again." Small farmers usually held heavily mortgaged land which, in the words of another historian, "made for the eventual consolidation of farming lands in the hands of a few absentee landlords and receiverships like Trans-America Company of the Bank of America, who coordinated the bankrupt but productive farms, both large and small, into systems that have been aptly characterized as 'factories in the fields.'"[16]

Tulare County farms shrank in size from an average 460 acres in 1900 to 242 in 1910 to 159 in 1925 as irrigation agriculture supplanted dry farmed wheat and made land too valuable for free grazing. But by 1945, as intensively cultivated farms were consolidated, the trend reversed and the average farm size rose to 213 acres. Nor was the Tulare county experience unique. Statewide,

15. Frank Adams, "Are We Developing Our Irrigated Areas Too Rapidly?" *TCC* 20 (November 24, 1925): 375–388, 397–399; *Water Resources of California: A Report to the Legislature of 1923*, Department of Public Works, Division of Engineering and Irrigation Bulletin no. 4 (Sacramento, 1923), 17, 39, 45, 55.

16. William L. Preston, *Vanishing Landscapes: Land and Life in the Tulare Lake Basin* (Berkeley, 1981), 199; Beach, "The Transformation of California, 1900–1920," 127.

the average farm decreased from 397 acres in 1900 to 250 acres in 1920 to 224 acres in 1930. But by 1950 the average stood at 267 acres. The number and size of farms larger than 1,000 acres increased dramatically from 1920 to 1945. In 1920, the state contained 4,906 farms in this category covering 17,638,199 acres. By 1945, the number had increased to 5,939 farms encompassing 24,663,631 acres.[17]

The 1920s represented a turning point in the history of California agriculture. By then, few of the state's army of boosters spoke of using irrigation to restore the dominance of the yeoman farmer, provide homes for jobless city dwellers, or reinforce traditional American values. The West had entered a new era. Little public land remained, successful farming required much more knowledge as well as equipment, and the city offered new opportunities as well as temptations. The economic hard times of the 1920s and 1930s increased the gulf between landownership and the act of tilling the soil—a gulf symbolized by the plight of migrant farm workers. Both the Central Valley Project and State Water Project contributed to the growth of agriculture as a business and the disappearance of farming as a way of life. The C.V.P. allowed large farmers to strengthen their hold on the countryside in several ways. It provided interest-free loans under the Reclamation Act of 1902. Power sales, sales of water to towns and cities, and direct federal appropriations subsidized the cost of massive dams and canals. Until the late 1950s, little effort was made to enforce the 160-acre limitation on cheap federal irrigation water. Even then, this restriction could often be evaded. Most important, both the Central Valley Project and State Water Project of the 1960s and 1970s represented a shift away from local control as water bureaucracies in Washington and Sacramento assumed vast new powers. Local water agencies, including irrigation, storage, and water conservation districts, survived but lost much of their autonomy. And as more and more decisions concerning water resource development were made in Washington and Sacramento, those water users who were well organized and wealthy—able to maintain armies of lobbyists and mold public opinion—virtually dictated state and national water

17. Preston, *Vanishing Landscapes*, 170, 200; *United States Census of Agriculture: 1935*, vol. 1 (Washington, D.C., 1936), 944–947; Lawrence Jelinek, *Harvest Empire: A History of California Agriculture* (San Francisco, 1979), 63.

policies. The rise of agribusiness coincided not just with the decline of the family farm ideal, but also with the virtual disappearance of the hope that California agriculture could be built on relatively autonomous, middle-class rural communities.

Of course, demographic shifts also contributed to the rise of agribusiness and the decline of "rural California." In 1900, 47.7 percent of the state's population resided in communities of 2,500 or fewer. This number decreased to 38.2 percent in 1910, 32.1 percent in 1920, and 26.7 percent in 1930. Though the state's rural population increased from 708,233 in 1900 to 1,099,902 in 1920, migration from out of state accounted for less than half the increase. Put in a broader perspective, California's urban population grew by 89 percent from 1900 to 1910 while the rural population increased by only 28.4 percent. In the next decade, the growth rate was 58.5 percent urban and 21 percent rural. During the 1920s, the rate was 78.8 percent to 37.9 percent. In the same three decades, some farm states, including Illinois and Iowa, suffered *absolute* declines in rural population. But both these states had a proportionately larger farm population than California, which had always been far more urban than other western or midwestern states.[18]

California had contributed much to western economic development since statehood. It served as a laboratory where the engineers, publicists, and promoters of irrigation, such as William Ellsworth Smythe, George H. Maxwell, and Elwood Mead, tested their ideas. For better or worse, it provided a system of water law, the "California Doctrine," followed by many other states. It had introduced the multiple-use concept of water planning on the Sacramento River long before that idea found expression in the great western water projects of 1928 and after. And it provided the West with the irrigation district, an institution responsible for much of the region's agricultural growth in the early decades of the century. But urbanization, farm mechanization, the soaring price of land and water, and other trends could not be reversed. After 1930, irrigation became one of the foundation blocks of agribusiness. The dream of using it to reform California society was all but forgotten.

18. For an excellent overview of California's population patterns from 1850–1950, see Warren S. Thompson, *Growth and Changes in California's Population* (Los Angeles, 1955). The statistics cited are from pp. 11, 13, and 277.

Counties of California.

Rivers of California.

Canals, Aqueducts and Dams, 1930.

Bibliography

I. Bibliographies

Barnes, Cynthia, and Giefer, Gerald J. *Index to Periodical Literature on Aspects of Water in California.* Berkeley, 1963.

Davis, Richard C. *North American Forest History: A Guide to Archives and Manuscripts in the United States and Canada.* Santa Barbara, Calif., 1977.

Dodds, Gordon B. "Conservation and Reclamation in the Trans-Mississippi West: A Critical Bibliography." *Arizona and the West* 13(1971):143–171.

———. "The Historiography of American Conservation: Past and Prospects." *Pacific Northwest Quarterly* 56(1965):75–81.

Fahl, Ronald J. *North American Forest and Conservation History: A Bibliography.* Santa Barbara, Calif., 1977.

Lee, Lawrence B. *Reclaiming the Arid West: An Historiography and Guide.* Santa Barbara, Calif., 1980.

Orsi, Richard J. *A List of References for the History of Agriculture in California.* Davis, Calif., 1974.

II. Archival Materials

A. National Archives, Washington, D.C., and Federal Records Center, Suitland, Md.

Record Group 8: Records of the Bureau of Agricultural Engineering (Irrigation and Drainage Investigations Division), Suitland, Md.

457

Record Group 48: Records of the Department of the Interior.
Record Group 77: Records of the Office of Chief of Engineers, United States Army.
Record Group 115: Records of the Bureau of Reclamation.
B. California State Archives, Sacramento.
Washington Bartlett Papers.
Newton Booth Papers.
William Hammond Hall Files (1878–1881; 1882–1904).
Herminghaus v. Southern California Edison Company File.
Lux v. Haggin File.
George C. Perkins Papers.
C. California Department of Water Resources Archives: Records of the Department of Engineering and Irrigation, 1921–1933.

III. Manuscript Collections

Adams, Frank. Collection. University of California Water Resources Center Archives, Berkeley, Calif.
Bancroft, Hubert Howe. Collection. Bancroft Scrapbooks. Bancroft Library, University of California, Berkeley, Calif.
Bard, Thomas R. Collection. Huntington Library, San Marino, California.
Bidwell, John. Collection. California State Library, Sacramento.
Cole, Cornelius. Collection. University of California, Los Angeles, Research Library.
Davidson, George. Collection. Bancroft Library, Berkeley, Calif.
Davis, Arthur Powell. Collection. Hebard Library, University of Wyoming, Laramie, Wyo.
Grunsky, C. E. Collection. Bancroft Library, Berkeley, Calif.
Hall, William Hammond. Collection. California Historical Society, San Francisco.
Harding, Sidney T. Collection. California Water Resources Center Archives, Berkeley, Calif.
Hichborn, Franklin. Collection. University of California, Los Angeles, Research Library.
Hilgard, E. W. Family Papers. Bancroft Library, Berkeley, Calif.
Hyatt, Edward. Collection. California Water Resources Center Archives, Berkeley, Calif.

Johnson, Hiram. Collection. Bancroft Library, Berkeley, Calif.
Lippincott, Joseph Barlow. Collection. California Water Resources Center Archives, Berkeley, Calif.
Manson, Marsden. Collection. Bancroft Library, Berkeley, Calif.
Markham, H. H. Collection. Huntington Library, San Marino, Calif.
Marshall, Robert. Collection. Bancroft Library, Berkeley, Calif.
Marshall, Robert. Collection. California Water Resources Center Archives, Berkeley, Calif.
Mead, Elwood. Collection. California Water Resources Center Archives, Berkeley, Calif.
Means, Thomas H. Collection. California Water Resources Center Archives, Berkeley, Calif.
Newell, Frederick Haynes. Collection. Hebard Library, University of Wyoming, Laramie, Wyo.
Newell, Frederick Haynes. Collection. Library of Congress, Washington, D.C.
Newlands, Francis G. Collection. Sterling Library, Yale University, New Haven, Conn.
North, John W. Collection. Huntington Library, San Marino, Calif.
Pardee, George C. Collection. Bancroft Library, Berkeley, Calif.
Phelan, James D. Family Papers. Bancroft Library, Berkeley, Calif.
Pinchot, Gifford. Collection. Library of Congress, Washington, D.C.
Ralston, William. Collection. Bancroft Library, Berkeley, Calif.
Sharon, William. Family Papers. Bancroft Library, Berkeley, Calif.
Shorb, James De Barth. Collection. Huntington Library, San Marino, Calif.
Waterman, Robert W. Family Papers. Bancroft Library, Berkeley, Calif.
Wheeler, Benjamin Ide. Correspondence. University of California Archives, Bancroft Library, Berkeley, Calif.
Works, John D. Collection. Bancroft Library, Berkeley, Calif.

IV. Dissertations and Theses

Angel, Arthur T. "Political and Administrative Aspects of the Central Valley Project of California." Ph.D. dissertation, University of California, Los Angeles, 1944.
Beach, Frank. "The Transformation of California, 1900–1920: The

Effects of the Westward Movement on California's Growth and Development in the Progressive Period." Ph.D. dissertation, University of California, Berkeley, 1963.

Carothers, Alice L. "The History of the Southern Pacific Railroad in the San Joaquin Valley." M.A. thesis, University of Southern California, 1934.

Casey, Jack T. "Legislative History of the Central Valley Project, 1933–1945." Ph.D. dissertation, University of California, Berkeley, 1949.

Coate, Charles E. "Water, Power, and Politics in the Central Valley Project, 1933–1967." Ph.D. dissertation, University of California, Berkeley, 1969.

Cole, Chester F. "Rural Occupance Patterns in the Great Valley Portion of Fresno County, California." Ph.D. dissertation, University of Nebraska, 1950.

Davison, Stanley Roland. "The Leadership of the Reclamation Movement, 1875–1902." Ph.D. dissertation, University of California, Berkeley, 1951.

Fitzpatrick, John J. III. "Senator Hiram W. Johnson: A Life History, 1866–1945." Ph.D. dissertation, University of California, Berkeley, 1975.

Force, Edwin T. "The Use of the Colorado River in the United States, 1850–1933." Ph.D. dissertation, University of California, Berkeley, 1936.

Francis, Jessie Davies. "An Economic and Social History of Mexican California, 1822–1846." Ph.D. dissertation, University of California, Berkeley, 1935.

Gidney, Ray M. "The Wright Irrigation Act in California." M.A. thesis, University of California, Berkeley, 1912.

Hayes, William J. "The Influence of Public Policy on the Development of California Water Law." J.D. thesis, University of California, Berkeley, 1911.

Hinricksen, Kenneth C. "Pioneers in Southern California Hydro-Electric Water Power: Chaffey and Baldwin." M.A. thesis, University of California, Berkeley, 1949.

Houston, Flora B. "The Mormons in California, 1846–1857." M.A. thesis, University of California, Berkeley, 1929.

Israelsen, Orson W. "A Discussion of the Irrigation District Movement." M.S. thesis, University of California, Berkeley, 1914.

Jewell, Marion Nielsen. "Agricultural Development in Tulare

County, 1870–1900." M.A. thesis, University of Southern California, 1950.

Kluger, J. R. "Elwood Mead: Irrigation Engineer and Social Planner." Ph.D. dissertation, University of Arizona, 1970.

Kratka, Genevieve. "The Upper San Joaquin Valley, 1772–1870." M.A. thesis, University of Southern California, 1937.

Lawrence, William D. "Henry Miller and the San Joaquin Valley." M.A. thesis, University of California, Berkeley, 1933.

Lee, Beatrice Paxson. "The History and Development of the Ontario Colony." M.A. thesis, University of Southern California, 1929.

Loofbourow, James R. "Conflict in Jurisdiction Between the Federal Power Commission and State Water Commissions." J.D. thesis, University of California, Berkeley, 1926.

Malone, Thomas E. "The California Irrigation Crisis of 1886: Origins of the Wright Act." Ph.D. dissertation, Stanford University, 1965.

Marten, Effie E. "The Development of Wheat Culture in the San Joaquin Valley, 1846–1900." M.A. thesis, University of California, Berkeley, 1924.

Metcalf, Barbara A. "Oliver M. Wozencraft in California, 1849–1887." M.A. thesis, University of Southern California, 1963.

More, Rosemary McDonald. "The Influence of Water-Rights Litigation upon Irrigation Farming in Yolo County, California." M.A. thesis, University of California, Berkeley, 1960.

Mott, Orra A. "The History of the Imperial Valley." M.A. thesis, University of California, Berkeley, 1922.

Mulkey, Marion J. "The Division of Water Rights of the Board of Public Works: Forfeiture of Riparian Rights by Non-User." J.D. thesis, University of California, Berkeley, 1923.

Olson, Keith W. "Franklin K. Lane: A Biography." Ph.D. dissertation, University of Wisconsin, 1964.

Packard, Walter E. "The Development and Present Status of Irrigation in Kings and Tulare Counties." M.S. thesis, University of California, Berkeley, 1909.

Paterson, Alan M. "Rivers and Tides: The Story of Water Policy and Management in California's Sacramento-San Joaquin Delta, 1920–1977." Ph.D. dissertation, University of California, Davis, 1978.

Paule, Dorothea J. "The German Settlement at Anaheim." M.A. thesis, University of Southern California, 1952.

Pisani, Donald J. "Storm over the Sierra: A Study in Western Water Use." Ph.D. dissertation, University of California, Davis, 1975.

Quastler, Imre E. "American Images of California Agriculture, 1800–1890." Ph.D. dissertation, University of Kansas, 1971.

Rhodes, Benjamin F. "Thirsty Land: The Modesto Irrigation District, a Case Study of Irrigation Under the Wright Law." Ph.D. dissertation, University of California, Berkeley, 1943.

Staniford, Edward F. "Governor in the Middle: The Administration of George C. Pardee, Governor of California, 1903–1907." Ph.D. dissertation, University of California, Berkeley, 1955.

Strebel, George L. "Irrigation as a Factor in Western History, 1847–1890." Ph.D. dissertation, University of California, Berkeley, 1965.

Thickens, Virginia E. "Pioneer Colonies of Fresno County." M.A. thesis, University of California, Berkeley, 1939.

Wallace, Thomas D. "The Status of Irrigation Companies in California." J.D. thesis, University of California, Berkeley, 1917.

Walters, Donald E. "Populism in California, 1889–1900." Ph.D. dissertation, University of California, Berkeley, 1952.

Wheaton, Donald W. "The Political History of California, 1887–1898." Ph.D. dissertation, University of California, Berkeley, 1924.

Wright, Doris Marion. "A Yankee in Mexican California: Abel Stearns, 1798–1848." Ph.D. dissertation, Claremont Graduate School, 1954.

V. Oral Histories

Adams, Frank. Transcript. Bancroft Library, Berkeley, Calif.

Bartlett, Louis. Transcript. Bancroft Library, Berkeley, Calif.

Bensley, John. Dictation. Hubert Howe Bancroft Collection, Bancroft Library, Berkeley, Calif.

Church, Moses J. Dictation. Bancroft Library, Berkeley, Calif.

Downey, Stephen W. Transcript. Bancroft Library, Berkeley, Calif.

Durbrow, William. Transcript. Bancroft Library, Berkeley, Calif.

Harding, Sidney T. Transcript. Bancroft Library, Berkeley, Calif.

Lambert, Charles F. Transcript. Bancroft Library, Berkeley, Calif.

Lux, Charles. Dictation. Hubert Howe Bancroft Collection, Bancroft Library, Berkeley, Calif.

Mason, J. Rupert. Transcript. Bancroft Library, Berkeley, Calif.
Miller, Henry. Dictation. Hubert Howe Bancroft Collection, Bancroft Library, Berkeley, Calif.
Shorb, J. De Barth. Dictation. Hubert Howe Bancroft Collection, Bancroft Library, Berkeley, Calif.

VI. Published Government Documents: United States

Congressional Globe.
Congressional Record.
Congress, House. *Colorado Desert.* 37th Cong., 2d sess., 1862. H. Rept. 87. Serial 1145.
———— . *U.S. Board of Commissioners on the Irrigation of the San Joaquin, Tulare and Sacramento Valleys of the State of California.* 43d Cong., 1st sess., 1874. H. Ex. Doc. 290. Serial 1615.
———— . *Geographical Surveys West of the One Hundredth Meridian in California, Nevada, Utah, Colorado, Wyoming, New Mexico, Arizona, and Montana.* 44th Cong., 2d sess., 1876. H. Ex. Doc. 1, pt. 2. Serial 1745.
———— . *Ceding the Arid Lands to the States and Territories.* 51st Cong., 2d sess., 1891. H. Rept. 3767. Serial 2888.
———— . *Report on the Climate of California and Nevada with Particular Reference to Questions of Irrigation and Water Storage in the Arid Region.* 51st Cong., 2d sess., 1891. H. Ex. Doc 287. 1891. Serial 2868.
———— . *Report on Agriculture by Irrigation in the Western Part of the United States at the Eleventh Census, 1890.* By F. H. Newell. 52d Cong., 1st sess., 1896. H. Misc. Doc. 340. Serial 3021.
———— . *Preliminary Report on Examination of Reservoir Sites in Wyoming and Colorado.* By Hiram Martin Chittenden. 55th Cong., 2d sess., 1897. H. Doc. 141. Serial 3666.
———— . *Fund for Reclamation of Arid Lands.* 61st Cong., 3d sess., 1911. H. Doc. 1262. Serial 6022.
———— . *Central Valley Project Documents.* 84th Cong., 2d sess., 1956. H. Doc. 416. Serial 11931.
Congress. Senate. *Report [on the Colorado Desert].* 36th Cong., 1st sess., 1860. S. Rept. 276. Serial 1040.
———— . *Irrigation: Its Evils, the Remedies, and the Compensations.* 43d Cong., 1st sess., 1874. S. Misc. Doc. 55. Serial 1584.

————— . *Irrigation and Reclamation of Land for Agricultural Purposes as Now Practiced in India, Egypt, Italy, Etc.* 44th Cong., 1st sess., 1875. S. Ex. Doc. 94. Serial 1664.

————— . *Irrigation in the United States.* 49th Cong., 2d sess., 1887. S. Misc. Doc. 15. Serial 2450.

————— . *Report of the Special [Stewart] Committee of the United States Senate on the Irrigation and Reclamation of Arid Lands.* 51st Cong., 1st sess., 1890. S. Rept. 928. Serial 2707.

————— . *Irrigation in the United States: Progress Report for 1890.* 51st Cong., 2d sess., 1890. S. Ex. Doc. 53. Serial 2818.

————— . *Report on Irrigation, 1891.* 52d Cong., 1st sess., 1892. S. Ex. Doc. 41, pt. 1. Serial 2899.

————— . *Cession of the Public Lands, etc.* 55th Cong., 1st sess., 1897. S. Doc. 130. Serial 3562.

————— . *Surveys of Reservoir Sites.* 55th Cong., 3d sess., 1899. S. Doc. 116. 1899. Serial 3735.

————— . *Irrigation Investigations in California.* 56th Cong., 2d sess., 1901. S. Doc. 108. Serial 4033.

————— . *Water From Sacramento River, California, For Irrigation Purposes.* 59th Cong., 1st sess., 1906. S. Rept. 2900. Serial 4905.

————— . *Report of the National Conservation Commission.* 60th Cong., 2d sess., 1909. S. Doc. 676. Serial 5398.

————— . *Report of the Committee on Irrigation and Reclamation of Arid Lands on the Investigation of Irrigation Projects.* 61st Cong., 3d sess., 1911. S. Rept. 1281. Serial 5846.

————— . *Problems of Imperial Valley and Vicinity.* 67th Cong., 2d sess., 1922. S. Doc. 142. Serial 7977.

Department of Agriculture. *Annual Reports* of the Commissioner and Secretary.

————— . *Irrigation in Arizona.* By R. H. Forbes. Office of Experiment Stations Bulletin no. 235. Washington, D.C., 1911.

————— . *Irrigation in California.* By F. W. Roeding. Office of Experiment Stations Bulletin no. 237. Washington, D.C., 1911.

————— . *Irrigation Districts: Their Organization, Operation, and Financing.* By Wells A. Hutchins. Technical Bulletin no. 254. Washington, D.C., 1931.

————— . *Irrigation in Fruit Growing.* By E. J. Wickson. Farmers Bulletin no. 116. Washington, D.C. 1900.

————— . *Irrigation Resources of California and Their Utilization.* By

Frank Adams. Office of Experiment Stations Bulletin no. 254. Washington D.C., 1913.

————— . *Irrigation in the San Joaquin Valley, California.* By Victor Cone. Office of Experiment Stations Bulletin no. 239. Washington, D.C., 1911.

————— . *Mutual Water Companies.* By Wells A. Hutchins. Technical Bulletin no. 82. Washington, D.C., 1929.

————— . *Report on the Climatic and Agricultural Features and the Agricultural Practice and Needs of the Arid Regions of the Pacific Slope, with Notes on Arizona and New Mexico.* By E. W. Hilgard, T. C. Jones, and R. W. Furnas. Report no. 20. Washington, D.C., 1882.

————— . *Report of Irrigation Investigations in California.* Office of Experiment Stations Bulletin no. 100. Washington, D.C., 1902.

————— . *Report of Irrigation Investigations, 1900.* Office of Experiment Stations Bulletin no. 104. Washington, D.C., 1902.

————— . *Report of Irrigation Investigations, 1901.* Office of Experiment Stations Bulletin no. 119. Washington, D.C., 1902.

————— . *Report of Irrigation Investigations, 1902.* Office of Experiment Stations Bulletin no. 133. Washington, D.C., 1903.

————— . *Summary of Irrigation-District Statutes of Western States.* By Wells A. Hutchins. Miscellaneous Publication no. 103. Washington, D.C., 1931.

Department of Commerce. Bureau of the Census. *Censuses of the United States.*

————— . *Irrigation: California* (Washington, D.C., 1920).

————— . *Irrigation of Agricultural Lands, 1930* (Washington, D.C., 1932).

————— . *Statistical Abstracts of the United States.*

Department of the Interior. *Annual Reports* of the Secretary.

————— . General Land Office. *Annual Reports.*

————— . Geological Survey. *Annual Reports.*

————— . Patent Office. *Annual Reports* of the Commissioner.

————— . Reclamation Bureau. *Annual Reports.*

————— . *California Hydrography.* By J. B. Lippincott. Geological Survey Water Supply and Irrigation Paper no. 81. Washington, D.C., 1903.

————— . *Development and Application of Water Near San Bernardino, Colton, and Riverside, California.* By J. B. Lippincott. Geological

Survey Water Supply and Irrigation Papers nos. 59 and 60. Washington, D.C., 1902.

————— . *Gazeteer of Surface Waters of California.* By B. D. Wood. Geological Survey Water Supply and Irrigation Papers no. 295, 296, and 297. Washington, D.C., 1912.

————— . *Irrigation Near Bakersfield, California.* By C. E. Grunsky. Geological Survey Water Supply Paper no. 17. Washington, D.C., 1898.

————— . *Irrigation Near Fresno, California.* By C. E. Grunsky. Geological Survey Water Supply and Irrigation Paper no. 18. Washington, D.C. 1898.

————— . *Irrigation Near Merced, California.* By C. E. Grunsky. Geological Survey Water Supply and Irrigation Paper no. 19. Washington, D.C., 1899.

————— . *Report on the Iron Canyon Project.* Washington, D.C., 1915.

————— . *Report on the Iron Canyon Project, California.* By Homer J. Gault and W. F. McClure. Washington, D.C., 1921.

————— . *Storage of Water on Kings River, California.* By J. B. Lippincott. Geological Survey Water Supply and Irrigation Paper no. 58. Washington, D.C., 1902.

————— . *The Transportation of Debris by Running Water.* By G. K. Gilbert. Geological Survey Professional Paper no. 86. Washington, D.C., 1914.

————— . *Water Resources of California: Stream Measurements in the Great Basin and Pacific Coast River Basins.* By H. D. McGlashan and F. F. Henshaw. Geological Survey Water Supply and Irrigation Paper no. 300. Washington, D.C., 1913.

————— . *Water Resources of California: Stream Measurement in Sacramento River Basin.* By H. D. McGlashan and F. F. Henshaw. Geological Survey Water Supply and Irrigation Paper no. 298. Washington, D.C., 1912.

————— . *Water Resources of California: Stream Measurements in San Joaquin River Basin.* By H. D. McGlashan and F. F. Henshaw. Geological Survey Water Supply and Irrigation Paper no. 299. Washington, D.C., 1912.

————— . *Water Storage on Cache Creek, California.* By Alfred E. Chandler. Geological Survey Water Supply and Irrigation Paper no. 45. Washington, D.C., 1901.

VII. Published Government Documents: California

Agricultural Society. *Transactions.*
Board of Agriculture. *Reports.*
Board of Directors of Drainage District No. 1. *Reports.*
Board of Equalization. *Reports.*
Board of Forestry. *Reports.*
Board of Health. *Reports.*
Board of Horticulture. *Reports.*
Bureau of Labor Statistics. *Reports.*
California Constitutional Convention of 1879–1880. *Debates and Proceedings of the Constitutional Convention of the State of California Convened at the City of Sacramento, Saturday, September 28, 1879.* 3 vols. Sacramento, 1880.
Commission on Land Colonization and Rural Credits. *Reports.*
Conservation Commission. *Report* (1913).
Department of Engineering. *Reports.*
Department of Public Works. Division of Engineering and Irrigation. *Reports.*
———— . *California Irrigation District Laws, as Amended, 1921.* Bulletin no 1. Sacramento, 1922.
———— . *The Control of Floods by Reservoirs.* Bulletin no. 14. Sacramento, 1928.
———— . *Coordinated Plan of Water Development in the Sacramento Valley.* Bulletin no. 15. Sacramento, 1928.
———— . *Coordinated Plan of Water Development in the San Joaquin Valley.* Bulletin no. 16. Sacramento, 1928.
———— . *Coordinated Plan of Water Development in Southern California.* Bulletin no. 17. Sacramento, 1928.
———— . *The Development of the Upper Sacramento River.* Bulletin no. 13. Sacramento, 1928.
———— . *Flow in California Streams.* Bulletin no. 5. Sacramento, 1923.
———— . *Ground Water Resources of the Southern San Joaquin Valley.* By S. T. Harding. Bulletin no. 11. Sacramento, 1927.
———— . *Industrial Survey of Upper San Francisco Bay Area with Special Reference to a Salt Water Barrier.* Sacramento, 1932.

————— . *Irrigation Districts in California, 1887–1915.* By Frank Adams. Bulletin no. 2. Sacramento, 1917.

————— . *Irrigation Districts in California.* By Frank Adams. Bulletin no. 21. Sacramento, 1930.

————— . *Irrigation Requirements of California Lands.* Bulletin no. 6. Sacramento, 1923.

————— . *A Proposed Major Development on American River.* By A. D. Edmonston. Bulletin no. 24. Sacramento, 1930.

————— . *Report on Kennett Reservoir Development: An Analysis of Methods and Extent of Financing by Electric Power Revenue.* By Lester S. Ready. Bulletin no. 20. Sacramento, 1929.

————— . *Report to the Legislature of 1931 on State Water Plan, 1930.* Bulletin no. 30. Sacramento, 1931.

————— . *Report on Salt Water Barrier.* By Walker R. Young. Bulletin no. 22. Sacramento, 1929.

————— . *Summary Report of the Water Resources of California and a Coordinated Plan for their Development.* Bulletin no. 12. Sacramento, 1927.

————— . *Supplemental Report on Water Resources of California.* Bulletin no. 9. Sacramento, 1925.

————— . *Water Resources of California: A Report to the Legislature of 1923.* Bulletin no. 4. Sacramento, 1923.

————— . *Water Resources of Tulare County and their Utilization, 1922.* Bulletin no. 3. Sacramento, 1922.

————— . Division of Water Rights. *Reports.*

————— . *Proceedings of the Sacramento River Problems Conference Held under the Auspices of the Sacramento Chamber of Commerce and the Division of Water Rights, State Department of Public Works, at Sacramento, California on January 25 and 26, 1924.* Sacramento, 1924.

————— . *Proceedings of the Second Sacramento-San Joaquin River Problems Conference and Water Supervisors Report, 1924.* Sacramento, 1925.

Senate and Assembly *Journals of the California Assembly and Senate*

————— . *Appendixes to the Journals.*

————— . *Assembly and Senate Bills, 1850–1930.* California State Law Library, Sacramento, Calif.

————— . *California Civil Code, 1872.*

_____ . *California Statutes,* 1850–1930.

_____ . *Majority Report of Special Committee to Whom was Referred Assembly Bill no. 84.* 7th sess. Sacramento, 1856.

_____ . *Report of Committee on Internal Improvements of the Assembly on Assembly Bill no. 321.* 16th sess. Sacramento, 1866.

_____ . *Report of the Senate Committee on Assembly Bill no. 321.* 16th sess. Sacramento, 1866.

_____ . *Report of the Engineer of the Sacramento Valley Irrigation and Navigation Canal.* 17th sess. Sacramento, 1868.

_____ . *Report of the Commissioner Appointed to Examine into the Practicability of Making a New Outlet for the Flood Waters of the Sacramento Valley.* 18th sess. Sacramento, 1870.

_____ . *Memorial Asking Aid from Congress for Irrigating Purposes.* 19th sess. Sacramento, 1872.

_____ . *Memorial and Report of the California Immigrant Union to the Legislature of the State of California.* 19th sess. Sacramento, 1872.

_____ . *Report of the Special Committee on Land Monopoly.* 20th sess. Sacramento, 1974.

_____ . *Reports of the Joint Committees of the Assembly on Mines and Mining Interests, and Agriculture, Relative to the Injury now being done to Lands and Streams in this State by the Deposit of Detritus from the Gravel Mines.* 21st sess. Sacramento, 1876.

_____ . *Majority Report of the [Assembly] Committee on Mining Debris.* 22d sess. Sacramento, 1878.

_____ . *Annual Message of George C. Perkins to the Legislature, January, 1881.* 24th sess. Sacramento, 1881.

_____ . *Biennial Message of George C. Perkins.* 25th sess. Sacramento, 1883.

_____ . *Report of the Assembly Committee on Claims, Twenty-Fifth Session, on Assembly Bill no. 207.* 25th sess. Sacramento, 1885.

_____ . *First Biennial Message of Governor George Stoneman, January 5, 1885.* 26th sess. Sacramento, 1885.

_____ . *Inaugural Address of Governor H. H. Markham, Delivered January 8, 1891.* 29th sess. Sacramento, 1891.

_____ . *First Biennial Message of Governor James H. Budd to the Legislature of the State of California, 1897.* 32d sess. Sacramento, 1897.

_____ . *Second Biennial Message of Governor James H. Budd to the*

Legislature of the State of California, 1899. 33d sess. Sacramento, 1899.

————— . *Statistical Summary of the Production and Resources of California, 1911.* 40th sess. Sacramento, 1913.

————— . *Report of the State Water Problems Conference [of 1916].* 42d sess. Sacramento, 1917.

————— . *Report of the Joint Committee of the Senate and Assembly Dealing with the Water Problems of the State, January 18, 1929.* 48th sess. Sacramento, 1929.

————— . *Report of the California Joint Federal-State Water Resources Commission, 1930.* 49th sess. Sacramento, 1932.

Railroad Commission. *Reports.*

State Controller. *Reports.*

State Engineer. *Reports.*

————— . *The Irrigation Question in California: Appendix to the Report of the State Engineer to His Excellency George C. Perkins, Governor of California.* By William Hammond Hall. Sacramento, 1881.

————— . *Irrigation Development: History, Customs, Laws, and Administrative Systems Relating to Irrigation, Water-Courses, and Waters in France, Italy, and Spain.* By William Hammond Hall. Sacramento, 1886.

————— . *Irrigation in [Southern] California.* By William Hammond Hall. Sacramento, 1888.

————— . *Memorandum Concerning the Improvement of the Sacramento River Addressed to James B. Eads and George H. Mendell.* By William Hammond Hall. Sacramento, 1880.

State Mineralogist. *Reports.*

State Surveyor-General. *Reports.*

University of California, College of Agriculture. *Reports to the Board of Regents.*

————— . *Alkali Lands, Irrigation and Drainage in their Mutual Relations.* By E. W. Hilgard. Sacramento, 1886.

————— . *Report on the Physical and Agricultural Features of the State of California, With a Discussion of the Present and Future of Cotton Production in the State.* By E. W. Hilgard. Sacramento, 1884.

————— . *Selected List of References Relating to Irrigation in California.* Agricultural Experiment Station Circular no. 260. Sacramento, 1923.

Water Commission. *Reports.*

VIII. Court Cases

Anaheim Water Company v. Semi-Tropic Water Company. 64 Cal.
    185 (1883).
Antioch v. Williams Irrigation District. 188 Cal. 451 (1922).
Benton v. Johncox. 17 Wash. 277 (1897).
Boggs v. Merced Mining Co. 14 Cal. 279 (1859).
Bradley v. Fallbrook Irrigation District. 68 Fed. 948 (1895).
Butte Canal and Ditch Company v. Vaughn. 11 Cal. 143 (1858).
Central Irrigation District v. De Lappe. 79 Cal. 351 (1889).
City of Los Angeles v. Baldwin. 53 Cal. 469 (1879).
City of Los Angeles v. Pomeroy. 124 Cal. 597 (1899).
Conger v. Weaver. 6 Cal. 548 (1856).
Crall v. Poso Irrigation District. 87 Cal. 140 (1890).
Crandall v. Woods. 8 Cal. 136 (1857).
Crawford v. Hathaway. 67 Neb. 365 (1903).
Creighton v. Evans. 53 Cal. 55 (1878).
Eddy v. Simpson. 3 Cal. 249 (1853).
Elms v. City of Los Angeles. 58 Cal. 80 (1881).
Fallbrook Irrigation District v. Bradley. 164 U.S. 112 (1896).
Feliz v. City of Los Angeles. 58 Cal. 73 (1881).
Ferrea v. Knipe. 28 Cal. 340 (1865).
Herminghaus v. Southern California Edison Company. 200 Cal. 81
    (1926.)
Hill v. King. 8 Cal. 336 (1857).
Hill v. Newman. 5 Cal. 445 (1855).
Hill v. Smith. 27 Cal. 476 (1865).
Irwin v. Phillips. 5 Cal 140 (1855).
Katz v. Walkinshaw. 141 Cal. 116 (1903).
Kelly v. Natoma Water Company. 6 Cal 105 (1856).
Lux v. Haggin. 69 Cal. 255 (1886); 4 Pac. 919 (1884); 10 Pac. 674
    (1886).
Maeris v. Bicknell. 7 Cal. 261 (1857).
In the Matter of the Bonds of the Madera Irrigation District. 92 Cal.
    296 (1891).
In the Matter of the Organization and Bonds of the Central Irri-
    gation District 117 Cal. 382 (1897).
Miller & Lux v. Enterprise Canal and Irrigation Company. 47 Cal.
    Dec. 1 (1913).

Miller & Lux v. Madera Canal and Irrigation Company. 155 Cal. 59 (1909).

Ortman v. Dixon. 13 Cal. 33 (1859).

Palmer v. Railroad Commission. 167 Cal. 163 (1914).

Pope v. Kinman. 54 Cal 3 (1879).

Rupley v. Welch. 23 Cal. 452 (1863).

Tartar v. Spring Creek Water and Mining Company. 5 Cal 395 (1855).

Thompson v. Lee. 8 Cal. 275 (1857).

Thorp v. Freed. 1 Mont. 651 (1872).

Tulare Irrigation District v. Strathmore Irrigation District. 3 Cal. (2d) 489 (1935).

Tulare Water Company v. State Water Commission. 187 Cal. 533 (1921).

Turlock Irrigation District v. Williams. 76 Cal. 360 (1888).

IX. Newspapers and Periodicals

American Agriculturalist (New York City)
Bakersfield Californian
California Advocate (San Francisco)
California Cultivator (Los Angeles)
California Culturalist (San Francisco)
California Development Board. *Annual Reports.*
California Farmer (San Francisco)
California Granger
California Mail Bag (San Francisco)
California State Board of Trade. *Annual Reports.*
Central Pacific Railroad Company. *Annual Reports.*
Chico Record
Citrograph (Redlands, Calif.)
Country Gentleman (Philadelphia)
DeBow's Commercial Review (New Orleans)
Engineering News-Record (Chicago)
Fall River Tidings (Shasta, Calif.)
Forest and Stream (New York City)
Forestry and Irrigation (Washington, D.C.)
Fresno Bee

Fresno Daily Republican
Fresno Expositor
Harper's Magazine (New York City)
Humboldt Times (Eureka, Calif.)
International Irrigation Congresses, *Proceedings.*
Irrigation Age (Chicago, Salt Lake City)
Kern County Weekly Courier (Bakersfield)
Land of Sunshine (Los Angeles)
Lodi Sentinel
Los Angeles Daily Commercial
Los Angeles Herald
Los Angeles Star
Los Angeles Times
Los Angeles Evening Express
Martinez *Daily Gazette*
Martinez Standard
Marysville Daily Appeal
Maxwell's Talisman (Chicago, Washington, New Orleans)
Mining and Scientific Press (San Francisco)
Modesto Morning Herald
Modesto Tribune
Mountain Messenger (Downieville, Calif.)
National Advocate (San Francisco)
New York *World*
Oakland *Enquirer*
Orland Register
Oakland Tribune
Oroville Register
Out West (Los Angeles)
Overland Monthly (San Francisco)
Pacific Rural Press (San Francisco)
Philadelphia *North American*
Pittsburgh Dispatch (Pittsburgh, Calif.)
Pittsburgh Post (Pittsburgh, Calif.)
Red Bluff Morning Times
Riverside Daily Press
Rural Californian (Los Angeles)
Sacramento Bee
Sacramento Daily Union

San Diego Union
San Francisco *Alta California*
San Francisco *Argonaut*
San Francisco *Bulletin*
San Francisco *Call*
San Francisco Chronicle
San Francisco *Daily Evening News*
San Francisco *Evening Post*
San Francisco *Examiner*
San Francisco Pacific Builder
San Jose Herald
San Jose Tribune
Santa Barbara Press
Scientific American (New York City)
Southern Pacific Railroad Company. *Annual Reports.*
Stanislaus County News (Modesto, Calif.)
Stockton Daily Independent
Stockton *Daily Record*
Stockton Democrat
Sunset
Transactions of the Commonwealth Club of California (San Francisco)
Tulare County Times
Turlock Weekly Journal
Visalia Weekly Delta
Washington Post
Water and Forest (San Francisco)
Weekly Butte Record (Chico, Calif.)
Weekly Colusa Sun

X. Articles

Adams, Edward F. "California Irrigation District Bonds." *Sunset* 27 (1911): 324–327.

Adams, Frank. "The Historical Background of California Agriculture." In *California Agriculture,* ed. Claude B. Hutchison, 1–50. Berkeley, 1946.

———. "Irrigation Development Through Irrigation Districts."

American Society of Civil Engineers Transactions 90 (1927): 773–790.

————— . "Pending Irrigation and Water Legislation." *Pacific Rural Press* 105 (1923): 388.

————— . "The Water Situation in California." *California Cultivator* 70 (1928): 93, 107.

————— . "The Water Supply of Sacramento Valley." *California Cultivator* 66 (1926): 125.

Alexander, Thomas G. "John Wesley Powell, the Irrigation Survey, and the Inauguration of the Second Phase of Irrigation Development in Utah." *Utah Historical Quarterly* 37 (1969): 190–206.

————— . "The Powell Irrigation Survey and the People of the Mountain West." *Journal of the West* 7 (1968): 48–53.

Allen, R. H. "The Spanish Land-Grant System as an Influence in the Agricultural Development of California." *Agricultural History* 9 (1935): 127–142.

Anderson, Henry S. "The 'Little Landers' Land Colonies: A Unique Agricultural Experiment in California." *Agricultural History* 5 (1931): 139–150.

Arrington, Leonard J., and May, Dean. "'A Different Mode of Life': Irrigation and Society in Nineteenth-Century Utah." *Agricultural History* 49 (1975): 3–20.

Bakken, Gordon M. "The English Common Law in the Rocky Mountain West." *Arizona and the West* 2 (1969): 109–128.

Barker, Charles A. "Henry George and the California Background of Progress and Poverty." *California Historical Quarterly* 24 (1945): 97–115.

Barrows, H. D. "Water for Domestic Purposes versus Water for Irrigation." *Annual Publication of the Historical Society of Southern California, 1911* 8 (1911): 208–210.

Bateman, Richard Dale. "Anaheim Was an Oasis in a Wilderness." *Journal of the West* 4 (1965): 1–20.

————— . "Gospel Swamp . . . The Land of Hog and Hominy." *Journal of the West* 4 (1965): 231–257.

Bean, Walton. "Ideas of Reform in California." *California Historical Quarterly* 51 (1972): 213–226.

————— . "James Warren and the Beginnings of Agricultural Institutions in California." *Pacific Historical Review* 13 (1944): 361–375.

Beattie, George William. "San Bernardino Valley before the Americans Came." *California Historical Quarterly* 12 (1933): 111–124.

Bennett, John E. "The District Irrigation Movement in California." *Overland Monthly* 19 (1897): 252–253.

Bennett, M. K. "Climate and Agriculture in California." *Economic Geography* 15 (1939): 153–164.

Bernstein, Harry. "Spanish Influence in the United States: Economic Aspects." *The Hispanic American Historical Review* 18 (1938): 46–65.

Billington, Ray A. "The Origin of the Land Speculator as a Frontier Type." *Agricultural History* 19 (1945): 204–212.

Bishop, William H. "Southern California." *Harper's Magazine* 65 (1882): 713–728, 863–882.

Bogue, Allan G. and Bogue, Margaret B. "'Profits' and the Frontier Land Speculator." *Journal of Economic History* 17 (1957): 1–24.

Bowers, Nathan A. "California's Land Settlement at Delhi." *Engineering News-Record* 88 (1925): 143–145.

Browne, J. Ross. "Agricultural Capacity of California: Overflows and Droughts." *Overland Monthly* 10 (1873): 297–314.

Burcham, L. T. "Cattle and Range Forage in California: 1770–1880." *Agricultural History* 35 (1961): 140–149.

Callow, Alexander, Jr. "San Francisco's Blind Boss." *Pacific Historical Review* 25 (1956): 261–280.

Carlson, Martin E. "William E. Smythe: Irrigation Crusader." *Journal of the West* 7 (1968): 41–47.

Carosso, Vincent P. "Anaheim: A Nineteenth Century Experiment in Commerical Viniculture." *Bulletin of the Business Historical Society* 23 (1949): 78–86.

Carter, Charles F., trans. "Duhaut-Cilly's Account of California in the Years 1827–1828." *California Historical Quarterly* 8 (1929): 306–336.

Caughey, John W. "The Californian and His Environment." *California Historical Quarterly* 51 (1972): 195–204.

Chandler, Alfred E. "Appropriation of Water in California." *California Law Review* 4 (1916): 206–215.

―――― . "The 'Water Bill' Proposed by the Conservation Commission of California." *California Law Review* 1 (1913): 148–168.

Chittenden, Hiram Martin. "Government Construction of Reservoirs in Arid Regions." *North American Review* 174 (1902): 245–258.

Clements, Kendrick A. "Politics and the Park: San Francisco's Fight for Hetch Hetchy, 1908–1913." *Pacific Historical Review* 48 (1979): 185–216.

Clifford, Josephine. "Tropical California." *Overland Monthly* 7 (1871): 297–312.

Conkin, Paul K. "The Vision of Elwood Mead." *Agricultural History* 34 (1960): 88–97.

Cooke, W. Henry. "The Controversy over Water Rights in the Santa Margarita River." *Pacific Historical Review* 25 (1956): 1–28.

Crafts, H. A. "Back to the Land: California Pioneers the State Land Settlement Plan." *Scientific American* 121 (1919): 185.

Davidson, George. "The Application of Irrigation to California." In *Contemporary Biography of California's Representative Men*, by Alonzo Phelps. San Francisco, 1881.

Donohue, Joan Marie. "Agostin Haraszthy: A Study in Creativity." *California Historical Quarterly* 48 (1969): 153–163.

Dunbar, Robert G. "History of Agriculture." In *Colorado and Its People*, edited by LeRoy Hafen, vol. 2:120–157. New York, 1948.

————— . "The Origins of the Colorado System of Water Right Control." *Colorado Magazine* 27 (1950): 241–262.

————— . "Pioneering Groundwater Legislation in the United States: Mortgages, Land Banks, and Institution-Building in New Mexico." *Pacific Historical Review* 47 (November 1978): 565–584.

————— . "The Significance of the Colorado Agricultural Frontier." *Agricultural History* 34 (1960): 119–126.

————— . "Water Conflicts and Controls in Colorado." *Agricultural History* 22 (1948): 180–186.

Durst, John H. "Riparian Rights from Another Standpoint." *Overland Monthly*, 2d ser., 6 (1885): 10–14.

Erdman, H. E. "The Development and Significance of California Cooperatives, 1900–1915." *Agricultural History* 32 (1958): 179–184.

Evans, T. "Orange Culture in California." *Overland Monthly* 12 (1874): 235–244.

Foley, Doris, and Morley, S. Griswold. "The 1883 Flood on the Middle Yuba River." *California Historical Quarterly* 28 (1949): 233–242.

Forbes, Jack D. "Indian Horticulture West and Northwest of the Colorado River." *Journal of the West* 2 (1963): 1–14.

Fortier, Samuel. "Irrigated Agriculture Dominant in California." *Pacific Rural Press* 76 (1908): 49, 55.

Ganoe, John T. "The Beginnings of Irrigation in the United States." *Mississippi Valley Historical Review* 25 (1938): 59–78.

———. "The Desert Land Act in Operation, 1877–1891." *Agricultural History* 11 (1937): 142–157.

———. "The Desert Land Act Since 1891." *Agricultural History* 11 (1937): 266–277.

———. "The Origin of a National Reclamation Policy." *Mississippi Valley Historical Review* 28 (1931): 34–52.

Gates, Paul W. "Adjudication of Spanish-Mexican Land Claims in California." *Huntington Library Quarterly* 21 (1958): 213–236.

———. "California's Agricultural College Lands." *Pacific Historical Review* 30 (1961): 103–122.

———. "California's Embattled Settlers." *California Historical Quarterly* 41 (1962): 99–130.

———. "The California Land Act of 1851." *California Historical Quarterly* 50 (1971): 395–430.

———. "Pre-Henry George Land Warfare in California." *California Historical Quarterly* 46 (1967): 121–148.

———. "Public Land Disposal in California." *Agricultural History* 49 (1975): 158–178.

———. "The Role of the Land Speculator in Western Development." *Pennsylvania Magazine of History and Biography* 64 (1942): 314–333.

———. "The Suscol Principle, Preemption, and California Latifundia." *Pacific Historical Review* 39 (1970): 453–471.

Gentilcore, R. Louis. "Ontario, California and the Agricultural Boom of the 1880s." *Agricultural History* 34 (1960): 77–87.

Gopalakrishnan, Chennat. "The Doctrine of Prior Appropriation and Its Impact on Water Development: A Critical Survey." *American Journal of Economics and Sociology* 32 (1973): 61–72.

Greenleaf, Richard E. "Land and Water in Mexico and New Mexico."

New Mexico Historical Review 47 (1972): 85–112.

Gressley, Gene M. "Arthur Powell Davis, Reclamation and the West." *Agricultural History* 42 (1968): 241–257.

Griffiths, David B. "Anti-Monopoly Movements in California, 1873–1898." *Southern California Quarterly* 52 (1970): 93–121.

Griswold, F. H. "Home-builders Displace Bonanza Farmers." *Sunset* 26 (1911): 578–579.

Guest, Francis F. "Municipal Government in Spanish California." *California Historical Quarterly* 47 (1967): 307–336.

Guinn, J. M. "The Great Real Estate Boom of 1887." *Annual Publication of the Historical Society of Southern California, 1890* 1 (1890): 13–21.

———. "A History of California Floods and Drought." *Annual Publication of the Historical Society of Southern California, 1890* 1 (1890): 33–39.

———. "The Passing of the Cattle Barons of California." *Annual Publication of the Historical Society of Southern California, 1909–1910* 8 (1909–1910): 51–60.

———. "The Passing of the Rancho." *Annual Publication of the Historical Society of Southern California* 10 (1915–1916): 46–53.

Haight, George W. "Riparian Rights." *Overland Monthly*, 2d ser., 5 (1885): 561–569.

Hall, Sharlot M. "The Problem of the Colorado River." *Out West* 25 (1906): 305–322.

Harding, Sidney T. "Background of California Water and Power Problems." *California Law Review* 38 (1950): 547–571.

Hardy, Osgood. "Agricultural Changes in California, 1860–1900." *Proceedings of the American Historical Association, Pacific Coast Branch, 1929*, 216–230. Eugene, Oreg., 1930.

Hicks, John D. "California in History." *California Historical Quarterly* 24 (1945): 7–16.

Hoffman, Abraham. "Joseph Barlow Lippincott and the Owens Valley Controversy: Time for Revision." *Southern California Quarterly* 54 (1972): 239–254.

Holt, L. M. "How the Reclamation Service is Robbing the Settler." *Overland Monthly* 50 (1907): 510–512.

———. "Our Irrigation Law: Its Defects and Remedies." *Rural Californian* 17 (1894): 631–632; 18 (1895): 432–433.

House, Albert V., Jr. "Proposals of Government Aid to Agricultural Settlement During the Depression of 1873–1879." *Agricultural History* 12 (1938): 46–66.

Hudanick, Andrew, Jr. "George Hebard Maxwell: Reclamation's Militant Evangelist." *Journal of the West* 14 (1975): 108–119.

Hundley, Norris, Jr. "Clio Nods: *Arizona* v. *California* and the Boulder Canyon Act: A Reassessment." *Western Historical Quarterly* 3 (1972): 17–51.

————. "The Politics of Reclamation: California, the Federal Government, and the Origins of the Boulder Canyon Act: A Second Look." *California Historical Quarterly* 52 (1973): 292–325.

————. "The Politics of Water and Geography: California and the Mexican-American Treaty of 1944." *Pacific Historical Review* 36 (1967): 209–226.

Hutchins, Wells. "The Community Acequia: Its Origin and Development." *The Southwestern Historical Quarterly* 31 (1928): 261–284.

Hutchins, Wells A., and Steele, Harry A. "Basic Water Rights Doctrines and Their Implications for River Basin Development." *Law and Contemporary Problems* 22 (1957): 276–300.

Hyatt, Edward. "Control of Appropriations of Water in California." *Journal of the American Water Works Association* 13 (1925): 125–144.

Irving, W. G. "Water Rights in California." *American Water Works Association Journal* 23 (1931): 1089–1109.

Iwata, Masakazu. "The Japanese Immigrants in California Agriculture." *Agricultural History* 36 (1962): 25–37.

Jensen, James M. "Cattle Drives from the Ranchos to the Gold Fields of California." *Arizona and the West* 2 (1960): 341–352.

Johnson, Hiram. "The Boulder Canyon Project." *Annals of the American Academy of Political and Social Science* 135 (1928): 150–156.

Johnson, Kenneth M. "Progress and Poverty—A Paradox." *California Historical Quarterly* 42 (1963): 27–32.

Jones, William K. "Los Angeles Aqueduct: A Search for Water." *Journal of the West* 16 (1977): 5–21.

Kahrl, William. "The Politics of California Water: Owens Valley and Los Angeles Aqueduct, 1900–1927." *California Historical Quarterly* 55 (1976): 2–25, 98–120.

Kauer, Ralph. "The Workingmen's Party of California." *Pacific Historical Review* 13 (1944): 278–291.

Kelley, Robert. "Taming the Sacramento: Hamiltonianism in Action." *Pacific Historical Review* 34 (1965): 21–49.

Kershner, Frederick D., Jr. "George Chaffey and the Irrigation Frontier." *Agricultural History* 27 (1953): 115–122.

Kirkendall, Richard S. "Social Science in the Central Valley of California: An Episode." *California Historical Quarterly* 43 (1964): 195–218.

Koppes, Clayton R. "Public Water, Private Land: Origins of the Acreage Limitation Controversy, 1933–1953." *Pacific Historical Review* 47 (1978): 607–636.

Korr, Charles P. "William Hammond Hall: The Failure of Attempts at State Water Planning in California, 1878–1888." *Southern California Quarterly* 45 (1963): 305–318.

Kramer, Howard D. "The Scientist in the West, 1870–1880." *Pacific Historical Review* 12 (1943): 239–251.

Lamar, Howard. "Persistent Frontier: The West in the Twentieth Century." *Western Historical Quarterly* 4 (1973): 5–25.

Le Duc, Thomas. "State Disposal of Agricultural Scrip." *Agricultural History* 28 (1954): 99–107.

Lee, Lawrence B. "The Canadian-American Irrigation Frontier, 1884–1914." *Agricultural History* 40 (1965): 271–283.

————. "Environmental Implications of Governmental Reclamation in California." *Agricultural History* 49 (1975): 223–229.

————. "The Little Landers Colony of San Ysidro." *Journal of San Diego History* 21 (1975): 26–51.

————. "William Ellsworth Smythe and the Irrigation Movement: A Reconsideration." *Pacific Historical Review* 41 (1972): 289–311.

————. "William E. Smythe and San Diego, 1901–1908." *Journal of San Diego History* 19 (1973): 10–24.

Lilley, William III, and Gould, Lewis L. "The Western Irrigation Movement, 1878–1902: A Reappraisal." In *The American West: A Reorientation*, edited by Gene M. Gressley, 57–74. Laramie, Wyo., 1966.

Lippincott, J. B. "General Outlook for Reclamation Work in California." *Forestry and Irrigation* 11 (1905): 349–353.

————. "Recent Irrigation Development in California." *California*

Cultivator 62 (1924): 91, 107, 111.

————— . "The Reclamation Service in California." *Forestry and Irrigation* 10 (1904): 162–169.

Maass, Arthur, and Zobel, Hiller B. "Anglo-American Water Law: Who Appropriated the Riparian Doctrine?" *Public Policy* 10 (1960): 109–156.

McAfee, Ward. "Local Interests and Railroad Regulation in California During the Granger Decade." *Pacific Historical Review* 37 (1968): 51–66.

McConnell, Grant. "The Conservation Movement—Past and Present." *Western Political Quarterly* 7 (1954): 463–478.

McDow, Roberta M. "State Separation Schemes, 1907–1921." *California Historical Quarterly* 49 (1970): 39–46.

————— . "To Divide or Not to Divide?" *The Pacific Historian* 10 (1966): 22–23.

McGinty, Brian. "The Legacy of Buena Vista: Agoston Haraszthy and the Development of California Viticulture." *The American West* 10 (1973): 17–23.

McKee, Irving. "John Paul Vignes, California's First Professional Winegrower." *Agricultural History* 22 (1948): 176–179.

Marks, Bernard. "The Riparian Decision in Interior California." *Overland Monthly,* 2d ser., 9 (1887): 145–162.

Maxwell, George H. "The Irrigation District: The Inherent Defects Which Have Caused Its Failure Can Only Be Remedied By a State System." *Irrigation Age* 12 (1898): 250–253.

————— . "National Irrigation: Work and Progress under the Reclamation Act." *Maxwell's Talisman* 6 (1906): 27.

————— . "Reclamation of Arid America." *Irrigation Age* 13 (1899): 407–409.

Mead, Elwood. "Farm Settlement on a New Plan." *American Review of Reviews* 59 (1919): 270–277.

————— . "The Irrigation Investigations in California of the Office of Experiment Stations." *Forestry and Irrigation* 11 (1905): 367–369.

Meinig, Donald W. "The Growth of Agricultural Regions in the Far West: 1850–1910." *The Journal of Geography* 54 (1955): 223–232.

Mendenhall, Walter C. "Studies of California Ground Waters." *Forestry and Irrigation* 11 (1905): 382–384.

Miles, Nelson A. "Our Unwatered Empire." *North American Review* 150 (1890): 370–381.

Miller, Gordon R. "Shaping California Water Law, 1781 to 1928." *Southern California Quarterly* 55 (1973): 9–42.

Montgomery, Mary. "Central Valley Project: Highlights of Its History." *Land Policy Review* 9 (1946): 18–21.

Moorhead, Dudley T. "Sectionalism and the California Constitution of 1879." *Pacific Historical Review* 12 (1943): 287–293.

Mosk, Sanford A. "Price-Fixing in Spanish California." *California Historical Quarterly* 17 (1938): 118–122.

Nash, Gerald D. "The California State Land Office, 1858–1898." *Huntington Library Quarterly* 27 (1964): 347–356.

————. "Henry George Reexamined: William S. Chapman's Views on Land Speculation in Nineteenth Century California." *Agricultural History* 33 (1959): 133–137.

————. "The Influence of Labor on State Policy, 1860–1920." *California Historical Quarterly* 42 (1963): 241–257.

————. "Problems and Projects in the History of Nineteenth-Century California Land Policy." *Arizona and the West* 2 (1960): 327–340.

————. "Stages of California's Economic Growth, 1870–1970: An Interpretation." *California Historical Quarterly* 51 (1972): 315–330.

Nash, Roderick. "John Muir, William Kent, and the Conservation Schism." *Pacific Historical Review* 36 (1967): 423–433.

Netz, Joseph. "The Great Los Angeles Real Estate Boom of 1887." *Annual Publication of the Historical Society of Southern California* 10 (1915–1916): 54–68.

Newell, Frederick H. "Irrigation." In *U.S. Smithsonian Institution, Annual Report, 1901,* 407–428. Washington, D.C., 1902.

————. "National Efforts at Home Making." In *U.S. Smithsonian Institution, Annual Report, 1922,* 517–531. Washington, D.C., 1922.

————. "The Reclamation Service and the Owens Valley." *Out West* 23 (1905): 454–461.

————. "The Reclamation of the West." In *U.S. Smithsonian Institution, Annual Report, 1903,* 827–841. Washington, D.C., 1904.

Olney, Warren. "The Present Status of the Irrigation Problem."

Overland Monthly, 2d ser., 9 (1887): 40–50.

Orsi, Richard J. "*The Octopus* Reconsidered: The Southern Pacific and Agricultural Modernization in California, 1865–1915." *California Historical Quarterly* 54 (1975): 197–200.

Parker, Edna Monck. "The Southern Pacific Railroad and Settlement in Southern California." *Pacific Historical Review* 6 (1937): 103–119.

Parsons, James J. "The Uniqueness of California." *American Quarterly* 7 (1955): 45–55.

Paul, Rodman. "The Beginnings of Agriculture in California: Innovation vs. Continuity." *California Historical Quarterly* 41 (1973): 16–27.

———. "The Great California Grain War: The Grangers Challenge the Wheat King." *Pacific Historical Review* 27 (1958): 331–350.

———. "The Wheat Trade between California and the United Kingdom." *Mississippi Valley Historical Review* 45 (1958): 391–412.

Peffer, E. Louise. "Memorial to Congress on an Agricultural College for California, 1853." *Agricultural History* 40 (1966): 53–56.

Petersen, Erik F. "The End of an Era: California's Gubernatorial Election of 1894." *Pacific Historical Review* 38 (1969): 141–156.

Peterson, Richard H. "The Failure to Reclaim: California State Swamp Land Policy and the Sacramento Valley, 1850–1866." *Southern California Quarterly* 56 (1974): 45–60.

Pisani, Donald J. "Conflict over Conservation: The Reclamation Service and the Tahoe Contract." *Western Historical Quarterly* 10 (1979): 167–190.

———. "Reclamation and Social Engineering in the Progressive Era." *Agricultural History* 57 (1983): 46–63.

———. "State vs. Nation: Federal Reclamation and Water Rights in the Progressive Era." *Pacific Historical Review* 51 (1982): 265–282.

———. "Water Law Reform in California, 1900–1913." *Agricultural History* 54 (1980): 295–317.

———. "Western Nevada's Water Crisis, 1915–1935." *Nevada Historical Society Quarterly* 22 (1979): 3–20.

———. "'Why Shouldn't California Have the Grandest Aqueduct in the World?': Alexis Von Schmidt's Lake Tahoe Scheme."

California Historical Quarterly 53 (1974): 347–360.

Poli, Adon, and Engstrand, W. M. "Japanese Agriculture on the Pacific Coast." *Journal of Land and Public Utility Economics* 21 (1945): 355–364.

Pomeroy, Earl. "California, 1846–1860: Politics of a Representative Frontier State." *California Historical Quarterly* 32 (1953): 291–302.

————. "Toward a Reorientation of Western History: Continuity and Environment." *Mississippi Valley Historical Review* 41 (1955): 579–599.

Powell, John Wesley. "The New Lake in the Desert." *Scribner's Magazine* 12 (1891): 463–468.

Prescott, Gerald L. "Farm Gentry vs. the Grangers: Conflict in Rural America." *California Historical Quarterly* 56 (1977): 328–345.

Putnam, Jackson K. "The Persistence of Progressivism in the 1920s: The Case of California." *Pacific Historical Review* 35 (1966): 395–411.

Raup, H. F., *San Bernardino, California: Settlement and Growth of a Pass-Site City.* University of California Publications in Geography 8 (1940–1962).

————. "Transformation of Southern California to a Cultivated Land." *Annals of the Association of American Geographers* 49 (1959): 58–78.

Reed, Howard S. "Major Trends in California Agriculture." *Agricultural History* 20 (1946): 252–255.

Reid, Bill G. "Franklin K. Lane's Idea for Veteran's Colonization, 1918–1921." *Pacific Historical Review* 33 (1964): 447–461.

Ressler, John B. "Indian and Spanish Water-Control on New Spain's Northwest Frontier." *Journal of the West* 7 (1968): 10–17.

Richardson, Elmo R. "Hetch Hetchy." In *California Water Atlas.* edited by William Kahrl, 29–31. Sacramento, 1978.

————. "The Struggle for the Valley: California's Hetch Hetchy Controversy, 1905–1913." California Historical Quarterly 38 (1959): 249–258.

Ritchie, Robert Welles. "The Rural Democracy at Delhi." *Country Gentleman* 85 (1920): 8, 36.

Sageser, A. Bower. "Los Angeles Hosts an International Irrigation Congress." *Journal of the West* 4 (1965): 411–424.

Sargent, A. A. "Irrigation and Drainage." *Overland Monthly,* 2d ser., 8 (1886): 19–32.

Shaw, John A. "Railroads, Irrigation, and Economic Growth: The San Joaquin Valley of California." *Explorations in Economic History* 10 (1973): 211–227.

Shaw, Lucien. "The Development of the Law of Waters in the West." *California Law Review* 10 (1922): 443–460.

————. "The Development of Water Law in California." *California Bar Association, Proceedings of the Thirteenth Annual Convention,* 13 (1913): 154–173.

Shelby, Gertrude Matthews. "Nobody Wants to Leave this Town." *Collier's* 75 (May 30, 1925): 26, 34.

Simmons, Marc. "Spanish Irrigation Practices in New Mexico." *New Mexico Historical Review* 47 (1972): 135–150.

Smith, Henry Nash. "Rain Follows the Plow: The Notion of Increased Rainfall for the Great Plains, 1844–1860." *Huntington Library Quarterly* 10 (1947): 169–193.

Smith, Herbert A. "The Early Forestry Movement in the United States." *Agricultural History* 12 (1938): 326–346.

Smith, Ray J. "The California Land Settlement Board at Durham and Delhi." *Hilgardia* 4 (1943): 399–492.

Smythe, William Ellsworth. "The Failure of the Water and Forest Commission." *Out West* 17 (1902): 751–757.

————. "Irrigation Principles." *Irrigation Age* 8 (1895): 10–12.

————. "A Study of Two Modern Instances." *Irrigation Age* 5 (1893): 112–113.

————. "A Success of Two Centuries." *Out West* 22 (1905): 72–76.

Stephenson, W. A. "Appropriation of Water in Arid Regions." *Southwestern Social Science Quarterly* 18 (1937): 215–226.

Sterling, Everett W. "The Powell Irrigation Survey, 1888–1893." *Mississippi Valley Historical Review* 27 (1940): 421–434.

Strong, Douglas H. "The Sierra Forest Reserve: The Movement to Preserve the San Joaquin Valley Watershed." *California Historical Quarterly* 45 (1967): 3–18.

Sunseri, Alvin R. "Agricultural Techniques in New Mexico at the Time of the Anglo-American Conquest." *Agricultural History* 47 (1973): 329–337.

Swain, Donald C. "The Bureau of Reclamation and the New Deal, 1933–1944." *Pacific Northwest Quarterly* 61 (1970): 137–146.

Taggart, H. F. "Thomas Vincent Cator: Populist Leader of California." *California Historical Quarterly* 27 (1948): 311–318; 28 (1949): 47–55.

Taylor, Paul S. "Central Valley Project: Water and Land." *Western Political Quarterly* 2 (1949): 228–253.

————. "Foundations of California Rural Society." *California Historical Quarterly* 24 (1945): 193–228.

————. "Reclamation." *The American West* 7 (1970): 27–33, 63.

————. "Water, Land, and People in the Great Valley." *The American West* 5 (1968): 24–29, 68–72.

Teilman, Hendrick. "The Role of Irrigation Districts in California's Water Development." *American Journal of Economics and Sociology* 22 (1963): 409–415.

Thickens, Virginia E. "Pioneer Agricultural Colonies of Fresno County." *California Historical Quarterly* 25 (1946): 17–38, 169–177.

Thompson, Kenneth, and Eigenheer, Richard A. "The Agricultural Promise of the Sacramento Valley: Some Early Views," *Journal of the West* 18 (1979): 33–41.

Thompson, Kenneth. "Historic Flooding in the Sacramento Valley." *Pacific Historical Review* 29 (1960): 349–360.

————. "Insalubrious California: Perception and Reality." *Annals of the Association of American Geographers* 59 (1969): 50–64.

————. "Irrigation as a Menace to Health in California: A Nineteenth Century View." *The Geographical Review* 59 (1969): 195–214.

————. "The Notions of Air Purity in Early California." *Southern California Quarterly* 54 (1972): 203–210.

Tomlinson, F. L. "Land Reclamation and Settlement in the United States." *International Review of Agricultural Economics* 4 (1926): 225–272.

Treadwell, Edward F. "Developing a New Philosophy of Water Rights." *California Law Review* 38 (1950): 572–587.

Vance, James E., Jr. "California and the Search for the Ideal." *Annals of the Association of American Geographers* 62 (1972): 185–210.

Von Geldern, Otto. "Reminiscences of the Pioneer Engineers of California." *Western Construction News* 4 (1929): 555–556.

Waldo, C. T. "Evaluation of California Water Rights Law." *Southern California Law Review* 18 (1945): 267–273.

Whitaker, Arthur P. "The Spanish Contribution to American Agriculture." *Agricultural History* 3 (1929): 1–14.

Wiel, Samuel C. "Fifty Years of Water Law." *Harvard Law Review* 50 (1936–1937): 252–304.

―――― . "Origin and Comparative Development of the Law of Watercourses in the Common Law and Civil Law." *California Law Review* 6 (1918): 245–267, 342–371.

―――― . "The Pending Water Amendment to the California Constitution, and Possible Legislation." *California Law Review* 16 (1928): 169–207, 259–280.

―――― . "Public Policy in Western Water Decisions." *California Law Review* 1 (1912): 11–31.

―――― . "The Recent Attorneys' Conference on Water Legislation." *California Law Review* 17 (1929): 197–213.

―――― . "Waters: American Law and French Authority." *Harvard Law Review* 33 (1919): 133–167.

Winther, Oscar O. "The Colony System of Southern California." *Agricultural History* 27 (1953): 94–102.

Woehke, W. V. "Food First: Work of the California Land Settlement Board." *Sunset* 45 (October 1920): 35–38.

Wood, H. J. "Water Plan for the Great Valley of California." *Economic Geography* 14 (1938): 354–362.

Wooster, Clarence M. "Building the Railroad down the San Joaquin in 1871." *California Historical Quarterly* 18 (1939): 22–31.

Works, John D. "Irrigation Laws and Decisions of California." In *History of the Bench and Bar in California*, by Oscar T. Shuck, 101–172. Los Angeles, 1901.

XI. Books and Pamphlets

Ackerman, E. A., and Lof, George O. G. *Technology in American Water Development*. Baltimore, Md. 1959.

Adams, Frank. *Edward F. Adams: 1839–1929*. Berkeley, Calif., 1966.

Alexander, J. A. *The Life of George Chaffey*. Melbourne, Australia, 1928.

Alexander, Thomas G. *A Clash of Interests: Interior Department and Mountain West, 1863–1896*. Provo, Utah, 1977.

Anderson, David B. *Riparian Water Rights in California*. Sacramento, Calif., 1977.

Archibald, Marybelle D. *Appropriative Water Rights in California.* Sacramento, Calif., 1977.

Arrington, Leonard. *Great Basin Kingdom: An Economic History of the Latter-Day Saints, 1830–1900.* Cambridge, Mass., 1958.

Bain, Joe S. et al. *Northern California's Water Industry.* Baltimore, Md., 1966.

Bancroft, Hubert Howe. *California Pastoral, 1769–1848.* San Francisco, 1888.

————. *History of Arizona and New Mexico, 1530–1888.* San Francisco, 1889.

————. *History of California, 1860–1890.* San Francisco, 1890.

Barker, Charles A. *Henry George.* New York, 1955.

Bean, Walton. *California: An Interpretive History.* New York, 1973.

Beattie, George William, and Beattie, Helen P. *Heritage of the Valley: San Bernardino's First Century.* Pasadena, Calif., 1939.

Berg, Norman. *A History of Kern County Land Company.* Bakersfield, Calif., 1971.

Blackford, Mansell B. *The Politics of Business in California, 1890–1920.* Columbus, Ohio, 1977.

Bowles, Samuel. *Our New West.* Hartford, Conn., 1869.

Boyd, David. *A History: Greeley and the Union Colony of Colorado.* Greeley, Colo., 1890.

————. *Irrigation Near Greeley, Colorado.* Washington, D.C., 1897.

Boyle, Robert H. et al., *The Water Hustlers.* San Francisco, 1971.

Brace, Charles Loring. *The New West: Or, California in 1867–1868.* New York, 1869.

Brereton, Robert M. *Reminiscences of Irrigation-Enterprise in California.* Portland, Oreg., 1903.

————. *Reminiscences of an Old English Civil Engineer, 1858–1908.* Portland, Oreg., 1908.

Brown, John Jr., and Boyd, James. *History of San Bernardino and Riverside Counties.* Chicago, 1922.

Browne, J. Ross. *Resources of the Pacific Slope.* San Francisco, 1869.

Bryant, Edwin. *What I Saw in California.* New York, 1848.

Buck, Solon J. *The Granger Movement: A Study of Agricultural Organization and Its Political, Economic and Social Manifestations, 1870–1880.* Cambridge, Mass., 1913.

Burnley, James. *Millionaires and Kings of Enterprise.* Philadelphia, 1901.

California Development Association. *Report on Problems of Agricultural Development in California*. San Francisco, 1924.

Carosso, Vincent P. *The California Wine Industry: A Study of the Formative Years*. Berkeley, 1951.

Carr, Ezra S. *The Patrons of Husbandry on the Pacific Coast*. San Francisco, 1875.

Caughey, John Walton. *The California Gold Rush*. Berkeley, 1948.

———. *California: A Remarkable State's Life History*. Englewood Cliffs, N.J., 1970.

Chambers, Clarke A. *California Farm Organizations*. Berkeley, 1952.

Chandler, Alfred E. *Elements in Western Water Law*. San Francisco, 1913.

Cleland, Robert G. *California in Our Time, 1900–1940*. New York, 1947.

———. *The Cattle on a Thousand Hills: Southern California, 1850–1880*. San Marino, Calif., 1951.

———. *The March of Industry*. Los Angeles, 1929.

———. *From Wilderness to Empire: A History of California, 1542–1900*. New York, 1944.

Cole, Cornelius. *Memoirs of Cornelius Cole*. New York, 1908.

Comfort, Herbert G. *Where Rolls the Kern*. Moorpark, Calif., 1934.

Conkling, Roscoe P., and Margaret B. *The Butterfield Overland Mail, 1857–1869*. Glendale, Calif., 1947.

Cooper, Erwin. *Aqueduct Empire*. Glendale, Calif., 1968.

Cory, H. T. *The Imperial Valley and the Salton Sink*. San Francisco, 1915.

Crane, Lauren E., ed. *Newton Booth of California: His Speeches and Addresses*. New York, 1894.

Crissey, Forrest. *Where Opportunity Knocks Twice*. Chicago, 1914.

Cronise, Titus Fey. *The Agricultural and other Resources of California*. San Francisco, 1870.

———. *The Natural Wealth of California*. San Francisco, 1868.

Cross, Ira B. *A History of the Labor Movement in California*. Berkeley, 1935.

Daggett, Stuart. *Chapters on the History of the Southern Pacific*. New York, 1922.

Dana, Julian. *The Sacramento: River of Gold*. New York, 1939.

Daniels, Roger. *The Politics of Prejudice: The Anti-Japanese Movement in California and the Struggle for Japanese Exclusion*. Berkeley, 1962.

Davis, Arthur P. *Irrigation Works Constructed by the United States Government.* New York, 1917.

Davis, Winfield J. *History of Political Conventions in California, 1849–1892.* Sacramento, 1893.

Deakin, Alfred. *Irrigation in Western America.* Melbourne, 1885.

De Pue and Company, *Illustrated Atlas and History of Yolo County, California.* San Francisco, 1879.

Dumke, Glenn. *The Boom of the Eighties in Southern California.* San Marino, Calif., 1944.

Dupree, A. Hunter. *Science in the Federal Government: A History of Policies and Activities to 1940.* Cambridge, Mass., 1957.

Dwinelle, John W. *The Colonial History, City of San Francisco.* San Francisco, 1867.

Ekirch, Arthur A. *Man and Nature in America.* New York, 1963.

Elliott, Wallace W. *History of San Bernardino and San Diego Counties.* San Francisco, 1883.

————. *History of Tulare County.* San Francisco, 1883.

Ellison, Joseph. *California and the Nation, 1850–1869: A Study of the Relations of a Frontier Community with the Federal Government.* Berkeley, 1927.

Ellison, William E. *A Self-Governing Dominion: California, 1849–1860.* Berkeley, 1950.

Engelhardt, Zephyrin. *The Missions and Missionaries of California.* San Francisco, 1908.

Fabian, Bentham. *The Agricultural Lands of California.* San Francisco, 1869.

Farnham, T. J. *Life, Adventures, and Travels in California: To Which are Added the Conquest of California, Travels in Oregon, and History of the Gold Regions.* New York, 1849.

Fisher, Lloyd, and Nielson, Ralph L. *The Japanese in California Agriculture.* Berkeley, 1942.

Fisher, Walter M. *The Californians,* San Francisco, 1876.

Fite, Gilbert C. *The Farmers' Frontier, 1865–1900.* New York, 1966.

Fogelson, Robert M. *The Fragmented Metropolis: Los Angeles, 1850–1930.* Cambridge, Mass., 1967.

Forbes, Alexander. *A History of Upper and Lower California.* London, 1839.

Gates, Paul W., ed. *California Ranchos and Farms, 1846–1862.* Madison, 1967.

————. *The Farmers' Age: Agriculture, 1815–1860.* New York, 1960.

————— . *History of Public Land Law Development.* Washington, D.C., 1968.

————— . *Land Policies in Kern County.* Bakersfield, Calif., 1978.

George, Henry. *Our Land and Land Policy.* San Francisco, 1871.

Golze, Alfred R. *Reclamation in the United States.* New York, 1952.

Gregg, Josiah. *Commerce of the Prairies: or the Journal of a Santa Fe Trader.* New York, 1844.

Gregory, Tom, et al. *History of Yolo County, California.* Los Angeles, 1913.

Gressley, Gene, ed. *The American West: A Reorientation.* Laramie, Wyo., 1966.

Guinn, J. M. *Historical and Biographical Record of Los Angeles and Vicinity.* Chicago, 1901.

Haber, Samuel. *Efficiency and Uplift: Scientific Management in the Progressive Era, 1890–1920.* Chicago, 1964.

Haggin, James B., et al. *Desert Lands of Kern County, California: Affadavits of Various Residents of Said County.* San Francisco, 1877.

Hamilton, Leonidas. *Hamilton's Mexican Law.* San Francisco, 1882.

Haraszthy, Agoston. *Grape Culture, Wines, and Wine-Making.* New York, 1862.

Harding, Sidney T. *Water in California.* Palo Alto, Calif., 1960.

Hart, James D. *American Images of Spanish California.* Berkeley, 1960.

Hastings, Lansford W. *The Emigrants' Guide to Oregon and California.* Cincinnati, 1845.

Hays, Samuel P. *Conservation and the Gospel of Efficiency: The Progressive Conservation Movement, 1890–1920.* Cambridge, Mass., 1959.

Helper, Hinton R. *The Land of Gold: Reality Versus Fiction.* Baltimore, 1855.

Hibbard, Benjamin H. *A History of Public Land Law Policies.* New York, 1924.

Hichborn, Franklin. *Story of the California Legislature of 1911.* San Francisco, 1911.

————— . *Story of the California Legislature of 1913.* San Francisco, 1913.

Hine, Robert V. *California's Utopian Colonies.* New Haven, 1966.

Hittell, John S. *Commerce and Industries of the Pacific Coast.* San Francisco, 1882.

————— . *The Resources of California.* San Francisco, 1863.

Hittell, Theodore H. *History of California.* San Francisco, 1897.

Hoffman, Abraham. *Vision or Villainy: Origins of the Owens Valley-Los Angeles Water Controversy.* College Station, Tex., 1981.

Hopkins, C. T. *Common Sense Applied to the Immigrant Question.* San Francisco, 1870.

Huffman, Roy E. *Irrigation Development and Public Water Policy.* New York, 1953.

Hundley, Norris, Jr., *Dividing the Waters: A Century of Controversy Between the United States and Mexico.* Berkeley, 1966.

———— . *Water and the West: The Colorado River Compact and the Politics of Water in the American West.* Berkeley, 1975.

Hutchins, Wells A. *The California Law of Water Rights.* Sacramento, 1956.

Hutchinson, W. H. *California: Two Centuries of Man, Land, and Growth in the Golden State.* Palo Alto, Calif., 1969.

———— . *Oil, Land, and Politics: The California Career of Thomas Robert Bard.* Norman, Okla., 1965.

Hutchinson, Claude B. *California Agriculture.* Berkeley, 1946.

Ingersoll, L. A. *Ingersoll's Century Annals of San Bernardino County, 1769 to 1904.* Los Angeles, 1904.

———— . *Ingersoll's Century History: Santa Monica Bay Cities.* Los Angeles, 1908.

Jackson, W. Turrentine, and Pisani, Donald J. *A Case Study in Interstate Resource Management: The California-Nevada Water Controversy, 1865–1955.* Davis, Calif., 1973.

Jackson, W. Turrentine. *Wagon Roads West: A Study of Federal Road Surveys and Construction in the Trans-Mississippi West, 1846–1869.* New Haven, 1952.

James, George Wharton. *Heroes of California.* Boston, 1910.

———— . *Reclaiming the Arid West: The Story of the United States Reclamation Service.* New York, 1917.

Jelinek, Lawrence J. *Harvest Empire: A History of California Agriculture.* San Francisco, 1979.

Kahrl, William, et al. *The California Water Atlas.* Sacramento, 1978.

———— . *Water and Power: The Conflict over Los Angeles' Water Supply in the Owens Valley.* Berkeley, 1982.

Kelley, Robert L. *Gold vs. Grain: The Hydraulic Mining Controversy in California's Sacramento Valley, A Chapter in the Decline of Laissez-Faire.* Glendale, Calif., 1959.

Kennan, George, *E. H. Harriman.* Boston, 1922.

Kerwin, Jerome G. *Federal Water Power Legislation.* New York, 1926.

King, Judson. *The Conservation Fight: From Theodore Roosevelt to the Tennessee Valley Authority.* Washington, D.C., 1959.

King, T. Butler. *Report of Hon. T. Butler King on California.* Washington, D.C., 1850.

Kinney, Abbot. *Forest and Water.* Los Angeles, 1900.

Kinney, Clesson S. *A Treatise on the Law of Irrigation and Water Rights and the Arid Region Doctrine of Appropriation of Waters.* San Francisco, 1912.

Kleinsorge, Paul L. *The Boulder Canyon Project.* Stanford, Calif., 1941.

Kuhn, James S., and Kuhn, W. S. *The New California.* 1910.

LaFuze, Pauliena B. *Saga of the San Bernardinos.* San Bernardino, Calif., 1971.

Lamar, Howard, ed. *The Reader's Encyclopedia of the American West.* New York, 1977.

Lampen, Dorothy. *Economic and Social Aspects of Federal Reclamation.* Baltimore, 1930.

Lavender, David, *Nothing Seemed Impossible: William C. Ralston and Early San Francisco.* Palo Alto, Calif., 1975.

Layne, J. Gregg. *Annals of Los Angeles.* San Francisco, 1935.

Layton, Edwin T., Jr. *The Revolt of the Engineers: Social Responsibility and the American Engineering Profession.* Cleveland, 1971.

Lewis, A. D. *Irrigation and Settlement in America.* Pretoria, South Africa, 1915.

Lewis, Oscar. *George Davidson: Pioneer West Coast Scientist.* Berkeley, 1954.

Lewis Publishing Company. *An Illustrated History of Los Angeles County, California.* Chicago, 1889.

Los Angeles Examiner. *Press Reference Library, 1915.* Los Angeles, 1915.

Lowenthal, David. *George Perkins Marsh: Versatile Vermonter.* New York, 1958.

Maass, Arthur, and Anderson, Raymond L. *. . . and the Desert Shall Rejoice: Conflict, Growth, and Justice in Arid Environments.* Cambridge, Mass., 1978.

Maass, Arthur. *Muddy Waters: The Army Engineers and the Nation's Rivers.* Cambridge, Mass., 1951.

MacArthur, Mildred Yorba. *Anaheim: "The Mother Colony."* Los Angeles, 1959.

McCarthy, Michael G. *Hour of Trial: The Conservation Conflict in Colorado and the West, 1891–1907*. Norman, Okla., 1977.

McCollum, William. *California As I Saw It*. Los Gatos, Calif., 1960.

McConnell, Grant. *Private Power and American Democracy*. New York, 1966.

McDonald, J. R., et al. *Prospectus of the West Side Irrigation District, San Joaquin Valley, California*. N.p., n.d.

McGowan, Joseph A. *History of the Sacramento Valley*. New York, 1961.

McGroarty, John Steven, ed. *History of Los Angeles County*. Chicago, 1923.

McWilliams, Carey. *California: The Great Exception*. New York, 1949.

————— . *Factories in the Field: The Story of Migratory Farm Labor in California*. Rev. ed., Santa Barbara, Calif., 1971.

————— . *Southern California: An Island on the Land*. Santa Barbara, Calif., 1973.

Manning, Thomas G. *Government in Science*. Lexington, Ky., 1967.

Mead, Elwood. *Helping Men Own Farms*. New York, 1920.

————— . *Irrigation Institutions*. New York, 1903.

Menefee, Eugene L. *History of Tulare and Kings Counties, California*. Los Angeles, 1913.

de Mille, Anna George. *Henry George: Citizen of the World*. Chapel Hill, N. C., 1950.

Miller, Thelma B. *History of Kern County, California*. Chicago, 1929.

Mills, William H. *The Hydrography of the Sacramento Valley*. San Francisco, 1904.

Montgomery, Mary, and Clawson, Marion. *History of Legislation and Policy Formation of the Central Valley Project*. Berkeley, 1946.

Morgan, Arthur E. *Dams and Other Disasters: A Century of the Army Corps of Engineers in Civil Works*. Boston, 1971.

Morgan, Wallace T. *History of Kern County, California*. Los Angeles, 1914.

Mowry, George. *The California Progressives*. Berkeley, 1951.

Nadeau, Remi A. *City-Makers: The Men Who Transformed Los Angeles From Village to Metropolis During the First Great Boom, 1868–1876*. Garden City, N.Y., 1948.

————— . *The Water Seekers*. Rev. ed. Santa Barbara, Calif., 1974.

Nash, Gerald D. *The American West in the Twentieth Century: A Short History of An Urban Oasis*. Englewood Cliffs, N.J., 1973.

————— . *State Government and Economic Development: A History of Administrative Policies in California, 1849–1933.* Berkeley, 1964.

Newell, Frederick H. *Irrigation in the United States.* New York, 1902.

Nordhoff, Charles. *California: For Health, Pleasure, and Residence.* New York, 1875.

Olin, Spencer C., Jr. *California's Prodigal Sons.* Berkeley, 1968.

Ostrom, Vincent. *Water and Politics: A Study of Water Policies and Administration in the Development of Los Angeles.* Los Angeles, 1953.

Paterson, Alan M., and Jackson, W. Turrentine. *The Sacramento-San Joaquin Delta: The Evolution and Implementation of Water Policy, An Historical Perspective.* Davis, Calif., 1977.

Pattie, James O. *The Personal Narrative of James O. Pattie of Kentucky.* Cincinnati, 1831.

Paul, Rodman W. *California Gold: The Beginning of Mining in the Far West.* Cambridge, Mass., 1947.

Peffer, E. Louise. *The Closing of the Public Domain.* Stanford, Calif., 1951.

Penick, James L. *Progressive Politics and Conservation: The Ballinger-Pinchot Affair.* Chicago, 1968.

Phelps, Alonzo. *Contemporary Biography of California's Representative Men.* San Francisco, 1881.

Phillips, Catherine C. *Cornelius Cole: California Pioneer and United States Senator.* San Francisco, 1929.

Pomeroy, Earl. *The Pacific Slope.* New York, 1965.

Pomeroy, John Norton. *A Treatise on the Law of Water Rights.* St. Paul, Minn., 1893.

Powers, Stephen. *Alone and Afoot: A Walk from Sea to Sea.* Hartford, Conn., 1872.

Preston, William L. *Vanishing Landscapes: Land and Life in the Tulare Lake Basin.* Berkeley, 1981.

Quaife, Milo M., ed. *Narrative of the Adventures of Zenas Leonard.* Chicago, 1934.

Richardson, Elmo R. *The Politics of Conservation: Crusades and Controversies, 1897–1913.* Berkeley, 1962.

Richman, Irving B. *California Under Spain and Mexico, 1535–1847.* Boston, 1911.

Robbins, Roy. *Our Landed Heritage: The Public Domain, 1776–1970.* Lincoln, Neb., 1976.

Robinson, Michael C. *Water for the West: The Bureau of Reclamation,*

1902–1977. Chicago, 1979.

Robinson, William W. *Land in California: The Story of Mission Lands, Ranchos, Squatters, Mining Claims, Railroad Grants, Land Scrip [and] Homesteads.* Berkeley, 1948.

Rockwell, John A., ed. *A Compilation of Spanish and Mexican Law in Relation to Mines, and Titles to Real Estate in Force in California, Texas and New Mexico.* New York, 1851.

de Roos, Robert. *The Thirsty Land: The Story of the Central Valley Project.* Stanford, Calif., 1948.

Roske, Ralph J. *Everyman's Eden: A History of California.* New York, 1968.

Royce, C. C. *John Bidwell: Pioneer, Statesman, Philanthropist.* Chico, Calif., 1906.

Sakolski, A. M. *The Great American Land Bubble.* New York, 1932.

Saunderson, Mont. *Western Land and Water Use.* Norman, Okla., 1950.

Seckler, David, ed. *California Water.* Berkeley, 1971.

Shuck, Oscar T. *Bench and Bar in California.* San Francisco, 1889.

————— . *History of the Bench and Bar of California.* Los Angeles, 1901.

Small, Kathleen Edwards. *History of Tulare County, California.* Chicago, 1926.

Smith, Henry Nash. *Virgin Land: The American West as Symbol and Myth.* Cambridge, Mass., 1950.

Smith, Truman. *On the Physical Character of the Northern States of Mexico.* Washington, D.C., 1848.

Smith, Wallace. *Garden of the Sun.* Los Angeles, 1939.

Smythe, William Ellsworth. *The Conquest of Arid America.* New York, 1900.

Stonehouse, Merlin. *John Wesley North and the Reform Frontier.* Minneapolis, 1965.

Swain, Donald C. *Federal Conservation Policy, 1921–1933.* Berkeley, 1963.

Swisher, Carl Brent. *Motivation and Political Technique in the California Constitutional Convention, 1878–79.* Claremont, Calif., 1930.

Taylor, Paul S. *Essays on Land, Water, and Law in California.* New York, 1979.

Teele, Ray P. *The Economics of Land Reclamation in the United States.* Chicago, 1927.

————— . *Irrigation in the United States.* New York, 1915.

Teilman, I., and Shafer, W. H. *The Historical Story of Irrigation in Central California*. Fresno, Calif., 1943.

Thomas, George. *The Development of Institutions Under Irrigation: With Special Reference to Early Utah Conditions*. New York, 1920.

——— . *Early Irrigation in the Western States*. Salt Lake City, 1948.

Thompson, Warren S. *Growth and Changes in California's Population*. Los Angeles, 1955.

Townley, John M. *Alfalfa Country: Nevada Land, Water and Politics in the 19th Century*. Reno, 1981.

Treadwell, Edward F. *The Cattle King*. Rev. ed. Boston, 1950.

Truman, Benjamin C. *Semi-Tropic California: Its Climate, Healthfulness, Productiveness, and Scenery*. San Francisco, 1874.

Vandor, Paul E. *History of Fresno County California*. Los Angeles, 1919.

Warne, William E. *The Bureau of Reclamation*. New York, 1973.

Warner, Juan Jose. *An Historical Sketch of Los Angeles County, California*. Los Angeles, 1876.

Webb, Walter Prescott. *The Great Plains*. Boston, 1931.

Wells, A. J. *Government Irrigation and the Settler*. San Francisco, 1910.

Werth, John J. *A Dissertation on the Resources and Policy of California*. Benicia, Calif., 1851.

Wicks and Phillips Investment House. *Irrigation Bonds: Their Security, Certainty, Desirability*. San Francisco, 1891.

Wickson, E. J. *California Nurserymen and the Plant Industry, 1850–1910*. Los Angeles, 1921.

——— . *Rural California*. New York, 1923.

——— . *One Thousand Questions in California Agriculture Answered*. San Francisco, 1914.

——— . *Second Thousand Questions in California Agriculture*. San Francisco, 1916.

Wiel, Samuel C. *Water Rights in the Western States*. San Francisco, 1905.

——— . *Waters: French Law and Common Law*. San Francisco, n.d.

Wilkes, Charles. *Narrative of the United States Exploring Expedition During the Years 1838, 1839, 1840, 1841, 1842*. Philadelphia, 1850.

Williams, R. Hal. *The Democratic Party and California Politics, 1880–1896*. Stanford, Calif., 1973.

Wilson, Iris Higbie. *William Wolfskill, 1798–1866: Frontier Trapper to California Ranchero*. Glendale, Calif., 1965.

Wilson, Warren. *History of San Bernardino County.* San Francisco, 1883.

Winchell, Lilbourne A. *History of Fresno County and the San Joaquin Valley.* Fresno, Calif., 1933.

Works, John D. *Answers to Objections Made to the Irrigation Bill Proposed by the Water and Forest Association.* N.p., n.d.

————. *Should the Irrigation Bill Pass?* N.p., n.d.

Zonlight, Margaret Aseman Cooper. *Land, Water and Settlement in Kern County California, 1850–1890.* New York, 1979.

Index

Designer: UC Press Staff
Compositor: Publisher's Typography
Printer: Braun-Brumfield
Binder: Braun-Brumfield
Text: 10/12 Palatino
Display: Palatino